The Molecular Biology of Enzyme Synthesis

The Molecular Biology of Enzyme Synthesis

Regulatory Mechanisms of Enzyme Adaptation

Roy Walker

Animal and Cell Physiology Group,
National Research Council of Canada, Ottawa

A Wiley-Interscience Publication
JOHN WILEY & SONS
New York • Chichester • Brisbane • Toronto • Singapore

Copyright © 1983 by John Wiley & Sons, Inc.

All rights reserved. Published simultaneously in Canada.

Reproduction or translation of any part of this work
beyond that permitted by Section 107 or 108 of the
1976 United States Copyright Act without the permission
of the copyright owner is unlawful. Requests for
permission or further information should be addressed to
the Permissions Department, John Wiley & Sons, Inc.

Library of Congress Cataloging in Publication Data:

Walker, P. Roy, 1945-
  The molecular biology of enzyme synthesis.

  "A Wiley-Interscience publication."
  Bibliography: p.
  Includes index.
  1. Enzyme synthesis. 2. Gene expression. 3. Molecular biology. I. Title

QP601.W22   574.19'25   82-10950
ISBN 0-471-06051-8

Printed in the United States of America

10 9 8 7 6 5 4 3 2 1

To my parents

PREFACE

Until comparatively recently, the synthesis of enzymes, especially in animal cells, was understood only descriptively. Moreover, the regulatory mechanisms that control enzyme synthesis were usually inferred from correlations between changes in enzyme synthesis and changes in the environment, rather than being based upon direct experimentation. In recent years, however, there have been considerable advances in our understanding of hormone action, second messenger systems, the biochemistry of protein and nucleic acid synthesis, as well as spectacular advances in the molecular biology of gene expression in both eukaryotes and prokaryotes. These areas of research have emerged as individual, specialized disciplines and as such tend to exist, and to be taught, separately. Consequently, the student is often left without an appreciation of the close interrelationships that exist between these fields and of the impact that each has upon our overall knowledge of enzyme synthesis. Moreover, a researcher working on the actions of steroid hormones, for example, is often unaware of the influence of the suprachiasmatic nucleus on enzyme synthesis or of developments in the structure of nucleosomes. In this book I have tried to bring together all these facets of enzyme synthesis to give both the student and the research worker a comprehensive view of this wide-ranging subject.

The book is organized into five sections. The first two sections deal in detail with the molecular biology of enzyme synthesis at the transcriptional and post-transcriptional levels, with particular emphasis on the regulatory mechanisms that control synthesis. Bacterial and animal cells are compared and contrasted to illustrate the considerable differences that have emerged between the two. It is now apparent that many of the mechanisms of gene expression that provide bacterial cells with their superb ability for rapid adaptation cannot operate in animal cells because of their greatly increased amounts of genetic material.

Section Three is devoted to a discussion on the regulation of enzyme synthesis in bacterial cells. The mechanisms for the control of both catabolic and biosynthetic pathways are described together with the mechanisms that operate to coordinate and integrate the overall metabolic activity of these cells. Section Four describes the regulation of enzyme synthesis in animal cells and is divided into two parts. Part One deals with the role of hormones and second messengers in the transfer of information in animal cells. It is the hormones that are principally involved in the regulation of enzyme synthesis, and a number of different mechanisms operate for the transfer of their information into the nucleus of target cells. In Part Two I describe the adaptive enzymes that are currently being studied at the molecular level and relate the changes in these enzymes to real physiological situations. This is particularly apparent in the discussion of diurnal rhythms of enzyme synthesis, which demonstrates that there are additional levels of regulatory complexity that are presently being overlooked.

Section Five deals with some of the more challenging aspects of enzyme adaptation—those occurring during tissue differentiation. An understanding of the molecular mechanisms that regulate enzyme synthesis during differentiation is the ultimate goal of molecular biology and is likely to form the basis of an understanding of neoplasia and other diseases of differentiation.

I have restricted the literature cited to the most recent reviews that are available for each topic, and these should provide a basis for further reading. I have only cited individual research papers for those areas that have not been adequately reviewed.

A project of this magnitude could not be completed without a great deal of cooperation from friends and colleagues. In particular, I would like to thank Jim Whitfield, head of the Animal and Cell Physiology Group, for his critical reading of the entire manuscript and his invaluable comments and advice. I would also like to thank Marianna Sikorska both for her encouragement and support and for her critical comments on the manuscript. Additionally, I would like to thank John MacManus for reading portions of the manuscript, and I am deeply indebted to Judy Walker, who prepared all the illustrations for the book. Furthermore, I wish to thank Denise Piché for typing the manuscript and Andrée McNeely and Huguette Blissett of the Canada Institute for Scientific and Technical Information for their help and expertise in dealing with the literature.

Roy Walker

Ottawa, Ontario
September 1982

CONTENTS

Chapter One Introduction to the Nature of Enzyme Synthesis 1

    The Concept of Enzyme Turnover, 1
    The Kinetics of Enzyme Turnover, 3
    Measurement of Enzyme Synthesis, 8
    The Role of Messenger RNA, 10
    References, 11

SECTION ONE **REGULATION OF ENZYME SYNTHESIS AT THE LEVEL OF TRANSCRIPTION** 13

Chapter Two Regulation of Gene Transcription in Bacterial Cells 15

    The Bacterial Genome, 15
        The Structure of DNA in the Nucleoid, 15
        Protein–Nucleic Acid Interactions, 18
        Sequence-Specific Protein Binding, 20
    The Mechanism of Transcription, 21
        RNA Polymerase, 22
        The DNA Template and the Initiation of Transcription, 22
        Transcription Termination, 25
    Regulation of Transcription, 27
        Promoter Activation by CAP, 27
        Promoter Obstruction by Repressor, 28
        Attenuation, 29
        Other Proteins That May Regulate Transcription, 34
    References, 35

CONTENTS

Chapter Three   **Structural Organization of DNA in the Nucleus of the Eukaryotic Cell**   38

Chromosomes, Chromatin, and the Complexity of DNA, 38
- Chromosomes, 38
- Chromatin, 39
- DNA Sequence Complexity, 40
- The Relationship between Sequence Complexity and Structure, 42

The Biochemical Composition of the Nucleus, 44
- The Proteins of the Nucleus, 44
- The Nuclear Matrix, 46
- Other Structural Proteins of the Nucleus, 47
- The Nuclear Membrane, 48

The Structural Organization of DNA, 48
- The Histone Core of the Nucleosome, 49
- Organization of DNA on the Histone Core, 52
- Polynucleosomes and the Arrangement of Histone H1, 52
- Higher Orders of Chromatin Structure, 54
- Other Structural Proteins Associated with Nucleosomes, 57

References, 59

Chapter Four   **Regulation of Gene Expression at the Transcriptional Level in Eukaryotes**   62

The Nature of Active Genes, 62
- The Nuclease Sensitivity of Active Gene Sequences, 62
- Nuclease Sensitivity and Nucleosome Phasing, 63
- Structural Correlates of Nuclease Sensitivity, 65
- The Binding of RNA Polymerase to Active Genes, 66

The Transcription Process, 67
- Multiplicity of Mammalian RNA Polymerases, 67
- Nucleotide Sequences Important for the Transcription of Structural Genes, 69
- Other RNA Polymerases Use Different "Promoter" Sequences, 72

The Regulation of Transcription, 77
  Activators of RNA Polymerases, 77
  Histone Modifications and Changes in Template Availability, 77
  The Role of Nonhistone Proteins, 82
  General Comments on the Regulation of Transcription in Eukaryotes, 85
References, 87

SECTION TWO	**REGULATION OF ENZYME SYNTHESIS BY POST-TRANSCRIPTIONAL MECHANISMS**	91
Chapter Five	**Regulation by Post-Transcriptional Modification of RNA Transcripts**	93

Intervening Sequences and the Nature of the Primary RNA Transcript, 93
  Split Genes, 94
  Sequence Complexity of HnRNA, 96
  The Structural Organization of HnRNP Particles, 97
RNA Processing in Eukaryotic Cells, 98
  Modification of the 5′ End—Capping, 98
  Methylation of Internal Adenyl Residues of HnRNA and mRNA, 100
  Modification of the 3′ End—Polyadenylation, 100
  Splicing and Joining of Coding Sequences to Generate a Contiguous mRNA Sequence, 103
  Splicing and the Regulation of Gene Expression, 106
  Summary of the Processing of HnRNA → mRNA, 110
  Processing of Other RNA Species, 112
RNA Processing in Bacterial Cells, 113
  Processing of Phage-T7 Early mRNAs, 113
  Processing of Bacterial-Cell tRNA Molecules, 114
  Processing of Bacterial-Cell rRNA Molecules, 114
The Transport of mRNA from Nucleus to Cytoplasm, 116
  Transport, 116
  mRNP Particles in the Cytoplasm, 117
  Degradation of mRNA, 118
References, 119

| Chapter Six | **Control of the Translation of Messenger RNA in Eukaryotes and Prokaryotes** | **122** |

The Protein-Synthesizing Machinery, 123
    Ribosomes: The Sites of mRNA Translation, 123
    Transfer RNA, 125
    Messenger RNA, 126
    Soluble Factors, 129

The Mechanism of Protein Synthesis, 129
    Initiation of Translation, 129
    Elongation and Translocation, 133
    Termination, 134

Regulation of Enzyme Synthesis at the Translational Level in Prokaryotes, 134
    Translational Control in Phage-Infected Cells, 134
    Tight Coupling between Transcription and Translation, 135

Regulation of Enzyme Synthesis at the Translational Level in Eukaryotes, 136
    Regulation of Translation at the Initiation Step, 137
    eIF-2 Cycling and the Regulation of Initiation, 138
    Regulation of Translation at the Elongation Step, 139
    mRNA Structure and the Regulation of Translation, 139
    Summary of the Role of mRNA in Determining its Own Rate of Translation, 140

References, 141

| Chapter Seven | **Regulation of Enzyme Activity by Post-Translational Modification** | **144** |

Post-Translational Modifications Involving Proteolysis, 144
    Activation of Enzymes by Proteolytic Cleavage, 144
    Translocation of Enzymes across Membranes, 145

Post-Translational Covalent Modification of Amino Residues, 147
    Phosphorylation, 147
    Adenylylation and Uridylylation, 151

ADP-Ribosylation, 151
Methylation, Acetylation, and Changes in
Sulfydryl Redox State, 152
Covalent Modification and Enzyme
Cascades, 154

Enzyme Degradation, 156
The Mechanism of Enzyme Degradation, 157
Specificity of the Process, 157

References, 158

SECTION THREE	**REGULATION OF ENZYME SYNTHESIS IN BACTERIAL CELLS**	**161**
Chapter Eight	**Catabolic Pathways**	**163**

Inducible and Noninducible Pathways, 165
Operons of Catabolic Pathways, 166
  The *lac* Operon, 166
  The *gal* Operon, 168
  The Catabolism of Arabinose, 169
  Histidine Utilization, 171

Integration of the Pathways of Catabolism, 173
  Glucose as a Preferred Source of Carbon and Energy and the Phenomenon of Catabolite Repression, 173
  cAMP as a Mediator of Catabolite Repression, 173
  The Catabolic Gene Activator Protein, 174
  CAP–cAMP and Auxic Growth, 176
  cAMP and Catabolite Repression, 177

Pathways of Nitrogen Source Catabolism, 178
  Escape from Catabolite Repression, 178
  The Central Role of Glutamine Synthetase in Cellular Nitrogen Metabolism, 178

References, 180

Chapter Nine	**Biosynthetic Pathways**	**182**

Pathways of Amino Acid Biosynthesis, 182
Operons for Pathways of Amino Acid Biosynthesis, 184
  Tryptophan Biosynthesis, 184
  The Phenylalanine Operon, 186
  Arginine Biosynthesis, 187

The Biosynthesis of Threonine, 189
Biosynthesis of the Branched-Chain Amino Acids—Leucine, Isoleucine, and Valine, 190
Histidine Biosynthesis, 194
Other Biosynthetic Operons, 196
Biotin Biosynthesis, 196
A General Model for the Mechanism of Attenuation, 196
Role of the DNA Sequence, 196
Role of the RNA Transcript of the Leader DNA, 198
Role of the Ribosome, 199
Role of the Amino Acyl tRNA Molecule, 200
Attenuation and the Regulation of Other Operons, 200
Regulation of the Synthesis of the Enzymes of Catabolism and Biosynthesis: A Summary, 201
References, 203

SECTION FOUR REGULATION OF ENZYME SYNTHESIS IN ANIMAL CELLS   205

PART ONE HORMONES, SECOND MESSENGERS, AND INFORMATION TRANSFER   207

Chapter Ten Hormones—The Inducers of Enzyme Synthesis   209

The Peptides of the Hypothalamus, 212
Hormones of the Pituitary, 213
Adrenocorticotrophin, 213
Thyroid-Stimulating Hormone, 215
Follicle-Stimulating Hormone and Luteinizing Hormone, 215
Prolactin, 215
Growth Hormone, 217
Melanocyte-Stimulating Hormone, 217
Oxytocin and Vasopressin, 217
Hormones of the Hypothalamus and Pituitary: A Summary, 218
The Steroid Hormones, 218
Chemistry of the Steroids, 218
The Glucocorticoids, 218
Androgens and Estrogens, 221

CONTENTS

    Thyroid Hormones, 222
        The Chemistry of T_3 and T_4, 222
        Secretion and Function of Thyroid Hormone, 222
    Hormones Not Directly under the Control of the Hypothalamic–Pituitary Axis, 223
        Insulin and Glucagon, 223
        Adrenaline and Noradrenaline, 224
    References, 225

Chapter Eleven   **Transfer of Information from Hormones to the Nucleus**   227

    Thyroid Hormones Transmit Information Directly to the Nucleus, 227
        Nuclear Receptors, 229
        Mechanism of Transcriptional Activation, 230
    Steroid Hormones Transmit Information via a Cytoplasmic Receptor, 232
        Cytoplasmic Receptors, 232
        Receptor Activation and Translocation to the Nucleus, 234
        Nuclear Acceptor Sites, 236
        The Mechanism of Transcriptional Activation, 237
    Polypeptide Hormones Transmit Information to the Plasma Membrane, 240
        Plasma-Membrane Receptors, 241
        Cyclic AMP as an Intracellular Second Messenger, 242
        Activation of Plasma-membrane Adenylate Cyclase, 244
        cAMP, Protein Kinases, and the Transfer of Information to the Nucleus, 246
        Phosphorylation and the Mechanism of Transcriptional Activation, 247
        Some Polypeptide Hormones Do Not Use cAMP as Second Messenger, 248
        Other Possible Hormone Second Messengers—cGMP, 249
        Other Possible Hormone Second Messengers-Calcium Ions, 250
        Prostaglandins: Hormones or Second Messengers, or What?, 253
    Catecholamines and Neurotransmitters Also Only

Transmit Information to the Cell Membrane, 253
Regulation of Receptor Concentration and the
Phenomenon of Down Regulation, 254
References, 256

PART TWO PHYSIOLOGICAL CONTROL OF ENZYME
ADAPTATION   261

Chapter Twelve **Induced Enzymes**   263

Hormonal Basis of Nutritional Adaptation, 266
  Insulin, Glucagon, and the Concentration of
    Hepatic cAMP, 266
  Changes in Other Hormones, 266
Hormonal Control of the Enzymes of Glucose
Production, 268
  General Pathways, 268
  Enzymes of Amino Acid Catabolism—TAT
    and TO, 268
  The Urea Cycle Enzymes, 273
  Glutamine Synthetase, 274
  Phosphoenolpyruvate Carboxykinase—A Key
    Enzyme of Gluconeogenesis, 275
  Summary of the Regulation of the Synthesis of
    Gluconeogenic Enzymes, 277
Hormonal Control of the Enzymes of Glucose
Conversion to Lipid, 279
  General Pathways, 279
  Glucokinase: The First Enzyme of Glycolysis, 280
  Fatty Acid Synthetase and AcetylCoA
    Carboxylase, 282
  The Enzymes of NADPH Production, 283
  Summary of the Regulation of the Synthesis
    of Lipogenic Enzymes, 284
Glucocorticoids and the Synthesis of Other
Enzymes and Proteins, 286
  Glucocorticoid Induction of Other Liver
    Enzymes, 286
  Regulation of Albumin and Transferrin Synthesis
    in Liver, 286
  Glucocorticoids and the Induction of Growth
    Hormone in Pituitary Cells, 287
Thyroid Hormones and the Synthesis of Other
  Proteins, 287

cAMP and the Regulation of the Synthesis of
Enzymes in Other Tissues, 288
  Lactate Dehydrogenase in Rat Glial Tumor
  Cells, 288
  Enzymes of Catecholamine Biosynthesis in
  Adrenal Cells, 289
  Other Proteins Controlled by cAMP, 290
The Mechanism of Action of cAMP in the
Regulation of Enzyme Synthesis, 290
References, 292

Chapter Thirteen   **Regulation of the Diurnal Rhythms of Enzyme Synthesis**   **296**

Diurnal Rhythms of Enzyme Synthesis, 297
  Hepatic Tyrosine Aminotransferase, 297
  Glucokinase, Glycogen Synthetase,
  Phosphorylase and Glycogen Rhythms, 300
  Rhythms of Other Hepatic Enzymes, 302
  The Rhythm of N-Acetyltransferase in
  the Pineal Gland, 303
  Diurnal Rhythms in Other Tissues, 306
Diurnal Rhythms of Other Biochemical
Functions, 307
  Chromatin Template Activity, RNA Synthesis,
  DNA Synthesis, and Mitosis, 307
  Glycogen and Lipid Deposition, 308
  Amino Acid Uptake and Protein Synthesis, 309
Diurnal Rhythms of Hormone Secretion, 309
  Corticosterone and ACTH, 309
  Melatonin, 310
  Pituitary Hormones, 311
  Rhythms of Insulin, Glucagon, and Hepatic
  cAMP, 312
  Thyroid Hormones, 312
Relationships between Diurnal Rhythms of
Hormones and Enzyme Synthesis, 312
  Glucocorticoids, Template Activity, and Enzyme
  Rhythms, 313
  Insulin and Carbon Storage, 313
  Glucagon, cAMP, and Rhythms of Enzyme
  Synthesis, 314
  Melatonin, Indoleamines, and Enzyme
  Rhythms, 315

Evidence for the Involvement of Higher Centers of the Brain in Enzyme Rhythms, 315
The Biochemical Nature of the Biological Clock, 316
  The Suprachiasmatic Nucleus, 316
  The Pineal Gland, 318
Diurnal Rhythms and Experimental Protocols, 318
References, 320

SECTION FIVE   REGULATION OF ENZYME SYNTHESIS DURING DIFFERENTIATION   325

Chapter Fourteen   Regulation of Enzyme Synthesis During Development   327

Enzyme Synthesis during Development is Discontinuous, 327
  Physiological Considerations, 327
  Patterns of Enzyme Synthesis in Liver, 328
Enzymes of the Late-Fetal Cluster in Liver, 330
  Phosphoenolpyruvate Carboxykinase, 330
  Tyrosine Aminotransferase, 334
  Other Enzymes in the Late-Fetal Cluster, 335
Enzymes Already Present in Maximal Amounts at Birth, 335
Enzymes of the Late-Suckling Period in Liver, 336
  Glucokinase, 336
  Malic Enzyme, 339
  Tryptophan Oxygenase, 340
Changes in the Secretion of Hormones During Development, 340
  Glucocorticoids, 340
  Insulin and Glucagon, 341
  Thyroid Hormones, 342
  Summary of Developmental Changes in Hormone Concentration, 343
Changes in the Development of Receptors and Second-Messenger Systems, 344
  Hormone Receptors, 344
  Adenylate Cyclase, cAMP, and Protein Kinases, 345
  Hormone-Responsive Calcium Transport, 346
  Summary of Changes in Information-Transfer Systems, 346

Development of Diurnal Rhythms in Liver Metabolism, 347
   The Pineal Gland, SCN, and Glucocorticoid Rhythms, 348
   Development of Diurnal Rhythms of Hepatic Enzymes, 349
References, 352

Chapter Fifteen   **Hormonal Control of Gene Expression During the Maturation of Reproductive Tissues**   355

Hormonal Control of the Maturation of the Chicken Oviduct, 355
   Changes in Morphology, 355
   Hormone Interactions at the Transcriptional Level, 357
Estrogen Regulation of Gene Expression in Avian and Amphibian Liver, 359
Hormonal Control of α_{2u}-Globulin Production in Male Rat Liver, 361
   α_{2u}-Globulin Synthesis during Development, 362
   Multihormonal Control in the Mature Animal, 362
Multihormonal Regulation of the Differentiation of the Mammary Gland, 364
   Morphological Changes, 364
   Changes in Hormone Levels During Pregnancy and Lactation in the Rat, 364
   Changes in Receptors and Second Messengers, 366
   Changes in the Synthesis of Milk Constituents, 366
   The Role of Prolactin, 368
   The Roles of Insulin and Glucocorticoids, 368
   The Role of Progesterone, 369
   The Role of cAMP, 370
   Summary of the Control of Mammary Gland Differentiation, 370
References, 371

Index   **373**

The Molecular Biology of Enzyme Synthesis

CHAPTER ONE

Introduction to the Nature of Enzyme Synthesis

Living cells have evolved complex regulatory mechanisms to control the concentration of their enzymes, particularly those catalyzing critical reactions. Enzymes are either expressed constitutively (i.e., synthesized continuously at an invariant rate) or are synthesized only in response to a specific stimulus. This latter group, called *adaptive enzymes*, is the subject of this book. Of all the levels of metabolic control the regulation of enzyme synthesis is the most important because it determines whether or not a reaction or pathway may operate.

THE CONCEPT OF ENZYME TURNOVER

In 1936 Karström (1) introduced the word *adaptive* to describe those enzymes that change in activity when bacterial cells are transferred from one environment to another and the word *constitutive* to describe those enzymes that do not. He defined *adaptive enzymes* as

> enzymes which are developed by a micro-organism only when the organism grows upon the specific substrate,

in contrast to *constitutive enzymes*,

> which are always developed by a given micro-organism irrespective of the composition of the medium on which the organism grows.

Since the nature of protein turnover was completely unknown at that time, these definitions carried no mechanistic implications.

Karström's definition of an adaptive enzyme proved to be too narrow when subsequent work revealed some enzymes that changed their activities in response to media constituents that were not substrates. For example, cells growing on leucine produced urease even though the amino acid is not its substrate. Dubos, in his 1940 review (2), broadened Karström's definition of an adaptive enzyme to an enzyme which appears as a specific response of the cell to the presence of a given substance in the medium.

This definition retained the concept of an enzyme that could respond specifically to the presence of a particular low-molecular-weight medium constituent, but did not restrict it to being the substrate of the enzyme. The terms *adaptive* and *constitutive* were widely accepted during the 1940s and 1950s with many examples of such enzymes being documented in both bacterial and animal cells.

The work on bacterial cells progressed rapidly, becoming more and more analytical. The question being asked—is there a fundamental difference between the synthesis of the two classes of enzymes?—provided impetus for the work that eventually led to the elucidation of both the biochemical mechanism of protein synthesis and the genetic basis of the regulation of enzyme adaptation.

Two different mechanisms of enzyme adaptation were discovered during this time in bacterial cells, and two new terms—enzyme *induction* and enzyme *repression*—were introduced. Enzyme induction was defined (3) as

a relative increase in the rate of synthesis of a specific apoenzyme resulting from exposure to a chemical substance.

This chemical substance was termed the *inducer*. In contrast, enzyme repression was defined (4) as

a relative decrease, resulting from the exposure of cells to a given substance, in the rate of synthesis of a particular apoenzyme.

The substance producing this effect was termed the *repressor*. These definitions have taken on very precise meanings as we have gained a better understanding of the biochemistry and genetics of these mechanisms in bacterial cells. Considerable confusion has been generated by using these terms to describe changes of enzyme synthesis in animal cells because they imply mechanisms that have not been demonstrated. Indeed, it is only very recently that apparent enzyme induction by hormones has been described in animal cells. Convincing demonstrations of enzyme repression have not yet been documented. Thus, considerable caution should be exercised in the use of these terms in animal cells.

Implicit in the concept of enzyme adaptation is the existence of enzyme turnover. Turnover was initially viewed at the cellular level. That is, in the presence of inducer, cells containing the enzyme were able to outgrow cells that had no enzyme. In the absence of inducer the enzyme-containing cells were at a disadvantage and were outgrown by another cell population. In other words, the enzyme complement of a cell was fixed, and only the fittest cells survived in any given environment. The demonstration that enzymes could be synthesized and degraded without death of the cell meant that turnover was an intracellular event. Moreover, enzymes expressed constitutively turn over even though their concentration is independent of the environment.

THE KINETICS OF ENZYME TURNOVER

To learn more about the nature of enzyme turnover let us examine some examples of enzyme adaptation in both bacterial and animal cells. Figures 1-1a and 1-1b show the induction of the enzyme β-galactosidase by methyl-β-D-thiogalactoside (a nonmetabolizable analogue of lactose, the physiological inducer of the enzyme) in *E. coli* cells. The response of hepatic tryptophan oxygenase to the administration of tryptophan is illustrated in Fig. 1-1c.

A thousandfold increase in the number of β-galactosidase molecules occurs within 1–2 h after the addition of the inducer to *E. coli* cells. This rate of increase is a function of both the synthesis of enzyme within cells and the exponential increase in the number of cells (since this is an actively growing culture). That the enzyme is preferentially induced is shown in Fig. 1-1b, where the activity of the enzyme is expressed as a function of total cell protein. In contrast, the administration of tryptophan to rats leads to an eightfold increase in enzyme over a period of 6 h, and the activity of the enzyme declines to the basal level by 12 h. These differences in both magnitude and time course are typical of the responses of bacterial and animal cells.

The shape of the tryptophan oxygenase curve gives some clues about the kinetics of enzyme turnover. The linearity of the increase suggests zero-order kinetics of synthesis, whereas the exponential decay indicates first-order kinetics of enzyme degradation.

Curves of similar shape have been observed for the adaptation of many mammalian enzymes, and, as a first approximation, the kinetics of enzyme synthesis may be described by the simple equation:

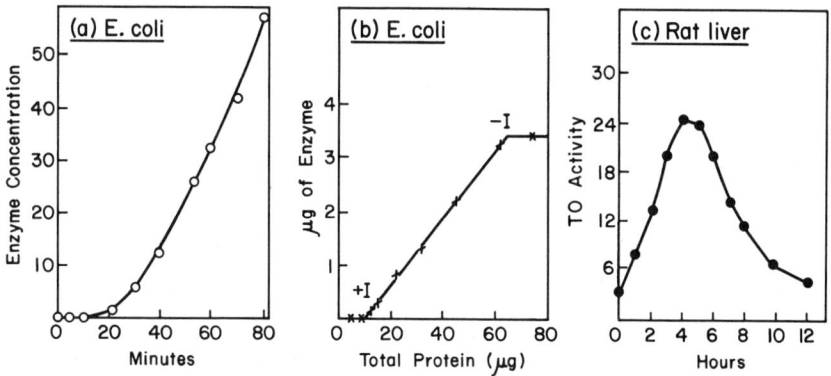

Fig. 1-1 Enzyme adaptation in bacterial and animal cells. The induction of β-galactosidase is expressed as total units (*a*) and as specific activity (*b*). The induction of tryptophan oxygenase is expressed as specific activity (*c*). [The data are reproduced from Benzer (5), with permission of Elsevier/North-Holland Biomedical Press, Cohn (6), with permission of the American Society for Microbiology, and Feigelson et al. (7), with permission of Dr. Feigelson.]

$$\frac{dE}{dt} = k_s - k_D E \tag{1}$$

where k_s is the rate constant for synthesis, k_D is the rate constant for degradation, and E is the concentration of enzyme at time t. [The review of Schimke and Doyle (8) should be consulted for more details on the kinetics of enzyme turnover.] Thus the concentration of an enzyme at any given time is a function if its rate of synthesis minus the fraction of enzyme molecules that have been degraded.

Physiologically, there are two important aspects to the change in the concentration of an enzyme; the magnitude of the change and the time course of the change, since the enzyme must adapt quickly and to a sufficient degree to deal with the environmental change. We can use Eq. (1) to learn more about the mechanisms by which enzymes change their concentration. Perhaps the simplest way to effect an increase in enzyme concentration is to completely stabilize the enzyme against degradation, reducing k_D to zero. Consider, for example, an enzyme with a steady-state basal level of 3.5 units/g of tissue and a k_D of 0.32. Since at steady state there is no net change in enzyme concentration, then dE/dt equals zero and $k_s = k_D E$. Thus $k_s = 1.1$ unit/h. If, in response to an external stimulus, k_D is reduced to zero, then $dE/dt = k_s = 1.1$ unit/h. That is, the concentration of enzyme will increase linearly by 1.1 unit/h until degradation resumes, as is shown in Fig. 1-2a (lower line). An alternative way in which to

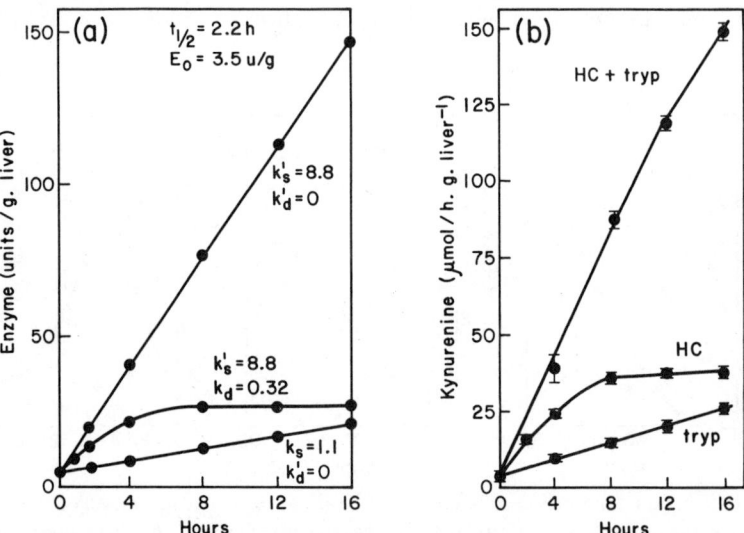

Fig. 1-2 The kinetics of enzyme adaptation. The curves in (*a*) are theoretical curves drawn from the formulas described in the text. The data in (*b*) show the increase in tryptophan oxygenase activity in response to injections of tryptophan, hydrocortisone, or a mixture of both. [Reproduced from Schimke et al. (9), with permission of the American Society of Biological Chemists, Inc.]

achieve an increase in enzyme concentration is to increase its rate of synthesis without changing the rate of degradation. Thus in the example quoted above, k_D remains at 0.32, whereas k_s becomes 8.8 unit/h if the rate of synthesis is increased eightfold. The time course calculated by substituting these figures in Eq. (1) is also shown in Fig. 1-2a (middle curve). It is nonlinear with the new steady-state level of enzyme being reached in about 6 h. If enzyme synthesis is increased eightfold and the enzyme is stabilized simultaneously, then there is a rapid linear increase in enzyme concentration (Fig. 1-2a, upper line).

All three mechanisms for increasing the concentration of an enzyme have been observed in physiological situations. For example, tryptophan oxygenase increases in the liver in response to either tryptophan or an injection of the steroid hormone hydrocortisone. The results of a typical experiment are shown in Fig. 1-2b. A comparison of the curves in Figs. 1-2a and 1-2b leads to the conclusion that tryptophan increases the enzyme by stabilizing the enzyme against degradation, whereas the steroid increases the rate of synthesis of the enzyme without affecting its rate of degradation. Moreover, when both are administered together there is a 50-fold increase in enzyme concentration in 16 h.

To explore further the roles of degradation and synthesis in determining enzyme responses, consider four enzymes, each having a different degradation constant, all being synthesized at an equal rate ($k_s = 0.35$). At steady-state each of the four enzymes has a different concentration, as shown in Table 1-1. Also included is the half-life of the enzyme ($t_{1/2}$) since it is easier to visualize the time taken for an enzyme level to fall by half than it is to visualize the fraction of enzyme molecules lost per unit time (k_D). Figure 1-3a shows the response of each enzyme to a tenfold stimulation of its rate of synthesis ($k_s' = 3.5$) with no change in rates of degradation ($k_D' = k_D$) using the following general solution of Eq. (1):

$$E_t = \frac{k_s'}{k_D'} - \left(\frac{k_s'}{k_D'} - E_0\right)\exp(-k_D' t)$$

as derived by Berlin and Schimke (10).

The time course by which each enzyme approaches the new, tenfold higher, steady-state is actually determined by the *degradation constant* of the enzyme. Enzyme A reaches this in approximately 2 h, whereas enzyme D has only increased 3.5-fold by 48 h (although it will eventually reach the new steady state). Thus, enzymes with large degradation constants (high rates of turnover) respond much more rapidly than enzymes with low degradation constants (low rates of turnover). Moreover, when the stimulus is removed the enzyme with the larger degradation constant returns to basal level much more rapidly (see enzymes B and C, dashed line in Fig. 1-3a). *Thus, adaptive enzymes are likely to be present at low basal levels and have high degradation constants (short half-lives).*

Constitutive enzymes, on the other hand, usually have low degradation constants to obviate the need for unnecessary enzyme turnover and to dampen the effects of general increases in protein synthesis as illustrated in Fig. 1-3b. In this

Table 1-1 Kinetic Constants for the Theoretical Enzymes Described in Fig. 1-3 as Discussed in the Text

Fig. 1-3a Enzyme	k_S (u/g/h)	k_D (h^{-1})	$t_{1/2}$* (h)	E_0 (u/g)	k_S' (u/g/h)
A	0.35	3.47	0.20	0.1	3.5
B	0.35	0.69	1.00	0.5	3.5
C	0.35	0.035	20	10	3.5
D	0.35	0.007	100	50	3.5

Fig. 1-3b Enzyme					
D$_a$	0.35	0.007	100	50	350
D$_b$	0.35	0.007	100	50	175
D$_c$	0.35	0.007	100	50	35
D$_d$	0.35	0.007	100	50	3.5

*$t_{1/2}$ is calculated from $t_{1/2} = \dfrac{\ln 2}{k_D}$

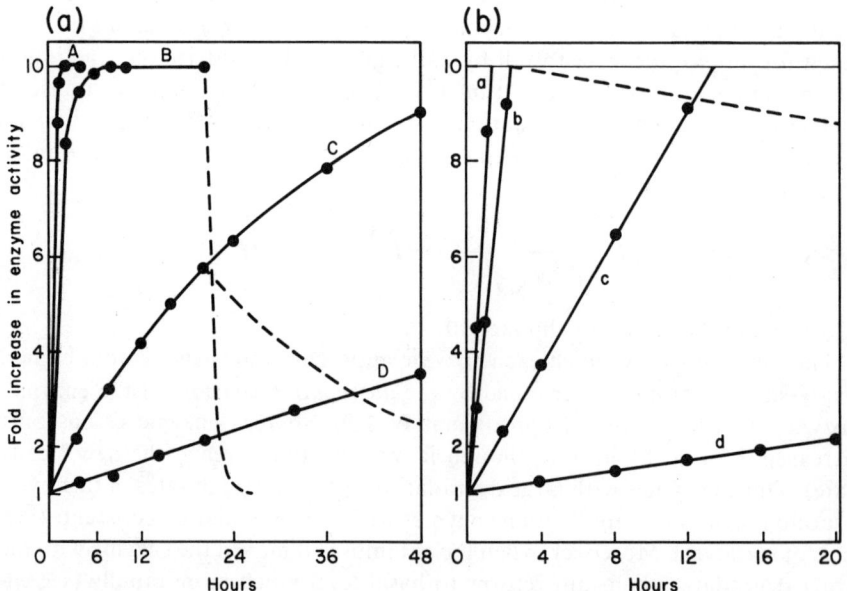

Fig. 1-3 Theoretical curves of enzyme adaptation using the kinetic constants from Table 1-1 as described in the text. (*a*) All four enzymes are induced tenfold at time zero. The dashed lines represent changes in the activities of B and C following withdrawal of the stimulus at 20 h. (*b*) Changes in the activity of enzyme D (long half-life) at four different rates of synthesis.

exercise the effect of changing the rate of synthesis of an enzyme with a low degradation constant (enzyme D from Fig. 1-3a) is examined. The curves are plotted to show the time taken to produce a tenfold increase in enzyme concentration with the various rates of synthesis given in Table 1-1. Curve c, for example, shows that the rate of enzyme synthesis must be increased 100-fold to produce a tenfold increase in enzyme concentration in 14 h. To produce a tenfold increase in concentration in 2 h the rate of synthesis must be increased 500–1000-fold! Such rate increases are not encountered in mammalian cells, although they are not uncommon in rapidly growing bacterial cells, where the rate of degradation is quite low. Mammalian cells can achieve the same increase in 2 h using an enzyme with a much higher degradation constant (enzyme A in Fig. 1-3a), and the rate of synthesis must be increased only tenfold. Clearly, it is also more economical to change the synthesis of an enzyme with a large k_D.

Moreover, because the activity of an enzyme with a low degradation constant returns to basal levels much more slowly following withdrawal of the stimulus (dashed line in Fig. 1-3b), an inhibitor would be required to prevent continued catalysis of that reaction. In bacterial cells this is less of a problem because once the substrate in the medium is depleted the reaction stops anyway, and the enzyme is rapidly diluted by the growth of cells without enzyme.

In tissues such as muscle that have little capacity for adaptation some enzymes have half-lives in excess of 100 days. In other cells such as hepatocytes there is not such a sharp distinction between the two classes of enzymes, and a whole spectrum of degradation constants is encountered, ranging from 3.78 ($t_{1/2}$ = 11 min) for ornithine decarboxylase to 0.0015 ($t_{1/2}$ = 20 days) for NAD glycohydrolase. Detailed lists of enzyme half-lives have been prepared by Rechcigl (11).

A kinetic analysis of changes in enzyme activity in response to exogenous stimuli can provide insight into mechanisms of enzyme adaptation in animal cells. For example, glucocorticoid hormones are known to have an effect on the levels of two enzymes of amino acid catabolism: tyrosine aminotransferase ($t_{1/2}$ = 2.0 h) and tryptophan oxygenase ($t_{1/2}$ = 2.2 h). In a study of this effect Berlin and Schimke (10) followed the changes in activity of these two enzymes in liver during a period of cortisone administration. They also included two other enzymes of amino acid catabolism, glutamate–alanine transaminase ($t_{1/2}$ = 48 h) and arginase ($t_{1/2}$ = 96 h), which have low k_D's. The changes in activity are shown in Fig. 1-4. The two enzymes with short half-lives responded very quickly, reaching maxima within a few hours, whereas during the same time period the changes in the other enzymes were negligible, leading one to the conclusion that the steroid only induced tryptophan oxygenase and tyrosine amino transferase. However, when the experiment was continued for a number of days both the transaminase and arginase increased. Analysis of these time courses revealed that the steroid was actually acting to immediately increase the rate of synthesis of *all four* enzymes by the same extent (four- to sevenfold). The time course of each enzyme was determined by its k_D since the steroid had no effect on degradation.

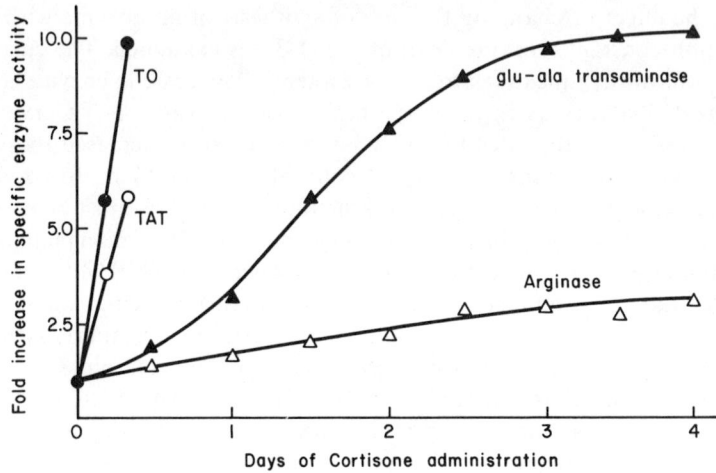

Fig. 1-4 The effect of cortisone on the enzymes of amino acid catabolism in liver. TO = tryptophan oxygenase, TAT = tyrosine aminotransferase, glu–ala = glutamine–alanine. [Reproduced from Fig. 2 of Berlin and Schimke (10), with permission.]

MEASUREMENT OF ENZYME SYNTHESIS

Estimates of k_D and k_s can be obtained from the time courses of the change in activity. The degradative phase of the time course is used to calculate k_D, and k_s is calculated from k_D and E_0. However, several assumptions are made with this approach. For example, degradation does not necessarily take place in the absence of synthesis once the inducer is removed. It is also assumed that the basal condition represents a true steady-state. More importantly, we have assumed that changes in activity measured by *in vitro* assay reflect changes in the number of enzyme molecules *in vivo*. Surprisingly accurate information has been obtained, however, despite these potentially serious assumptions.

It must be emphasized that a change in *activity* cannot be directly equated with a change in enzyme *concentration*. Of the seven ways in which an increase in catalytic activity can be achieved, only three require an increase in the synthesis of new enzyme molecules (Fig. 1-5). These are ①—a transcriptional increase in the synthesis of precursor mRNA, ②—an increase in the processing of precursor mRNA to mature mRNA (or perhaps an increase in the selection of a particular mRNA from a pool of rapidly turning-over RNA in the nucleus), or ③—an increase in the rate at which the mRNA is translated into enzyme molecules. The other four ways all increase enzyme activity without increasing the rate of enzyme synthesis. For example, the translation product—④—may be an inactive holoenzyme or zymogen requiring proteolysis or some other step to generate a fully active enzyme. Furthermore, the catalytic efficiency of the enzyme may be increased without a change in its concentration either by ⑤—increasing the affinity or accessibility of the active site for the substrate or

⑥—by facilitating cooperative interactions between subunits. Decreasing the rate of enzyme degradation—⑦—leads to an accumulation of enzyme molecules without changing the rate of synthesis.

A direct measure of the number of enzyme molecules is the unequivocal solution to this problem, and immunoprecipitation of the enzyme with a specific antibody is the method of choice. Moreover, the rate of synthesis of the enzyme can be estimated by carrying out immunoprecipitation when the enzyme molecules have been labeled with a radioactive precursor. This methodology is discussed extensively in the review of Schimke and Doyle (8), and can also be adapted to obtain true rates of enzyme degradation.

Satisfactory results can only be obtained if the pulse of labeling is very brief, particularly for those enzymes that turn over rapidly. A great deal of work can be saved by careful choice of precursor. If lysine or leucine is used the specific activity of the intracellular amino acid pool must be measured, a tedious procedure. However, if arginine is used it rapidly equilibrates with urea, whose specific activity is more easily determined. Moreover, arginine is not substantially reutilized, which obviates the need to correct the data. Other precursors commonly used are bicarbonate, which is rapidly converted to arginine, and the sulfur-containing amino acids, labeled with ^{35}S, since they equilibrate with inorganic sulfur in the cell, and sulfur's specific activity is also easily determined.

Only those enzymes for which these data are available are used as examples in subsequent chapters. It is unfortunate that of the many enzymes for which activity changes have been demonstrated, only a few have been adequately demonstrated to be actual changes in enzyme synthesis.

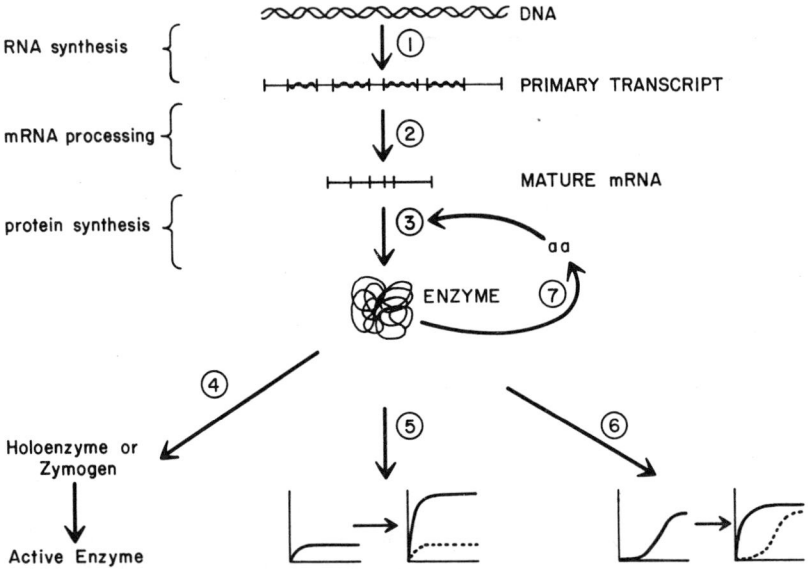

Fig. 1-5 The seven ways by which an increase in enzyme activity can be achieved. Each step is described in the text.

THE ROLE OF MESSENGER RNA

Enzyme proteins are synthesized by translation of their respective mRNAs and there are two ways to effect an increase in enzyme synthesis—either increase the rate of translation of preexisting mRNAs or increase the number of mRNA molecules available for translation (either by transcription or by processing). Consequently, a measure of mRNA concentration is an important factor in understanding the nature of enzyme synthesis.

There are two techniques for the measurement of mRNA concentration. The first involves isolation of an mRNA fraction that can be used in an *in vitro* protein translation assay. Antibody to the enzyme being studied is used to identify the amount synthesized, which is directly proportional to the concentration of its mRNA. The second technique involves the synthesis of a radioactively labeled DNA copy (cDNA) of the purified mRNA. Since the mRNA must first be isolated and purified, however, this approach is limited to only those mRNAs with sufficiently unusual characteristics (abundance, size, etc.) to be purified. Advances in the application of recombinant DNA techniques that permit individual cDNAs to be cloned and subsequently isolated are rapidly overcoming some of these problems (see Ref. 12, for example). Pure, radioactively labeled cDNA is then used to hybridize to cell extracts to measure the amount of its mRNA in the same way as antigen–antibody precipitation measures enzyme protein concentration.

In the last two or three years these methods have been used to measure the concentrations of the mRNAs of adaptive enzymes in animal cells. In every case so far, an increase in mRNA concentration precedes the increase in enzyme

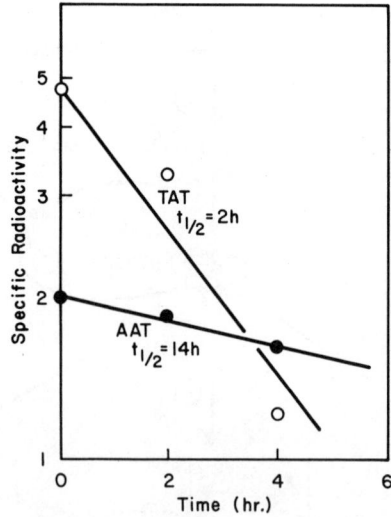

Fig. 1-6 Differences in the turnover of animal cell messenger RNAs. TAT = tyrosine aminotransferase, and AAT = alanine aminotransferase. Note that the ordinate is a log scale. [Reproduced from Stiles et al. (13), with permission.]

activity (see Fig. 12-6, for example). Thus a tenfold increase in the rate of enzyme synthesis requires a tenfold increase in the concentration of its mRNA on the polysomes. The turnover of mRNA has not yet been studied in sufficient detail to distinguish between increases in its rate of synthesis following stimulation or decreases in its rate of degradation. The synthesis of mRNA appears to be zero order, and degradation appears first order. Thus estimates of turnover can be made using Eq. (1) as shown in Fig. 1-6 for mRNA^{TAT} and mRNA^{AAT}. These measurements contribute greatly to our understanding of enzyme synthesis and its regulation.

Bacterial cell mRNAs are very unstable ($t_{1/2}$ = 1–2 min) and difficult to isolate, rendering measurements of their concentration almost impossible. However, since transcription and translation are so tightly coupled in bacterial cells it is usually the rate of transcription of the gene that determines the rate of enzyme synthesis. The kinetics of this system are usually examined *in vitro* with isolated gene transcription–mRNA translation-coupled systems (14).

REFERENCES

1 H. Karström (1936). The problem of adaptation in bacteriology. *Proceedings of the 2nd International Congress on Microbiology (London)*, pp. 473–474.
2 R. J. Dubos (1940). The adaptive production of enzymes by bacteria. *Bact. Rev.* **4**, 1–16.
3 M. Cohn, J. Monod, M. R. Pollock, S. Spiegelman, and R. Y. Stanier (1953). Terminology of enzyme formation. *Nature* **172**, 1096.
4 H. J. Vogel (1957). Repressed and induced enzyme formation: a unified hypothesis. *Proc. Natl. Acad. Sci. USA* **43**, 491–496.
5 S. Benzer (1953). Induced synthesis of enzymes in bacteria analysed at the cellular level. *Biochim. Biophys. Acta* **11**, 383–393.
6 M. Cohn (1957). Contributions of studies on the β-galactosidase of *Escherichia coli* to our understanding of enzyme synthesis. *Bact. Rev.* **21**, 140–168.
7 P. Feigelson, T. Dashman, and F. Margolis (1959). The half-lifetime of induced tryptophan peroxidase *in vivo*. *Arch. Biochem. Biophys.* **85**, 472–482.
8 R. T. Schimke and D. Doyle (1970). Control of enzyme levels in animal tissues. *Ann. Rev. Biochem.* **39**, 929–976.
9 R. T. Schimke, E. W. Sweeney, and C. M. Berlin (1965). The roles of synthesis and degradation in the control of rat liver tryptophan pyrrolase. *J. Biol. Chem.* **240**, 322–331.
10 C. M. Berlin and R. T. Schimke (1965). Influence of turnover rates on the response of enzymes to cortisone. *Molec. Pharmacol.* **1**, 149–156.
11 M. Rechcigl, Jr. (1970). Intracellular protein turnover and the roles of synthesis and degradation in regulation of enzyme levels. In M. Rechcigl, Jr., Ed., *Enzyme synthesis and degradation in mammalian systems*. University Park Press, Baltimore, pp. 236–310.
12 J. R. Parnes, B. Velan, A. Felsenfeld, L. Ramanathan, O. Ferrini, E. Appella, and J. G. Seidman (1981). Mouse $β_2$-microglobulin cDNA clones: A screening procedure for cDNA clones corresponding to rare mRNAs. *Proc. Natl. Acad. Sci. USA* **78**, 2253–2257.
13 C. D. Stiles, K. Lee, and F. T. Kenney (1976). Differential degradation of messenger RNAs in mammalian cells. *Proc. Natl. Acad. Sci. USA* **73**, 2634–2638.
14 G. Zubay (1973). In vitro synthesis of proteins in microbial systems. *Ann. Rev. Genet.* **7**, 267–287.

SECTION ONE

Regulation of Enzyme Synthesis at the Level of Transcription

The first stage of enzyme synthesis is the transcription of the gene coding for the polypeptide. In a bacterial cell this requires the activation of a single gene from a population of approximately 4000. In a mammalian cell one gene must be selected from enough DNA to code for one million genes. The selection and activation processes are directed by DNA binding proteins that regulate the binding to, and transcription of, the gene by RNA polymerase.

Research into the molecular biology of the mechanisms operating in bacterial cells has progressed rapidly in the past 20 years, and we now have a reasonable understanding of their general characteristics, although many fine details are still to be worked out. Until recently our knowledge of the mechanisms operating in mammalian cells was virtually nonexistent. Moreover, there seemed little hope of being able to gain any knowledge because the genetic techniques applied so successfully to bacterial cells were not suitable for eukaryotes. However, the past few years have seen the application of new techniques that are powerful enough to permit a biochemical and "genetic" analysis of the eukaryotic genome. These techniques include biochemical analysis of the genome using specific nucleases to cut the DNA into manageable pieces; recombinant DNA methods to isolate, purify, and amplify specific segments of DNA that contain active genes; "*in vitro*" or "surrogate" genetical analysis of these DNA fragments; the development of *in vitro* transcription systems to test the activity of these "mutated" genes; improvements in DNA sequencing methodology; and improvements in electron microscopy that permit an examination of chromatin structure. Many of these techniques are described in detail by Lewin (Ref. 1 of Chapter 3).

This section is divided into three chapters. Chapter 2 describes how transcription is regulated in bacterial cells. It deals with the proteins involved and how they interact with both the DNA of the genome and the nascent RNA transcript. Since the bacterial genome is so small, proteins are able to locate and bind to specific DNA sequences and activate individual genes or groups of genes, called *operons*. The molecular biology of these interactions is well understood in terms of the sequences that are recognized, but is less well understood in terms of the structural changes in the genome that must accompany gene activation.

Chapters 3 and 4 deal with the regulation of transcription in eukaryotic cells. DNA is organized into a specific organelle, the nucleus, and it is now apparent that the structural organization of DNA in the nucleus is a major factor affecting its transcriptional activity. Chapter 3 describes our current knowledge of chromatin structure as a prelude to the discussion of the regulation of transcription in Chapter 4. For a long time it was assumed that the process of transcription in animal cells was very similar to that in bacterial cells. This is no longer the case. The overwhelming amount of DNA that has accumulated in the higher eukaryotes has placed constraints on the way in which it can be expressed. We are only just now beginning to appreciate the mechanisms involved.

CHAPTER TWO

Regulation of Gene Transcription in Bacterial Cells

The genome of a bacterial cell like *Escherichia coli* is a single chromosome of approximately 5×10^6 nucleotide base pairs. It is approximately 1 mm in length, and in many bacteria the ends join to form a loop. Since a bacterial cell is about 2 μm in length, packing the DNA into the cell presents a considerable problem. It is folded extensively into a highly compacted body called the nucleoid, as shown in Fig. 2-1. The extent of this folding is readily apparent in cells from which the DNA has been "liberated" by gentle lysis of the cell membrane (1). Although the nucleoid appears restricted to a certain part of the cell, there is no nuclear membrane, and the DNA is free to interact with cytoplasmic proteins. The shape of the chromosome is determined by this interaction and changes continuously.

This single chromosome contains all the information required for the survival of the cell. Extensive genetic analysis of the *E. coli* chromosome has located the positions of more than 600 of the 4000 possible genes. A genetic linkage map of *E. coli* with the loci of some of the genes that are described in this textbook is shown in Fig. 2-2. A more complete map is given in the review by Bachmann et al. (2).

THE BACTERIAL GENOME

The Structure of DNA in the Nucleoid

The Watson–Crick double helix of DNA exists, under most physiological conditions, in the B-configuration (as shown in Fig. 2-3). In this state there are approximately 10 base pairs per helical turn, with the base pairs 3.4 Å apart. The helix has a major groove (22 Å wide) and a minor groove (12 Å wide) in each turn. DNA, however, is not a rigid, rodlike structure but, rather, is capable of taking up a variety of different structures. For example, it can exist in the A-configuration (Fig. 2-3a), in which the shapes of the grooves are greatly altered. Moreover, the entire helix can be wound into a superhelix (100–120 Å in diameter). This supercoiled structure of DNA is maintained *in vivo* by its

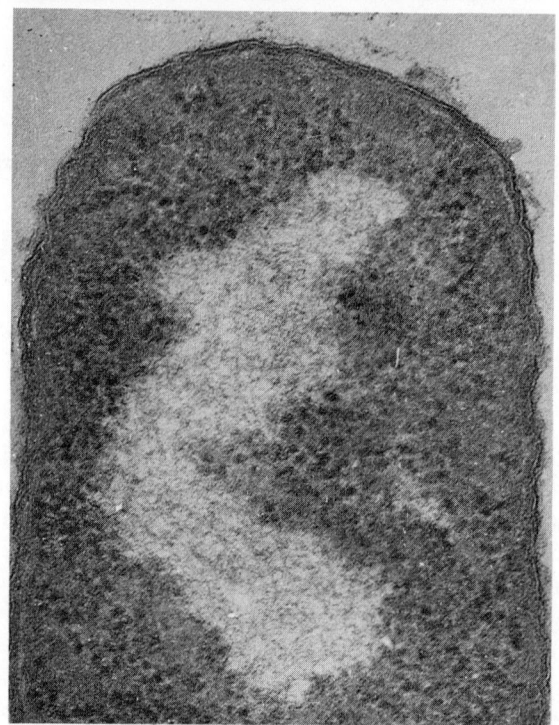

Fig. 2-1 The genome of *E. coli* compacted into a nucleoid structure (×90,000). Photograph kindly supplied by Dr. A. Ryter [Reproduced, with permission, from A. Ryter and A. Chang (1975). *J. Mol. Biol.* **98**, 797–810. Copyright by Academic Press Inc., London.]

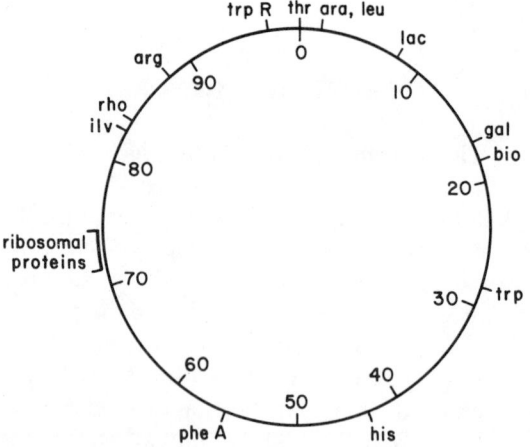

Fig. 2-2 Abbreviated linkage map of the *E. coli* genome.

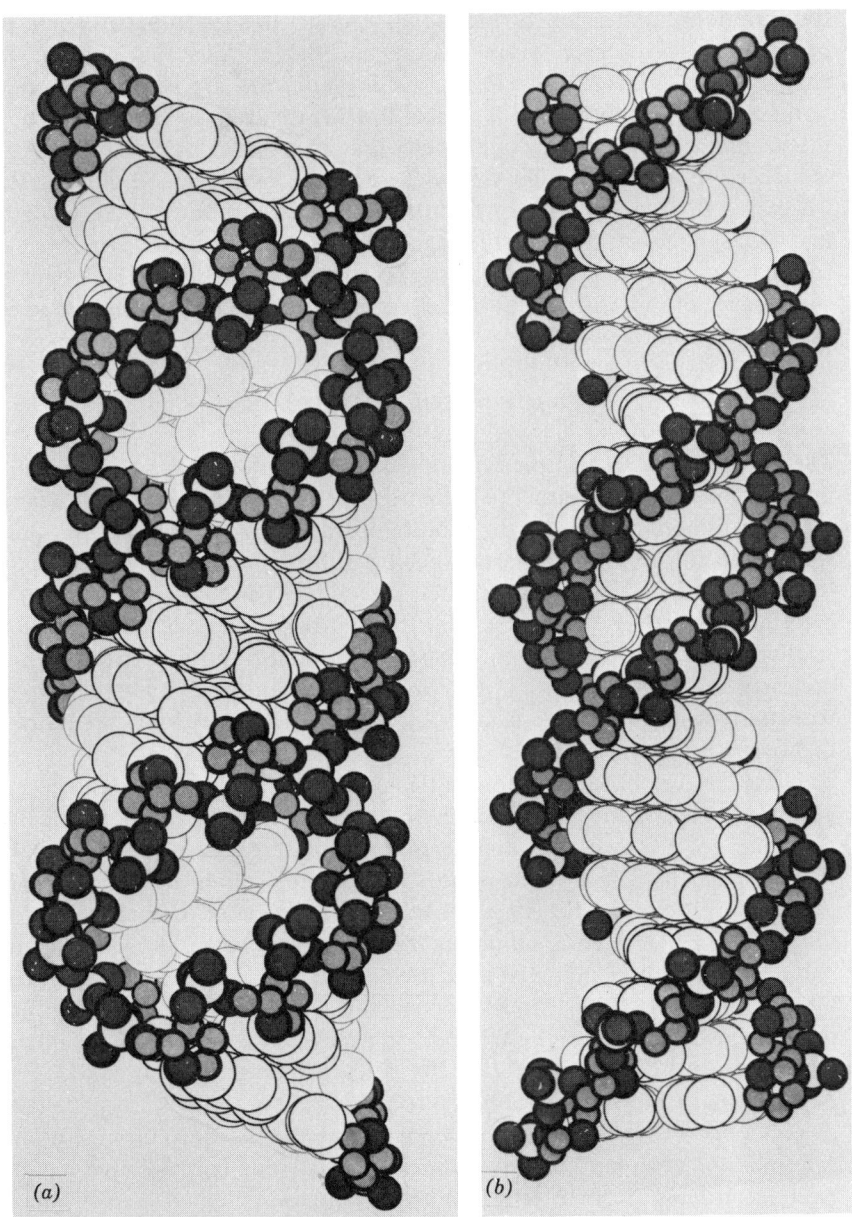

Fig 2-3 Two examples of the different conformations of the DNA double helix. (*a*) A-DNA; (*b*) B-DNA. Photographs kindly supplied by Dr. R. D. Wells. [Reprinted with permission from R. D. Wells (1977). *CRC Crit. Rev. Biochem.* **4**, 305–340. Copyright 1977 The Chemical Rubber Co., CRC Press, Inc.]

associated proteins and polyamines, giving rise to the characteristic "beads on a string" appearance of the bacterial chromosome (1).

It is becoming evident that the structure of DNA plays an important role in transcriptional activation of individual genes. Changes in supercoiling have been shown to influence transcription, and the enzymes that catalyze these changes, such as DNA gyrase and unwinding enzymes, are assuming regulatory significance (4, 5). Other levels of DNA structure are also important, such as the local deviations in the double helix that are created by certain DNA base sequences. For example, an A:T-rich sequence favors the B-configuration, whereas a G:C-rich sequence favors the A-configuration. These deviations may act as recognition sites for regulatory proteins that are otherwise confronted with a very uniform DNA sugar–phosphate backbone. The regulatory significance of these structural features of DNA in the nucleoid is discussed in the next section.

Protein–Nucleic Acid Interactions

The functions of DNA in the cell are to store information and to transmit this information accurately either to daughter cells, by the process of replication, or into a translatable form, by the process of transcription. The necessary precision of this information transfer is achieved through the pairing of the purine bases, adenine and guanine, with the pyrimidine bases, thymine (uracil) and cytosine, respectively. The bases pair by hydrogen bonding, which, although relatively weak in strength, only exists when donor and acceptor groups are precisely aligned. The transfer of information is controlled by regulatory proteins that must also be able to interact with these sequences of base pairs in a highly specific manner.

In the DNA double helix most of the hydrogen-bonding donor and acceptor groups are involved in interstrand base-pairing and, being on the inside of the molecule, are not available for interaction with proteins. However, each base has potential hydrogen-bond donors or acceptors along its "edges," which project out into either the major or the minor groove of the helix. Figure 2-4 illustrates the hydrogen-bonding capacity of the four bases, showing the donor and acceptor groups involved in helix base-pairing and the ones that project into the grooves. Other atoms, for example, those of the methyl group of thymine, also project into the grooves and may be involved in steric interactions that improve the recognition capacity of each base pair. Only the G:C and A:T base pairs are shown in the figure, but the T:A and C:G base pairs also project a unique configuration of donor and acceptor groups into the grooves. Thus all four base-pair combinations can be individually recognized by DNA-binding proteins. Seeman et al. (5) analyzed these potential recognition sites and concluded that specific interaction could take place provided that at least two side groups are hydrogen-bonded for each base pair.

DNA-binding proteins must possess amino acid residues capable of interacting with these edges of the base pairs. This issue was also addressed by Seeman

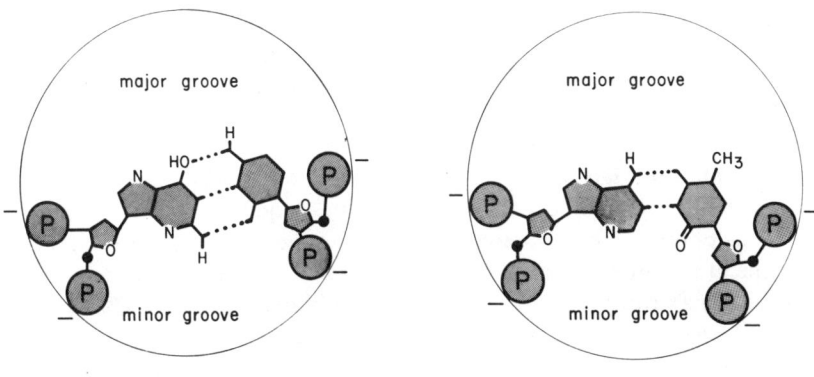

Fig. 2-4 The projection of potential hydrogen-bond donor/acceptor groups into the major and minor grooves of the DNA double helix. G:C and A:T base pairs are shown. The circle around each base pair is the diameter of the helix. Note the negatively charged phosphate groups, always on the outside of the helix. [Reprinted, with permission, from the November 1973 *Bioscience*, © American Institute of Biological Sciences (7).]

et al. (5), and they demonstrated that glutamine and asparagine could interact directly with, for example, adenine in the major groove and guanine in the minor groove, as shown in Fig. 2-5. Asparagine can also interact with guanine in the major groove, and arginine hydrogen-bonds in a similar way. Furthermore, by virtue of its tertiary structure, a protein may contribute suitable donors and acceptors from different amino acid residues into a configuration that can interact with the nucleotide bases.

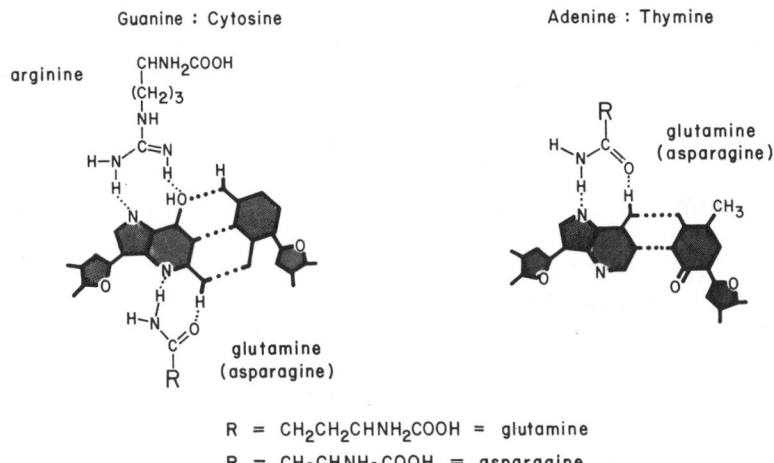

Fig. 2-5 Hydrogen bonding between amino acids and the "edges" of G:C and A:T base pairs. The dotted lines represent hydrogen bonds. The bases (shaded) are in the same configuration as in Fig. 2-4. [Adapted from Figs. 2 and 3 of Ref. 6, with permission.]

When considering the nature of protein–nucleic acid interactions it is essential to consider the influence of changes in DNA conformation. DNA is a long polymer that is in a dynamic state, continuously undergoing changes in shape that will alter its ability to interact with regulatory proteins. This phenomenon has received more attention recently, and there are reviews by Wells et al. (8), Sobell (9), and Von Hippel (10), as well as a book on molecular stereodynamics (11) devoted to this subject. For example, the sequence of nucleotides may change the polymer from the B-configuration to the A-configuration, changing the shape of the grooves of the helix and profoundly affecting protein binding. Thus sequences rich in G:C are likely to have binding characteristics different from A:T-rich sequences. Moreover, the binding of one protein to the DNA polymer may induce a conformational change in DNA that affects the binding of a second protein by altering the structure of its binding site. Such conformational changes may be "transmitted" for considerable distances along the DNA helix.

Hydrogen bonds are not the only types of molecular interaction between proteins and double-stranded DNA (or RNA). The aromatic amino acids tyrosine, tryptophan, phenylalanine, and histidine may intercalate into the hydrophobic environment between the base pairs of the helix. Like hydrogen bonds, however, hydrophobic interactions are relatively weak, although they may play an important role in the sequence-specific recognition of DNA by a protein.

Most of the binding energy between nucleic acids and proteins comes from charge–charge interactions between positively charged amino acid residues (arginine, lysine, histidine) on the protein and the negatively charged phosphates on the DNA backbone. Each bond provides up to 6 kcal/mol toward the 15–20 kcal/mol required for the high-affinity binding that characterizes the interactions between regulatory proteins and nucleic acids. Ionic bonding is sequence-independent, since the DNA backbone is quite uniform in character. It is possible, however, that a protein may recognize regions of local change in helix configuration.

Sequence-Specific Protein Binding

A protein that is required to interact with a unique nucleotide base sequence must recognize and bind to that site much more tightly than to other similar sequences of the chromosome. Von Hippel (10) has addressed the problem of the number of nucleotide pairs that are required to define a unique sequence. He calculates that a sequence of 12 nucleotides is the minimum number that will define a unique sequence on a genome of 10^7 base pairs. Since 12 nucleotides is about 40 Å in length, specific binding proteins must have active sites large enough to accommodate this sequence. Figure 2-6 shows that a tetrameric protein of $M_r = 150,000$ could accommodate such a sequence quite easily.

For each unique sequence of 12 nucleotides on the chromosome there are 12 other sequences in which only one nucleotide is different (together with many other sequences differing by 2, 3, 4, . . . nucleotides), and the protein will also

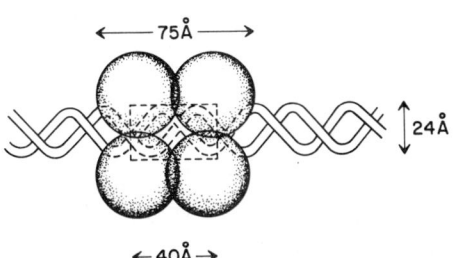 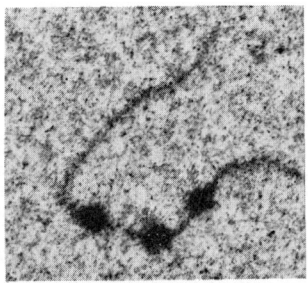

Fig. 2-6 The interaction between a tetrameric protein and the DNA double helix showing the size requirement of the active site and the relative size of the two macromolecules. The insert shows an electron micrograph (×200,000) of three RNA polymerase molecules bound to promoter DNA. [From Fig. 1 of C. Bordier and J. Dubochet, (1974). *Eur. J. Biochem.* **44**, 617, with permission.]

have a high affinity for these sequences. Thus the protein–nucleic acid interaction must be such that even if one base is incorrect the binding energy is lowered so drastically that it binds much less tightly. This can usually be achieved if one ionic bond is disrupted or prevented from forming.

The binding of a protein to nucleic acid is described by its association constant K; $K = k_a/k_b$, where k_a is the rate constant for association, and k_b is the rate constant for dissociation. For a tightly bound protein k_a is very high, with values of 10^{10} M^{-1} s^{-1}, whereas k_b is much lower (~10^{-3} s^{-1}, equivalent to a half-life of 20–30 min), giving an overall association constant of 10^{13} M^{-1}. Binding to nonspecific sites is also of high affinity, with association constants of 10^7–10^{10} M^{-1}. Therefore the protein spends much of its time bound to DNA rather than free in solution. This raises the question of how the protein finds its *unique* 12-base sequence on a chromosome that is 10^7 base pairs long. Studies on the rates at which DNA-binding proteins locate their specific sites (10) indicate that rate constants of 10^{10} M^{-1} s^{-1} are much faster than would be expected for a process mediated by simple diffusion. Thus the reaction must be facilitated in some way. Several models (reviewed in Ref. 10) have been proposed in which the protein is transferred between loops of the folded chromosome as they come into contact with each other, until the protein reaches its unique site (or binds near enough to slide along the DNA to the proper site). Such facilitated diffusion has rate constants similar to those observed experimentally for repressor binding.

THE MECHANISM OF TRANSCRIPTION

Transcription is the process that generates an RNA copy of a DNA template. It involves the binding of DNA-dependent RNA polymerase "upstream" from the structural gene, followed by the translocation of the enzyme along the gene, generating an RNA polymer complementary to the coding strand of the DNA double helix. The polymerase then disengages from the DNA, and the completed RNA copy is released. The basis of transcription is this interaction between the polymerase molecule and the DNA template—*gene transcription is regulated by changing the rate at which this reaction takes place*. The structure of the two components of the system, polymerase and the DNA template, will now be described.

RNA Polymerase

RNA polymerase catalyzes the stepwise addition of ribonucleoside monophosphates to a growing RNA chain. The chain is synthesized in the 5' to 3' direction complementary to the *3' to 5'* sense strand of DNA. The bacterial enzyme exists in two states; as a core enzyme that, although it can carry out chain elongation, cannot initiate transcription, or as a holoenzyme that has the ability to recognize and bind specifically to promoter sites and initiate RNA synthesis. The core enzyme has four subunits, $\alpha\alpha\beta\beta'$, and is converted into the holoenzyme by the binding of a fifth subunit, designated sigma (i.e., $\alpha\alpha\beta\beta'\sigma$).

The function of each subunit is not yet completely understood, and there are several reviews summarizing current knowledge (12–15). The β and β' subunits have M_r = 155,000 and 165,000, respectively, with the β subunit being responsible for binding the nucleoside triphosphates and the positively charged β' subunit being involved in DNA binding. However, binding at specific promoter sites only occurs when the σ subunit (M_r = 90,000) is bound to the enzyme. In the cell there are fewer molecules of σ subunit than there are of core enzyme (approximately 0.33:1), making the association between the two a key step in the initiation of transcription. The σ subunit alone has no DNA binding capacity. Presently, the function of the α subunits (M_r = 37,000) is not known.

The binding of polymerase holoenzyme nonspecifically to DNA is relatively weak (K = $10^7 M^{-1}$); however, upon location of a promoter, in the presence of σ, there is a rapid ($t_{1/2}$ = 0.2 s) initiation reaction, and transcription commences. Recent work has shown that as transcription proceeds the association constant of the enzyme increases to K = $10^{11} M^{-1}$, and after eight or nine nucleosides have been polymerized, σ is displaced (16). It is likely that the core enzyme makes additional ionic bond contacts with the DNA backbone, accounting for the increase in affinity, and supplying the energy for displacement of σ. The free σ subunit subsequently binds to another molecule of core enzyme, and another round of initiation of transcription takes place.

The DNA Template and the Initiation of Transcription

Until about 20 years ago, although the positions of many genes on the *E. coli* chromosome had been located, little was known about how the genetic information was used to produce proteins. In 1961 Jacob and Monod published papers (17, 18) that described a model for the expression of genetic information and its regulation, with particular reference to the *lac* operon. They introduced the concept of two kinds of gene: *structural genes* that coded for the actual protein and *regulatory genes* that controlled the expression of the structural genes. The regulatory gene governs the synthesis of an intracellular substance that inhibits information transfer from structural gene to protein. The intracellular substance, called the repressor, binds to a site near the structural gene and prevents its transcription. This site, called the operator, lies in between the structural gene and the site of RNA polymerase binding, which is called the promoter. Jacob and Monod also postulated that information transfer from gene to protein involves an unstable intermediate that they equated with the unstable RNA species previously discovered by Volkin and Astrachan (19). This unstable RNA was designated messenger RNA.

The concept of the operon was also introduced and defined as

two or more structural genes which behave as a *unit* in the transfer of information.

For example, lactose induces not only β-galactosidase, but also lactose permease and thiogalactoside transacetylase in a coordinate manner. The structure of the *lac* operon was proposed to be

where z, y, and a are the structural genes for the three enzymes, and p and o are the promoter and operator, respectively. Discussion of the biochemical and physiological aspects of the *lac* operon can be found in Chapter 8; the present discussion is confined to the interaction of RNA polymerase with the portion of the DNA template known as the promoter.

The application of sophisticated genetic fine-mapping techniques [reviewed by Lewin (20) and by Beckwith (21)] confirmed that the promoter was contiguous with the first structural gene, the z gene (coding for β-galactosidase). This entire region of the *lac* operon has now been sequenced [as have the promoter-operator regions of a number of other operons (10, 22–24)]. The nucleotide sequence of this regulatory region of the *lac* operon is shown in Fig. 2-7.

RNA polymerase is thought to make contact at three distinct sites on the promoter as shown in the following sequence (base pairs 44–88 from Fig. 2-7):

```
CAG GCTTTACATTT ATGCTTCCGGCTCG TATGTTGT TGTG GAATT
GTC CGAAATGTAAA TACGAAGGCCGAGC ATACAAC ACAC CTT TAA
```

The 11 base pairs boxed-in on the left are thought to be the binding site of the σ subunit of the holoenzyme. The center box is a site of attachment of the core enzyme, often referred to as the Pribnow box (22). When the enzyme is correctly aligned with these two binding sites the active site of the enzyme is brought into contact with the DNA about four or five bases downstream from the Pribnow box, and initiation of transcription takes place.

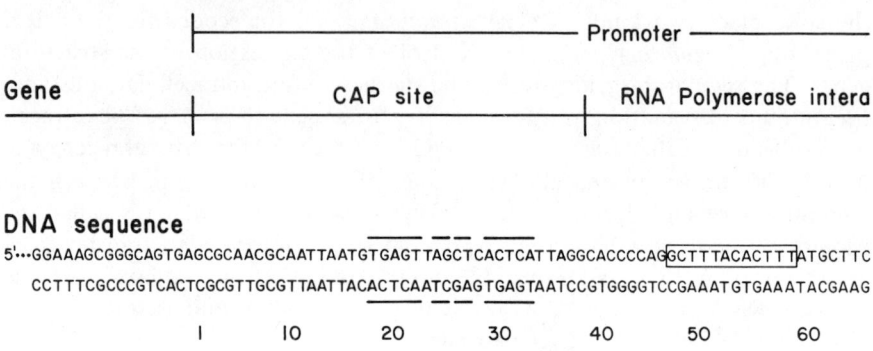

Fig. 2-7 The nucleotide sequence of the regulatory region of the *lac* operon of *E. coli*. Both strands of DNA are shown together with the sequence of mRNA and the first three amino acids of β-galactosidase. The horizontal lines above and below the DNA se-

The affinity of binding between polymerase and the promoter is determined by the precise nucleotide sequence of these three points of contact. Thus some promoters are very strong, ensuring efficient initiation of transcription, whereas other promoters have considerable sequence diversity in these areas, and polymerase molecules bind weakly.

The polymerase needs access to the nucleotide sequence of the sense strand so that it can make a complementary RNA copy. To achieve this the two strands of the helix are "melted" by the polymerase molecule (an expression used to describe strand separation *in vivo* that is equivalent to that achieved *in vitro* by heating), with approximately 11 base pairs being separated. The mechanism involves the formation of ionic contracts between the DNA backbone and the enzyme that are strong enough to pull the strands apart. The nucleotide sequence near the transcription-start site is usually rich in the relatively weak A:T base pairs to facilitate the initial melt. Once open, the first complementary ribonucleoside triphosphate (usually ATP or GTP) binds to polymerase, and transcription begins. The enzyme moves along the DNA template, and chain elongation occurs in a stepwise fashion. As the core enzyme translocates along the gene the strands are melted ahead of the active site of the enzyme, exposing the sense strand. The two strands reanneal after the enzyme has passed. Thus, although the initial melt occurs at an A:T-rich site, subsequent melting is sequence-independent.

When a polymerase molecule has translocated away from the promoter-operator region the site becomes available for another molecule of enzyme. Indeed, it is not uncommon to find several polymerase molecules actively transcribing the same gene, as illustrated in Fig. 2-8 (25). This electron micrograph

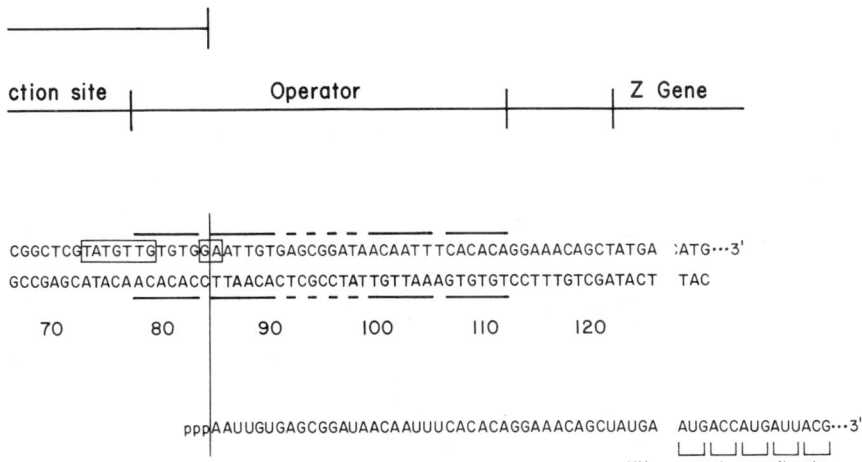

quence are the palindromic, protein-binding sites for CAP and for repressor as explained in the text. The boxed sequences are the sites of RNA polymerase binding. [Redrawn from Fig. 4 of Ref. 23, with permission.]

also shows that the mRNA transcript is often translated before transcription is complete. Polymerase molecules transcribe the gene until they reach a DNA sequence that signals the end of the gene, transcription stops, and the completed mRNA chain is released.

Transcription Termination

The DNA sequence at the site of termination of a number of genes has been found to show certain common characteristics, namely, a region of DNA that is rich in G:C base pairs and possesses dyad symmetry followed immediately by

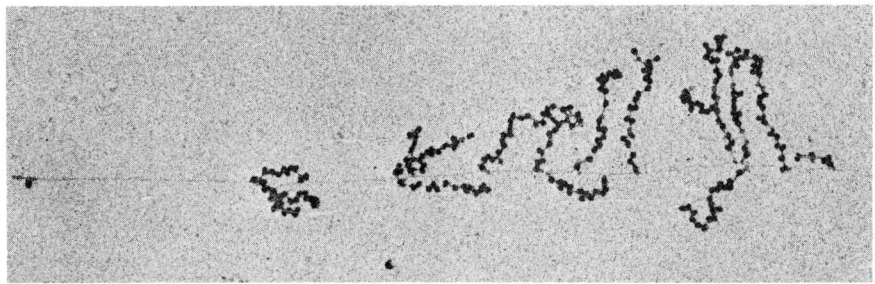

Fig. 2-8 Bacterial genes in action. The gene is being transcribed from left to right, generating longer and longer transcripts. The electron-dense particles on the transcripts are ribosomes actively engaged in translation. ×43,036. [Reprinted from Fig. 2 of Miller et al. (1970). *Science* **169**, 392–395, with permission. Copyright 1970 by the American Association for the Advancement of Science. The photograph was kindly supplied by Professor O. L. Miller, Jr.]

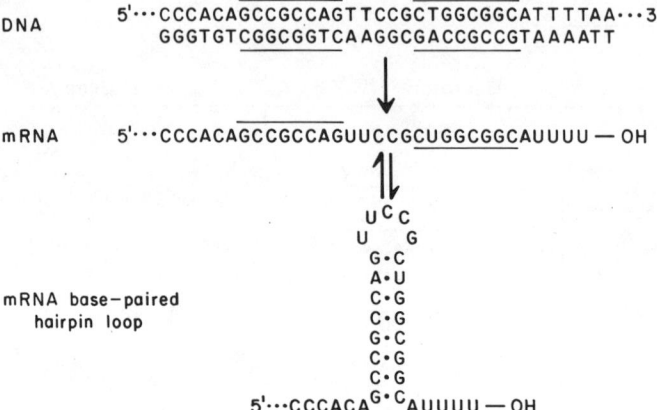

Fig. 2-9 The terminator of the tryptophan operon showing the sequence of DNA, the mRNA transcript, and the base-paired secondary structure of the mRNA transcript. The sequences with dyad symmetry are underlined.

a region of DNA containing a number of consecutive A:T base pairs as shown in Fig. 2-9 for the terminator of the tryptophan operon. Such a sequence confronts RNA polymerase with an unusual configuration of the helix, causing it to pause at this site. This provides enough time for the reactions that promote dissociation of the enzyme to take place. The mechanism of termination has been reviewed in detail by von Hippel (10), Adhya and Gottesman (26), and most recently by Platt (27).

Transcription of a region of DNA that possesses dyad symmetry yields an RNA copy that has a sequence capable of generating a base-paired hairpin loop, as shown in Fig. 2-9. It is thought that as the polymerase brakes at the terminator it allows this hairpin loop to be formed, which sterically hinders further transcription, causing dissociation of polymerase from the gene and release of the completed RNA transcript. The RNA transcript usually ends with a sequence of uridylic acid residues transcribed from the A:T base pairs of the terminator.

Some terminators are very strong, and others are very weak, depending on the G:C and A:T content of the terminator sequence and the degree of dyad symmetry. Some, but not all, weak terminators have been found to become very efficient in the presence of a certain protein. This protein, rho, has a subunit $M_r = 50,000$ in *E. coli* and 45,000 in *Bacillus subtilis*, although it appears to be physiologically active as a dimer or a tetramer. It does not bind to DNA, but has an affinity for single-stranded RNA and for the polymerase core enzyme. The mechanism by which rho effects termination is not known, but it probably involves the stabilization of the RNA hairpin-loop structure or facilitation of its interaction with core enzyme. The rho protein possesses ATPase activity, indicating that energy is required to disrupt the polymerase–DNA ionic bonds.

REGULATION OF TRANSCRIPTION

The last section described the frequency and rate of initiation of transcription in terms of two factors—the availability of sigma factor and the strength of the enzyme-promoter interaction. However, inducible and repressible catabolic operons must be switched on and off to meet the metabolic requirements of the cell. Additionally, biosynthetic operons must only be expressed when there is a demand for their end-product. A number of other proteins have been found to interact with the catabolic operons to regulate their activity, and a novel modification of the mechanism of termination has evolved to effectively regulate the expression of biosynthetic operons. These proteins and mechanisms will now be described, starting with a protein, the catabolite activator protein (CAP) that is required for the expression of a number of catabolic operons. Only mechanisms are discussed here; more physiological details of the operons mentioned can be found in Chapters 8 and 9.

Promoter Activation by CAP

The promoters for the operons of catabolic pathways fall into two categories: those that can initiate transcription unaided and those that require the presence of the catabolite activator protein. This protein serves to coordinate the catabolic activity of bacterial cells so that they utilize their carbon sources most efficiently (10, 28; for details see Chapter 8).

Glucose is the preferred carbon source for most cells, and other carbon sources, such as lactose, are only used when the glucose in the medium is depleted. The concentration of cAMP in the cell is inversely related to the concentration of glucose in the medium (see Fig. 8-9 in Chapter 8). cAMP binds to CAP (CAP is sometimes referred to in the older literature as CRP, the cAMP-receptor protein), and as its concentration increases the affinity of CAP for specific DNA binding is increased, and the CAP–cAMP complex becomes restricted to certain sites on the chromosome.

These binding sites for the CAP–cAMP complex have been found to be adjacent to the promoters of a number of catabolic operons (as shown in Fig. 2-7 for the *lac* operon). In this case the protein binds 12–15 base pairs from the σ-subunit recognition site. The binding site has the sequence symmetry typical of many sites on DNA that bind multimeric proteins. The site is palindromic with both chains having essentially the same sequence when read in their 5'–3' direction. Thus rotation of one strand 180° about an axis of symmetry generates the same sequence on the other strand. Such an arrangement is thought to facilitate the binding of identical protein subunits.

In the presence of cAMP, CAP forms a stable complex with DNA, and, in the case of the *lac* operon, for example, it promotes a 20–50-fold increase in transcription. The mechanism is not completely understood but is likely to involve protein:protein interactions between CAP–cAMP and the σ subunit of the holoenzyme. There may also be a CAP–cAMP-mediated conformational

Promoter Obstruction by Repressor

The *lac* operon is only expressed when there is lactose present in the medium. The repressor protein ensures that the operon is inactive at all other times. This protein is a tetramer of $M_r = 152,000$ and is synthesized constitutively from a gene, the i gene, located next to the regulatory region of the *lac* operon (see Fig. 8-4). Each subunit of the repressor contains 347 amino acid residues, and the entire sequence is known. A structure–function analysis of the subunit has revealed 60 or 90 residues at the N-terminal end of the molecule that are responsible for DNA binding, and in paticular a sequence of 10 amino acids (50 residues from the N-terminal) is very important:

NH$_2$ ... ASN–ARG–VAL–ALA–GLN–GLN–LEU–ALA–GLY–LYS–GLN ...

This sequence contains five residues that can bind specifically to the "edges" of DNA bases (asn, arg, gln; see Fig. 2-5) and a lysine residue that can form an ionic bond with a DNA-backbone phosphate group.

The repressor binds to a site between the promoter and the first structural gene. This site, termed the operator, is illustrated in Fig. 2-7. It overlaps the polymerase recognition site and the site of initiation of transcription. Thus the presence of repressor effectively blocks the binding of polymerase, preventing transcription. The interactions between repressor, operator, and polymerase have been discussed in a number of recent reviews (10, 23, 29, 30). The operator sequence is highly palindromic, as shown in Fig. 2-7, and is approximately 90 Å in length, which is close to the size of the repressor molecule (100 Å × 90 Å). Studies with pure repressor and synthetic operator have shown that two of the four subunits of the repressor make contact with this palindromic operator sequence.

In the absence of lactose the repressor forms a very stable complex at its operator sequence with an association constant of $10^{13}\ M^{-1}$. The presence of allolactose (the inducer of the *lac* operon; see Chapter 8) causes a conformational change in the repressor, decreasing its affinity to $10^9–10^{11}\ M^{-1}$, thereby freeing the site for polymerase binding and transcription.

Under normal conditions there are approximately 10 molecules of repressor per cell (equivalent to $2 \times 10^{-8}\ M$), and since its affinity even for nonspecific DNA binding is high ($10^9–10^{11}\ M^{-1}$), most of these molecules will be bound to DNA. The advantages of this have already been discussed in relation to a protein being able to locate its specific binding site(s). With most of the repressor molecules bound to DNA, its free concentration is reduced considerably. This is important because a high free-repressor concentration would still result in repressor occupation of its operator binding site even if inducer decreases its affinity to $10^9\ M^{-1}$, and operon depression could never occur. However, if there are only one or two free repressor molecules per cell its concentration is so low that, in the presence of inducer, the operator binding site would be vacant for a substantial amount of time (see Ref. 10 for a detailed computer analysis of operon derepression).

The regulation of the *lac* operon by CAP and by its repressor is an example of the elegant simplicity of operon control. Some other operons are controlled in an identical way, whereas yet others have somewhat more complex controls. For example, the *gal* operon is actually controlled by two different repressors acting at overlapping operators. These operons are discussed in more detail in Chapter 8. However, all of them are controlled by proteins that have two common characteristics: they can be allosterically modified by a small molecule, and they have the ability to bind tightly to DNA in only one of these conformations. These proteins are the key to regulation because they transmit information about the metabolic status of the cell (reflected by changes in cAMP, lactose, etc.) to the genome and can elicit a change in gene expression.

Experimentally, DNA sequences that represent important regulatory sites such as operators are usually located by selecting for mutants that have altered regulatory capacities. During studies of some operons, mutants were isolated that mapped at a locus downstream from that expected of promoter mutations. One such operon was that of tryptophan biosynthesis, and further work on these mutants led to the discovery of a major new mechanism for the regulation of gene transcription (31, 32). This is known as attenuation and has now been found to be the major, and in some cases the sole, mechanism for the regulation of the operons for the enzymes of amino acid biosynthetic pathways. Many of these operons have recently been reviewed by Platt (27), Yanofsky (32), and Keller and Calvo (33). The next section describes the role of attenuation in the regulation of the *trp* operon; a discussion of the regulation of other operons of amino acid biosynthesis can be found in Chapter 9.

Attenuation

The tryptophan operon regulates the expression of the five enzymes of tryptophan biosynthesis as shown in Fig. 2-10 and is only expressed during periods of tryptophan starvation. The operon has a promoter–operator region located about 160 base pairs upstream from the start of the first structural gene (*trp E* coding for anthranilate synthetase), and RNA polymerase molecules transcribe this leader segment of the operon. During mutation studies designed to identify the regulatory components of this operon, it was noticed that a number of constitutive mutations mapped at a locus within the leader segment. When this section of the leader was sequenced it was found to contain a termination sequence very similar to those normally found *at the end of operons* (compare Figs. 2-10 and 2-9). It contains a typical G:C-rich sequence followed by a string of successive thymine residues. The detection of RNA transcripts 140 nucleotides in length (equivalent to the distance between the *trp* promoter and

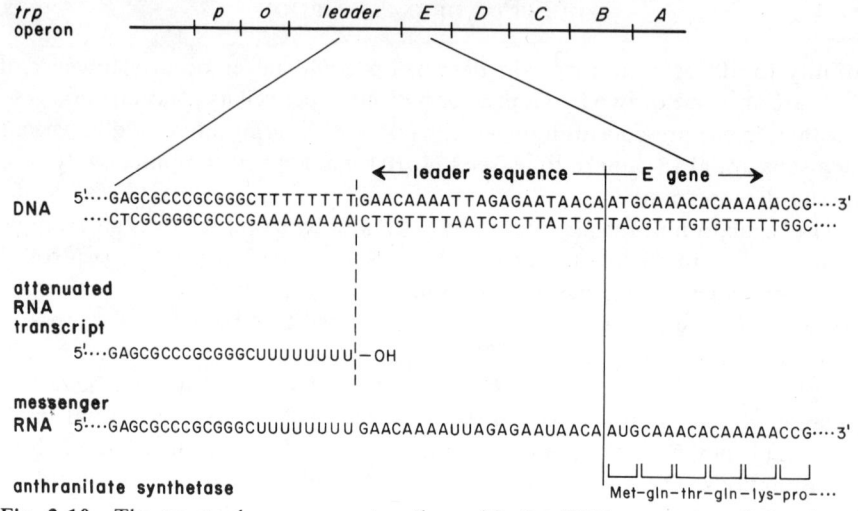

Fig. 2-10 The tryptophan operon together with the DNA sequence of the site of attenuation. The sequences of attenuated and nonattenuated (mRNA) transcripts are also shown. The E gene codes for the mRNA for the enzyme anthranilate synthetase.

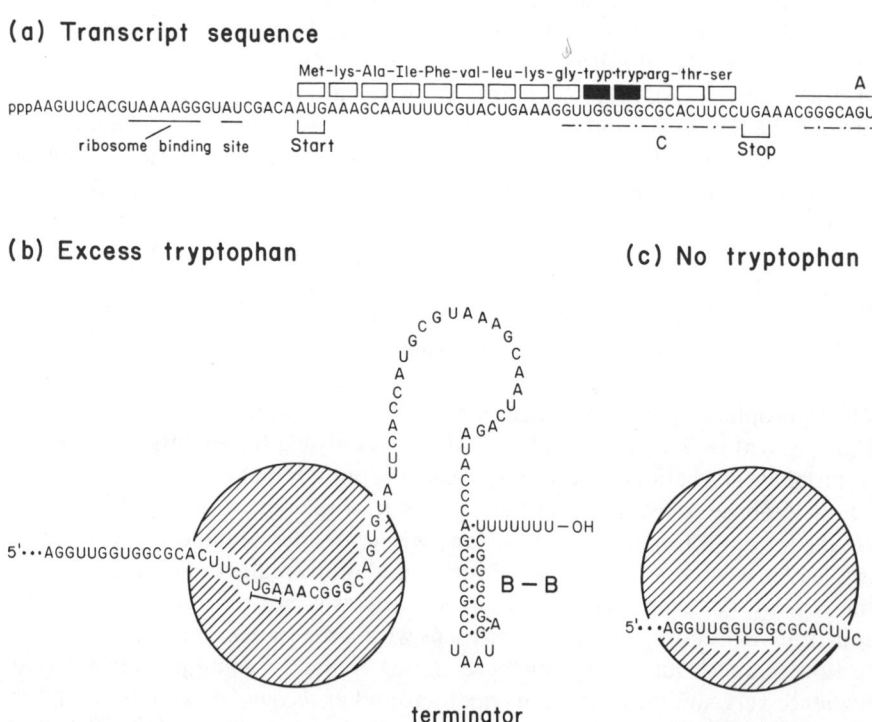

Fig. 2-11 Components of the mechanism for attenuation. The sequence of the leader transcript showing the leader peptide and the three sequences capable of generating base-paired loops is shown in (*a*). In (*b*), (*c*), and (*d*) the three possible secondary

the terminator sequence) provided evidence that this sequence does indeed function as a terminator (32).

RNA polymerase molecules transcribing the leader segment of the operon either stop transcription at this terminator and dissociate from the operon before reaching the structural genes, or read through this sequence and go on to transcribe the structural genes to express the operon (Fig. 2-10). This is the basis of attenuation. The term *attenuation* was introduced to describe transcription termination *within* an operon and to distinguish it from the termination that occurs at the end of operons.

The attenuated leader transcript has now been sequenced (Fig. 2-11a) and found to contain several interesting features. For example, immediately before the string of uridyl residues at the 3' end of the transcript is a G:C-rich sequence that could form a stable, stem-loop structure by base-pairing of the sequences designated B–B in Fig. 2-11a. Such a structure is directly analogous to the one that participates in conventional transcription termination (Fig. 2-9). In addition, there are two other segments of sequence symmetry [A–A and C–C in Fig. 2-11a, following the nomenclature of Oxender et al. (34)] that could also give rise to base-paired, stem-loop structures. Moreover, since there is sequence overlap the formation of one stem-loop may preclude the formation of another. For example, the formation of stem-loop structure A–A precludes the

structures are shown as explained in the text. The shaded circle in (*b*) and (*c*) represents the ribosome binding site. [Adapted from Ref. 34, with permission from Dr. C. Yanofsky.]

formation of B–B (the terminator structure). Thus it appears that the leader RNA transcript can exist in a number of different, *mutually exclusive* secondary structures.

The 5' end of the transcript has a sequence for ribosome binding (Fig. 2-11a, compare with Figs. 6-3a, 6-3b) followed by an AUG codon that could initiate the synthesis of a peptide of 14 residues before a translation-stop codon (UGA) is reached. Two adjacent tryptophan codons are found in this sequence, a very rare event considering the low tryptophan content of most proteins. If a ribosome were to bind to the transcript and commence translation of this peptide it could only complete the process if there were sufficient tryptophanyl tRNA^{trp} present in the cell to allow it to translocate past the *trp–trp* double codon. The concentration of tryptophanyl tRNA^{trp} is, in turn, dependent upon the concentration of tryptophan, the end-product of the biosynthetic pathway controlled by the *trp* operon.

The model for the regulation of transcription of the *trp* operon by attenuation proposes the following sequence of events as the RNA polymerase molecule moves from the promoter toward the first structural gene. As the leader transcript is being produced, a ribosomal subunit binds at the 5' end, moves along it, and when the AUG codon is reached translation is initiated. Thus translation of the leader peptide commences while RNA polymerase is still transcribing the leader segment. Moreover, during the time (approximately 0.6 s) it takes the ribosome to translate the leader coding sequence, RNA polymerase transcribes the segment of the leader that contains all three domains of sequence symmetry (A–A, B–B, and C–C) as shown in Fig. 2-12a, 2-12b, ①–③.

If there is sufficient tryptophan in the cell the *trp–trp* double codon is translated, and the ribosome moves to the end of the coding sequence (UGA) *just before* RNA polymerase reaches the terminator (attenuator) sequence (Fig. 2-12b, ③). At this point the ribosome blocks, by steric hindrance, the formation of stem-loop structures A–A and C–C (Figs. 2-11b; 2-12b, ④), and as sequence B–B is transcribed it forms a stable *terminator* structure that destabilizes RNA polymerase (Fig. 2-12b, ⑤). The polymerase molecule dissociates from the genome, and the structural genes are not transcribed under conditions when there is excess tryptophan (Fig. 2-12b, ⑥).

When there is no tryptophan in the cell the translating ribosome is arrested at the *trp–trp* double codon, and even though it blocks the formation of stem-loop structure C–C it does not block the formation of A–A (Figs. 2-11c; 2-12a, ④ ⑤). The formation of stem-loop structure A–A *preempts* the formation of B–B, there is no attenuation, polymerase can read through the remainder of the leader sequence to the structural genes, and the operon is expressed to satisfy the demand for tryptophan (Fig. 2-12a, ⑥). Stem-loop structure A–A is termed the preemptor (33).

Under conditions when there is no ribosome available to translate the leader the first symmetrical sequence to be transcribed (C–C) forms a stem loop. This is called the protector since it protects the terminator sequence (Fig. 2-11d), permitting the terminator structure to be formed, and the operon is not ex-

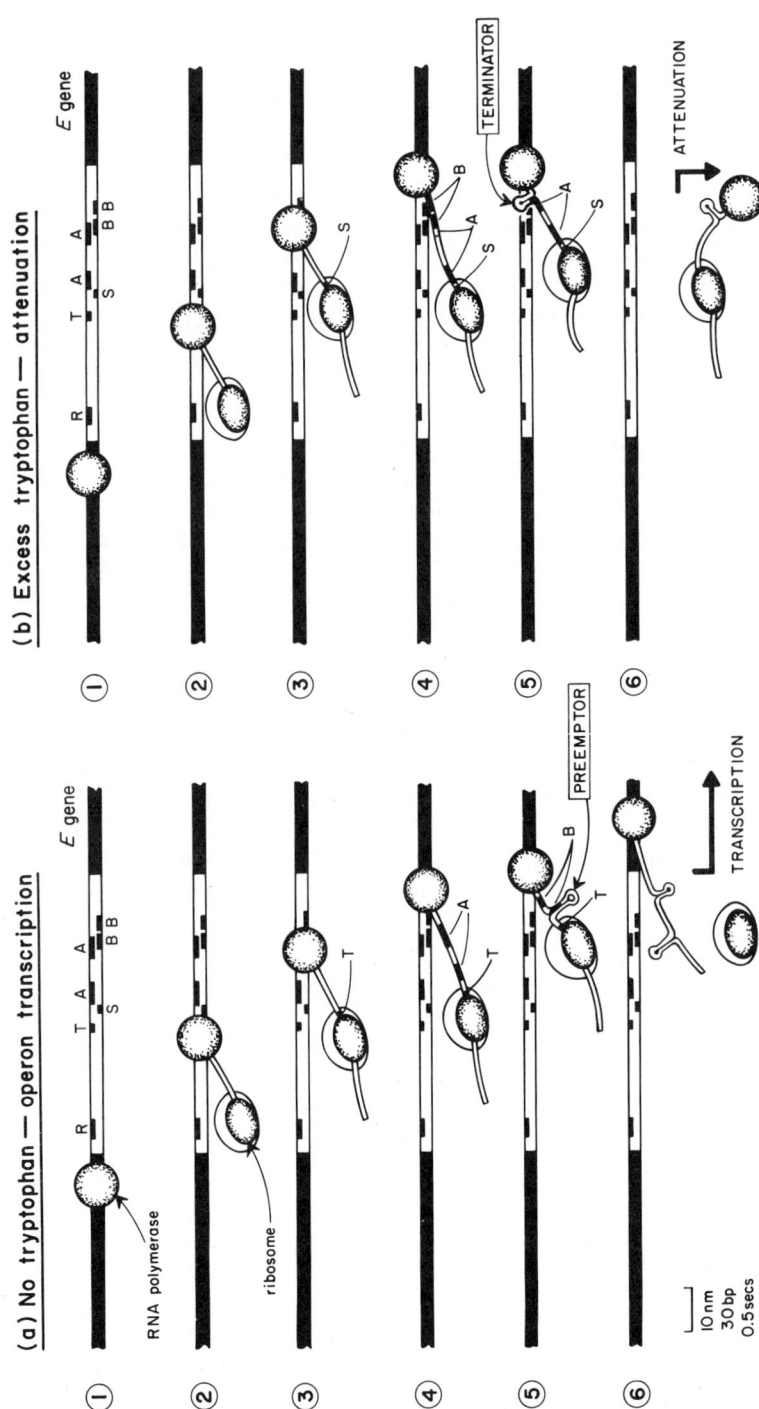

Fig. 2-12 A scale drawing of the processes of attenuation (b) and transcription read-through (a). In ① the following sequences are represented (compare Fig. 2-11): R = ribosome binding sequence, T = *trp–trp* double codon, S = stop codon, A–A = preemptor sequence, B–B = terminator sequence (C–C is omitted for clarity). The sequences of events following polymerase binding to DNA are described in the text. The scale bar shows distances in nanometers and base pairs as well as the approximate time taken for either polymerase or a ribosome to move that distance.

pressed. The protector is also formed when a ribosome, having reached the *trp–trp* double codon dissociates (Fig. 12-2a, ⑥). The formation of the protector under these circumstances prevents further translation of the attenuated leader RNA.

The above description of the mechanism of attenuation considers the two extremes—excess tryptophan and no tryptophan—which are not often encountered *in vivo*. Under more usual circumstances the rate of translation of the leader decreases as the concentration of tryptophanyl tRNA^{trp} falls. This, in turn, increases the number of preemptors formed, and the number of RNA polymerase molecules permitted to read through to the structural genes increases. Thus it is the *rate of translation of the leader peptide by the ribosome* that forms the basis of the mechanism of attenuation.

Although direct proof of this mechanism is still forthcoming, a number of experiments do support such a model. For example, mutations in the sequence that forms the terminator (weakening its stability) all result in a change in the degree of attenuation (32). Moreover, nuclease digestion studies indicate that those sequences predicted to be double standard are resistant to nucleases specific for single-stranded RNA. The 14-residue peptide that should be translated by the ribosome has not been isolated, but a mutation in the DNA that changes the AUG codon to AUA seriously impairs expression of the operon, indicating that translation initiation, of the leader peptide, is indeed part of the mechanism. More recently (35), studies on the kinetics of transcription of the leader segment by RNA polymerase indicate that the enzyme pauses during transcription at a site about 20 base pairs past the 3′ end of the coding portion of the leader. It is proposed that this serves to synchronize the movement of the ribosome and polymerase.

The protein rho has been found to interact with some attenuators by a mechanism similar to its role in termination, and this further increases the regulatory capacity of the system (36).

Other Proteins That May Regulate Transcription

A number of proteins have been isolated from bacterial cells that have some capacity to modulate the rate of gene transcription. In general these proteins are poorly characterized because they are available in only small quantities. Moreover, many of these proteins modulate gene expression by interacting with the protein components of the system rather than with the DNA template. This makes their function much more difficult to establish because the problem cannot be approached genetically (i.e., their binding sites cannot be identified by mutations). However, this should not detract from their potential regulatory importance. The properties of a number of these proteins have been recently reviewed (14, 37).

Some of these proteins, for example, σ', y, and γ, can replace the σ subunit of RNA polymerase and alter the enzyme's promoter specificity. There are also a number of heat-stable, low-molecular-weight DNA-binding proteins (H_1, H_2,

D) that can influence the rate of gene transcription. A number of the proteins involved in the mechanism of protein synthesis (see Chapter 6) such as IF-2, EF-G, EF-Tu and EF-Ts have also been shown to interact with polymerase and may coordinate the rate of transcription with the rate of translation.

The bacteriophage λ codes for a protein that functions as an antiterminator, permitting polymerase to transcribe across rho-dependent terminators. This protein is the product of the phage N gene and appears to bind to a specific site on DNA downstream from the promoter. When a translocating core enzyme encounters this protein the antiterminator becomes incorporated into the polymerase molecule, rendering it insensitive to rho-dependent terminators. This is an interesting protein since a knowledge of its mechanism of action will contribute greatly to our understanding of the stereochemistry of termination and, in particular, the role of secondary and possibly tertiary structures of RNA. There is also some evidence for the presence of a similar protein (factor L) in uninfected *E. coli* cells.

Much further work is needed to characterize these proteins and to establish their role in the regulation of gene transcription in bacterial cells.

REFERENCES

1. J. D. Griffith (1976). Visualization of prokaryotic DNA in a regularly condensed chromatin-like fiber, *Proc. Natl. Acad. Sci. USA* **73**, 563–567.
2. B. J. Bachmann, K. B. Low, and A. L. Taylor (1976). Recalibrated linkage map of *Escherichia coli* K-12. *Bact. Rev.* **40**, 116–167.
3. A. Ryter and A. Chang (1975). Localisation of transcribing genes in the bacterial cell by means of high resolution autoradiography. *J. Mol. Biol.* **98**, 797–810.
4. S. M. Mirkin, E. S. Bogdanova, Zh. M. Gorlenko, A. I. Gragerov, and O. A. Larionov (1979). DNA supercoiling and transcription in *Escherichia coli*: Influence of RNA polymerase mutations. *Molec. Gen. Genet.* **177**, 169–175.
5. J. J. Champoux (1978). Proteins that affect DNA conformation. *Ann. Rev. Biochem.* **47**, 449–479.
6. N. C. Seeman, J. M. Rosenburg, and A. Rich (1976). Sequence-specific recognition of double helical nucleic acids by proteins. *Proc. Natl. Acad. Sci. USA* **73**, 804–808.
7. W. Etkin (1973). A representation of the structure of DNA. *Bioscience* **23**, 652–653.
8. R. D. Wells, T. C. Goodman, W. Hillen, G. T. Horn, R. D. Klein, J. E. Larson, U. R. Müller, S. K. Nevendorf, N. Panayotatos, and S. M. Stirdivant (1980). DNA structure and gene regulation. *Prog. Nuc. Acid Res. Mol. Biol.* **24**, 167–267.
9. H. M. Sobell (1979). Importance of symmetry and conformational flexibility in DNA structure for understanding protein–DNA interactions. In R. F. Goldberger (Ed.). *Biological Regulation and Development*, Vol. 1: *Gene Expression*. Plenum Press, New York, pp. 171–199.
10. P. H. von Hippel (1979). On the molecular basis of the specificity of interaction of transcriptional proteins with genome DNA. In R. F. Goldberger (Ed.). *Biological Regulation and Development*, Vol. 1: *Gene Expression*. Plenum Press, New York, pp. 279–347.
11. R. H. Sarma (1979). *Stereodynamics of Molecular Systems*. Pergamon Press, New York.
12. M. J. Chamberlin (1976). RNA polymerase—an overview. In R. Losick and M. J. Chamberlin (Eds.). *RNA polymerase*. Cold Spring Harbor Laboratory, pp. 17–67.

13 M. J. Chamberlin (1976). Interaction of RNA polymerase with the DNA template. In R. Losick and M. J. Chamberlin (Eds.). *RNA polymerase*. Cold Spring Harbor Laboratory, pp. 159–191.
14 R. Lathe (1978). RNA polymerase of *Escherichia coli*. *Curr. Topics Microbiol. Immunol.* **83**, 37–91.
15 W. Zillig, R. Palm, and A. Heil (1976). Function and reassembly of subunits of DNA-dependent RNA polymerase. In R. Losick and M. J. Chamberlin (Eds.). *RNA polymerase*. Cold Spring Harbor Laboratory, pp. 101–125.
16 V. M. Hansen and W. R. McClure (1980). Role of the σ subunit of *Escherichia coli* RNA polymerase in initiation. *J. Biol. Chem.* **255**, 9564–9570.
17 F. Jacob and J. Monod (1961). Genetic regulatory mechanisms in the synthesis of proteins. *J. Mol. Biol.* **3**, 318–356.
18 F. Jacob and J. Monod (1961). On the regulation of gene activity. *Cold Spring Harbor Quant. Symp. Mol. Biol.* **26**, 193–211.
19 E. Volkin and L. Astrachan (1957). RNA metabolism in T2-infected *Escherichia coli*. In W. D. McElroy and B. Glass (Eds.). *The Chemical Basis of Heredity*. The Johns Hopkins Press, Baltimore, pp. 686–695.
20 B. Lewin (1977). Gene Expression, Vol. 1. John Wiley & Sons, London, pp. 272–309.
21 J. R. Beckwith (1978). *lac*: The genetic system. In J. H. Miller and W. S. Reznikoff (Eds.). *The Operon*. Cold Spring Harbor Laboratory, pp. 11–30.
22 D. Pribnow (1979). Genetic control signals in DNA. In R. F. Goldberger (Ed.). *Biological Regulation and Development*, Vol. 1: *Gene Expression*. Plenum Press, New York, pp. 219–277.
23 R. Wu, C. P. Bahl, and S. A. Narang (1978). Lactose operator–repressor interaction. *Curr. Topics, Cell. Reg.* **13**, 138–178.
24 G. Zubay (1980). Regulation of transcription in prokaryotes, their plasmids, and viruses. In L. Goldstein and D. M. Prescott (Eds.). *Cell Biology: A comprehensive treatise*, Vol. 3. Academic Press, New York, pp. 153–214.
25 O. L. Miller, B. A. Hamkalo, and C. A. Thomas Jr. (1970). Visualization of bacterial genes in action. *Science* **169**, 392–395.
26 S. Adhya and M. Gottesman (1978). Control of transcription termination. *Ann. Rev. Biochem.* **47**, 967–996.
27 T. Platt (1981). Termination of transcription and its regulation in the tryptophan operon of *E. coli*. *Cell* **24**, 10–23.
28 B. DeCrombrugghe and I. Pastan (1978). Cyclic AMP, the cyclic AMP receptor protein, and their dual control of the galactose operon. In J. H. Miller and W. S. Reznikoff (Eds.). *The Operon*. Cold Spring Harbor Laboratory, pp. 303–324.
29 S. Bourgeois and M. Pfahl (1976). Repressors. *Adv. Protein Chem.* **30**, 1–99.
30 M. D. Barkley and S. Bourgeois (1978). Repressor recognition of operator and effectors. In. J. H. Miller and W. S. Reznikoff (Eds.). *The Operon*. Cold Spring Harbor Laboratory, pp. 177–220.
31 K. Bertrand, L. Korn, F. Lee, T. Platt, C. L. Squires, C. Squires, and C. Yanofsky (1975). New features of the regulation of the tryptophan operon. *Science* **18**, 22–26.
32 C. Yanofsky (1981). Attenuation in the control of expression of bacterial operons. *Nature* **289**, 751–759.
33 E. B. Keller and J. M. Calvo (1979). Alternative secondary structures of leader RNAs and the regulation of the *trp, phe, his, thr,* and *leu* operons. *Proc. Natl. Acad. Sci. USA* **76**, 6186–6190.
34 D. L. Oxender, G. Zurawski, and C. Yanofsky (1979). Attenuation in the *Escherichia coli* tryptophan operon: role of RNA secondary structure involving the tryptophan coding region. *Proc. Natl. Acad. Sci. USA* **76**, 5524–5528.

REFERENCES

35 M. E. Winkler and C. Yanofsky (1981). Pausing of RNA polymerase during *in vitro* transcription of the tryptophan operon leader region. *Biochemistry* **20**, 3738-3744.

36 I. P. Crawford and G. V. Stauffer (1980). Regulation of tryptophan biosynthesis. *Ann. Rev. Biochem.* **49**, 163-195.

37 R. Losick and J. Pero (1976). Regulatory subunits of RNA polymerase. In R. Losick and M. J. Chamberlain (Eds.). *RNA polymerase*. Cold Spring Harbor Laboratory, pp. 227-246.

CHAPTER THREE

Structural Organization of DNA in the Nucleus of the Eukaryotic Cell

The structural organization of the nucleic acids in bacterial cells is relatively primitive. DNA is condensed into the nucleoid, and transcription and translation take place within, or in close proximity to, this structure. The nucleoid having no membrane, there is direct communication between the nucleic acid and the cytoplasm. In eukaryotic cells the situation is very different. The genetic material is segregated into a specific organelle, the nucleus. Until recently little was known about the structure of this organelle or the way in which DNA was organized within it. A clearer image of the organization of the vast amount of DNA is now emerging following the discovery of the nucleosome as the unit of chromatin structure. In this chapter we discuss the various levels of DNA organization and the nature of the proteins that interact with it to form chromatin. All of this serves as a basis for the discussion of the regulation of gene transcription in Chapter 4.

CHROMOSOMES, CHROMATIN, AND THE COMPLEXITY OF DNA

Chromosomes

The DNA content of the haploid genome of the higher eukaryotes is $(2-3) \times 10^9$ base pairs, or about 500 times more DNA than that of a bacterial cell. The genetic material is organized into a discrete number of chromosomes, with even the smallest chromosome having 10 times more DNA than *E. coli*. All of this nucleic acid (approximately 120 cm in length) is packaged into the nucleus. In the interphase cell much of the DNA of the chromosomes is dispersed, and it is difficult to establish which DNA belongs to each chromosome. Because of this, the genetic material of eukaryotic cells is not usually studied biochemically at the level of individual chromosomes; rather, all of the DNA is extracted as chromatin and then analyzed.

Chromatin

Chromatin is the term used to describe the complex of DNA and its associated proteins that stains histologically with basic dyes and can be isolated from cells and studied *in vitro*. Microscopists have long recognized two types of chromatin—heterochromatin and euchromatin [see reviews of Lewin (1), Ris and Korenberg (2), and the book by Bostock and Sumner (3)]. Heterochromatin is that part of the chromatin that remains in the condensed state in the interphase (nonmitotic) cell, whereas euchromatin is the part of chromatin that disperses during interphase. The condensed heterochromatin is considered to be in an inactive state, whereas autoradiographic studies have demonstrated that RNA synthesis (i.e., gene transcription) takes place on the dispersed euchromatin.

Heterochromatin can be further subdivided into *constitutive heterochromatin*, which is the chromatin that remains condensed in all cells of a given organism, and *facultative heterochromatin*, which is that chromatin condensed in some cells but not in others. Although transcription takes place on the dispersed euchromatin, not all of it is active at any one time. Thus we can distinguish five classes of DNA in the interphase nucleus:

1. DNA that remains highly condensed, has no information content, and is never transcribed.
2. DNA that carries information but in particular cells is never expressed.
3. DNA that is condensed but disperses under appropriate stimulation.
4. DNA that is dispersed but is not being actively transcribed.
5. DNA that is dispersed and is being actively transcribed.

The bacterial cell has virtually all of its DNA in categories 4 and 5; these categories of eukaryotic DNA, together with category 3, are the ones in which we are mainly interested.

The bacterial chromosome carries structural genes to code for all the proteins in the cell, and most of the genome is capable of being expressed, although only 4–10% of it may actually be transcribed at any given time. Each structural gene sequence is present in only one copy per chromosome (with the exception of rDNA, which is present in seven copies in *E. coli*, for example), and there are potentially about 4000 genes (structural and regulatory). In marked contrast, if all the DNA of a mammalian cell were in the form of structural genes (of approximately 2000 base pairs each), the genome would have the capacity to code for about *1 million* proteins. It is highly unlikely that there are so many genes for two reasons. First, this number far exceeds the number of known proteins in any one cell. Second, an organism carrying such a load of structural genes could not survive, because it would accumulate too many mutations, some of which would inevitably be lethal. It is likely that the total number of genes in the genome of higher mammals does not exceed 50,000 and is probably somewhere in the region of 20,000–50,000 (1).

This number of genes accounts for only 2–4% of the DNA content of the eukaryotic genome. Thus either each gene is represented many times on the genome or the remaining 96–98% of the genome is in a form that is never expressed in conventional structural gene products. Sequence complexity studies (see below) have shown that most structural genes are present in only one copy per haploid genome. Other studies have shown that, although the ribosomal and transfer RNA genes and even some structural genes (e.g., the histone genes) are greatly amplified, this can only account for a small additional percentage of DNA. Thus most of the DNA does not code for proteins, raising the question of whether this DNA is ever transcribed or whether it is completely inactive. Certainly, much more than 4% of the DNA is dispersed as euchromatin in the interphase nucleus. Studies on DNA sequence complexity and the structural organization of these sequences have provided a partial answer to this question.

DNA Sequence Complexity

During the 1960s a technique for estimating the relative amounts of DNA that are present as single copy and the amounts that are repeated many times was introduced by Britten and Kohne (Ref. 4, see also Refs. 1 and 5–7). The technique is based upon the ability of two complementary single strands of DNA to recognize each other in solution and form a base-paired double strand. If only one species of complementary strand is present the reaction occurs very quickly, whereas if there is a mixture of sequences it takes longer for each individual strand to find its own complementary sequence. At various times during the incubation the amount of DNA that is double-stranded is measured by UV spectrophotometry (double-stranded DNA absorbs less UV light than single-stranded DNA), chromatography (under appropriate conditions hydroxyapatite binds only double-stranded DNA), or enzymatically (with a nuclease that digests only single-stranded DNA). The reaction is followed to completion, and the results are plotted as a fraction of the DNA that is single-stranded versus the product of the amount of DNA used in the reaction (Co) and the reaction time (t). This is generally referred to as a Cot curve, and typical curves are shown in Fig. 3-1.

When two synthetic polynucleotides, such as poly-U and poly-A, are incubated together the rate of reassociation is very fast, and the reaction goes to completion rapidly at low Cot values. On the other hand, single strands of DNA from the *E. coli* genome, which consists of many unique sequences, take much longer to reassociate. Consequently the *E. coli* DNA Cot curve is displaced to the right. Similarly, DNA from a mammalian source reassociates even more slowly because the genome is even larger, and the concentration of each individual sequence is correspondingly lower. A Cot curve can be calibrated to indicate the total number of base pairs or the total complexity of the nucleic acid. The midpoint of the reassociation curve, designated $Cot_{1/2}$, is the value used to determine complexity. The value for poly-A:U is 1, since there is only

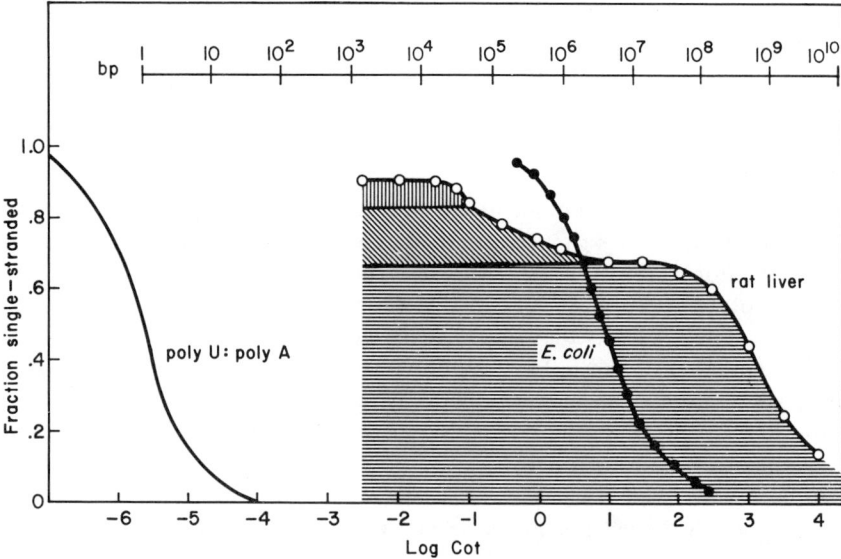

Fig. 3-1 DNA sequence complexity analyzed by hybridization techniques. The graph shows Cot curves for a synthetic homopolymer poly-(A:U), *E. coli* DNA, and rat liver DNA (all are sheared to the same length). The abscissa is plotted as a log scale, and the interpretation of the curves is given in the text. The upper scale gives the sequence complexity in base pairs, and the liver curve is shaded to illustrate the three complexity classes. [Reproduced from Fig. 1 of Britten and Kohne (1968). *Science* **161**, 529–540, with permission. Copyright 1968 by the American Association for the Advancement of Science.]

one type of base pair. The value obtained for *E. coli* is 5×10^6 base pairs, which is the size of the entire genome. The $Cot_{1/2}$ of the unique sequences (most slowly hybridized) of rat-liver DNA is equivalent to about 2×10^9 base pairs, and this is taken to be the size of the mammalian genome.

The shape of the *Cot* curve for mammalian cells is quite different from those of *E. coli* and the synthetic polynucleotides. The curve for the synthetic poly-A:U resembles very closely that of a theoretically drawn curve; it is symmetrical, and the reaction is completed over about 2 logs of *Cot*. Any significant deviation from this shape of curve or the number of logs required for completion indicates that the DNA consists not only of unique sequences but also contains sequences that are present in more than one copy on the genome. The more copies there are of a particular sequence the faster it finds a complementary sequence, and it becomes double-stranded at a lower *Cot* value. The shape of the *Cot* curve for *E. coli* DNA is both symmetrical and complete over 2 logs of *Cot*, indicating that it is made up almost entirely of unique sequences. In contrast, the liver-DNA curve spans about 6 logs of *Cot* and is markedly asymmetrical, indicating that mammalian genomes contain a mixture of unique sequences and sequences that are repeated many times. This complex curve is usually divided into three sections (shaded areas in Fig. 3-1) corresponding to the highly repetitive (hybridizing at a log Cot of about -1), moderately repetitive (hybridizing between -1 and $+1$ logs of Cot), and single-copy DNA sequences (hybridizing above $+1$ log of Cot). Usually about 50–75% of the genome is single copy, 10% is highly repetitive, and the remainder is moderately repetitive.

A knowledge of the amounts of DNA in each abundance class still does not solve the problem of why there is so much more DNA than appears to be necessary. For example, all the known structural genes account for at most 2–4% of total DNA or 4–8% of the single-copy DNA. Thus there is still a great deal of single-copy DNA that is never expressed as cytoplasmic mRNA. Moreover, the number of genes that are known to be amplified cannot account for the 15–40% of the DNA that is moderately repetitive. In addition, the function of the 10% of the DNA that is highly repetitive and presumably of sequences too simple to code for proteins is also not clear.

The Relationship between Sequence Complexity and Structure

Without a knowledge of the actual nucleotide sequence of very long (i.e., many thousands of base pairs) segments of DNA it is difficult to know how the various complexity classes are organized in relation to each other. Many of the techniques that have been employed to gain some understanding of DNA structural organization, without a knowledge of its sequence, have been plagued by artifacts. However, some general conclusions can be drawn, and these are summarized in the book of Bostock and Sumner (3), in Bradbury et al. (7), and in a more recent review by Bostock (8).

The most abundant class of DNA (i.e., the highly repetitive sequences) can be separated from the bulk of chromatin by cesium chloride density-gradient centrifugation. Isolation and characterization of this material, usually referred to as satellite DNA, has been achieved from a number of mammalian sources. It has been found to consist of very simple sequences repeated millions of times. The repeat may be of only a few bases such as —CCCTAA— repeated over and over, sometimes with slight variations that form the basis of larger 100–150 and 300–400 base-pair repeat patterns. Some of these sequences are characterized by the presence of the unusual nucleotide 5-hydroxymethylcytosine. Satellite DNA is never transcribed and appears to play a structural role, being located in the region of the centromeres. This accounts for much of the 10% of the DNA that is the most highly repetitive.

Single-copy sequences are not linked together but appear to be interspaced with DNA that is not transcribed. In fact, more than half of the genome is composed of stretches of unique sequences of about 2500 base pairs separated by repetitive sequences of about 300 base pairs. Much of the moderately repetitive class of DNA can be accounted for by these 300 base-pair spacer sequences. The organization of these spacer sequences has been recently studied with a novel approach by Moreau et al. (9). This study took advantage of the observation that the spacer DNA segments are A:T rich. Thus at an appropriate formamide concentration these weakly base-paired regions denature, whereas the more G:C-rich regions remain double-stranded. DNA isolated by gentle techniques in the presence of 75–85% formamide was examined under the microscope and found to be composed of characteristic loops of denatured DNA punctuating otherwise double-stranded DNA. Much of the DNA was organized into segments of approximately 2500 base pairs separated by A:T-rich spacer sequences of about 800 base pairs. On the other hand, some sequences of DNA that were uninterrupted for about 80,000 base pairs were flanked by several A:T-rich spacer regions arranged in tandem (each approximately 800 base pairs). Thus the DNA that must contain single-copy DNA, which may represent structural genes, is organized into a characteristic sequence pattern. At present it is not clear whether these spacer regions are involved solely in the structural organization of DNA or whether they play a functional role by delimiting transcription units and possibly being involved in the binding of RNA polymerase or other regulatory molecules.

Some of the moderately repetitive DNA is transcribed. The best example of this is the rRNA precursor as shown in Fig. 3-2. The rDNA genes are organized into transcription units separated by nontranscribed spacers of repetitive DNA. In some cells there may be as many as 500 copies of rDNA organized in this way. The RNA transcript of each gene is a large 45S molecule that is cleaved to give one each of the ribosomal RNA components 5.8S, 18S, and 28S. Each transcript contains some short sequences of repetitive nucleotides that are excised and degraded during the processing of the 45S precursor. The nontranscribed spacer sequences contain many repeats of simple sequences and also contain hydroxymethylcytosine. The histone genes also appear to be organized into an amplified transcription unit, with the genes for each histone being separated by repetitive DNA (see Fig. 4-1 in Chapter 4). At the present time the histones are the only proteins known to be transcribed from repetitive DNA (this will be discussed in greater detail in Chapter 4).

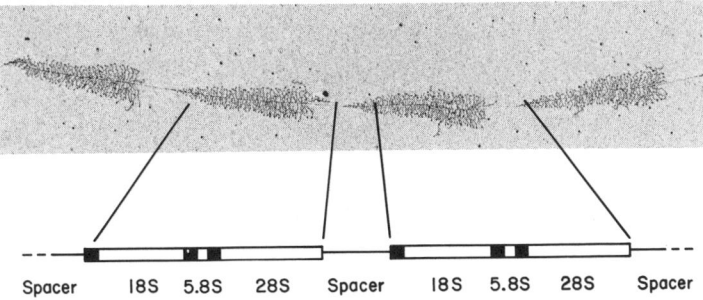

Fig. 3-2 The structural organization of ribosomal DNA. Each ribosomal RNA gene (coding for 18S, 5.8S, and 28S rRNAs) is separated by a nontranscribed spacer DNA. The genes are being transcribed from left to right by many polymerase molecules (×8000). [The photograph was kindly supplied by Professor Werner Franke and is reprinted from Ref. 58, with permission.]

DNA sequencing of these large transcription units is the only way to definitively interrelate all the classes of DNA and to establish their function. However, even this approach will only establish the function of a small amount of the total DNA of a mammalian cell. Thus the significance of the apparent vast excess of DNA is going to be difficult to elucidate in terms of a functional role. Indeed, there is some question whether all of this DNA is functional (in the sense that it contributes to the phenotype of the organism by coding for or regulating the synthesis of cellular proteins) or whether some DNA serves no function other than to ensure its own survival. This notion of "selfish" DNA has recently been discussed by Doolittle and Sapienza (10) and by Orgel and Crick (11). Certainly there seems to be much more nontranscribed DNA than can be accounted for as either regulatory sequences or sequences involved in chromatin structure.

The organization of DNA within the nucleus has also been studied from the tertiary structure point of view, that is, the way in which it folds to generate the various levels of dispersed and condensed chromatin seen in electron micrographs of nuclei. An appreciation of this aspect of DNA structure requires a knowledge of the other components in the nucleus, particularly the proteins.

THE BIOCHEMICAL COMPOSITION OF THE NUCLEUS

The nucleus occupies about 20% of the volume of mammalian cells and contains approximately 2.5 mg of DNA, 0.5 mg of RNA, and 3–5 mg of protein per gram of tissue, together with smaller amounts of phospholipid and other cell constituents. It is a highly complex structure, as illustrated in the diagrammatic representation of the nucleus shown in Fig. 3-3. DNA is shown as both condensed heterochromatin, often associated with the inner nuclear membrane, and as dispersed euchromatin. Each nucleus has at least one nucleolus—the site of ribosomal RNA synthesis. The outer nuclear membrane is contiguous with the endoplasmic reticulum and is usually covered with ribosomes. Nuclear pores span both the inner and outer membrane, permitting communication between the nucleus and the cytoplasm (although the pores are not the only means of nucleo–cytoplasmic communication). The structures within the nucleus are supported by the nuclear matrix, a fibrous skeleton contiguous with the inner nuclear membrane (particularly in the vicinity of the pore complex). The small, electron-dense granules associated with the matrix are the various species of ribonucleoprotein particles involved in the processing and transport of RNA.

The Proteins of the Nucleus

Historically the proteins of the nucleus have been divided into two classes, the histone and the nonhistone proteins, based on their gross physicochemical properties. The histones were found to be highly basic (positively charged), whereas the nonhistone proteins were generally found to be acidic (negatively

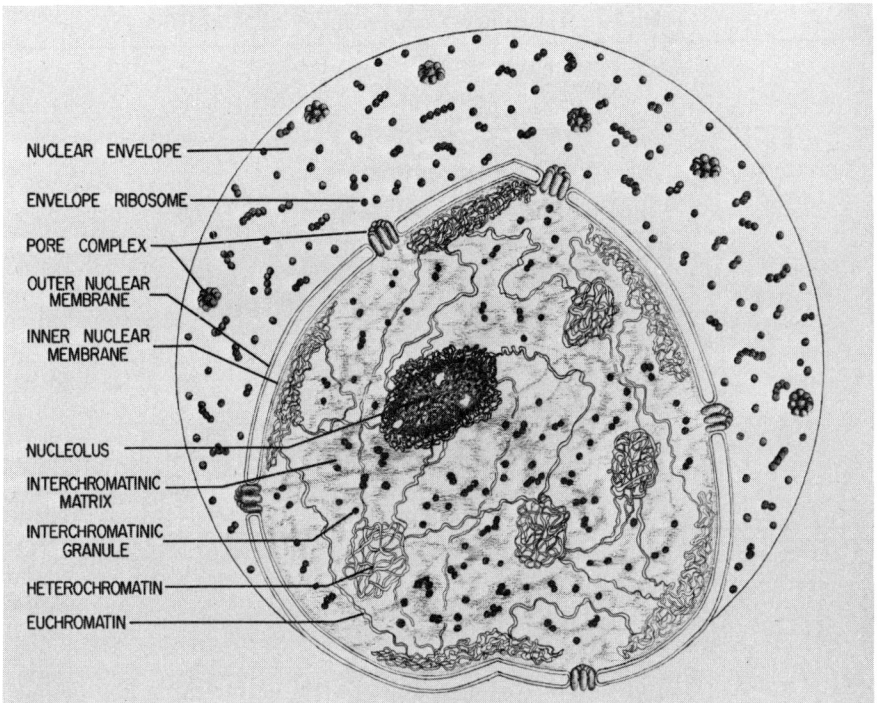

Fig. 3-3 Diagrammatic representation of the structure of an animal cell nucleus. [Reproduced from Ref. 16, with permission. The drawing was kindly supplied by Dr. R. Berezney.]

charged). The latter group of proteins is often referred to either as the acidic nuclear proteins or the nonhistone chromosomal proteins, but, since both of these terms are somewhat misleading, they will be referred to here as the nonhistone proteins (NHP). The properties of both classes of proteins have been studied extensively in the past; much of this early work is described in two excellent books edited by Cameron and Jeter (12) and by Fitzsimmons and Wolstenholme (13).

The ratio of DNA:histone:NHP is 1:1:0.5 on a weight basis. The nuclear proteins can be divided into several subclasses, as shown in Table 3-1, which also gives the approximate amount of each type. The histone class is made up of only five proteins; H1, H2A, H2B, H3, and H4 (in some cases H1 is replaced by H5), whereas the NHP class is now known to contain several hundred.

Only about half of the NHP are chromatin-associated, and of these only a small proportion bind tightly to DNA (1% of total nuclear proteins or 3% of NHP). These estimates are based on the number and intensity of spots on stained 2D polyacrylamide gels. This technique has revealed some 500 protein spots per nucleus (14). However, since the limit of detection of this method is about 5000 molecules of each protein per nucleus, many regulatory molecules may escape detection because their concentration must, necessarily, be very

Table 3-1 The Protein Composition of the Nucleus[a]

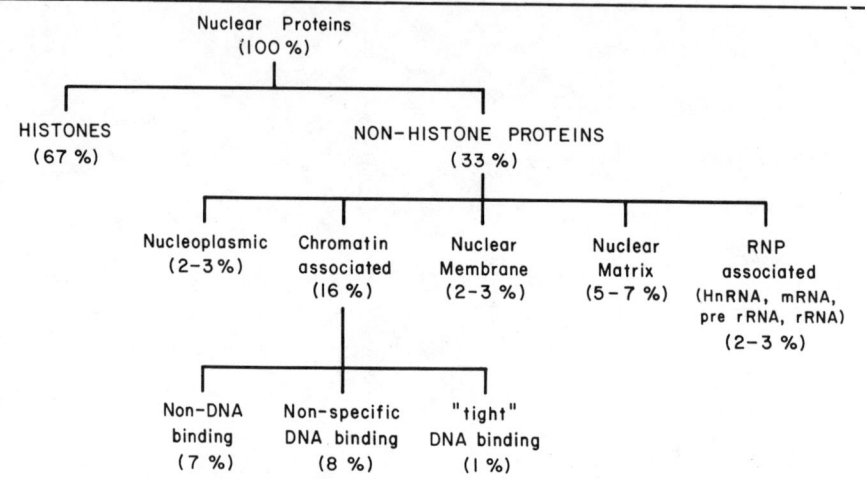

[a] The percentages are representative of the amounts of the various proteins found in a number of mammalian cell nuclei.

low. The other half of the NHP are associated with subnuclear structures such as the matrix and also with the various ribonucleoprotein particles.

There are major technical problems associated with the study of the NHP because of their insolubility in buffers of physiological ionic strength and pH. Moreover, they have a marked tendency to form aggregates with each other and with the histones. Consequently, the proteins must be extracted with high salt ($>2M$) or agents such as guanidine hydrocholoride, urea, or phenol, all of which alter or denature the proteins. However, despite these limitations, considerable progress has been made with some NHP, including the proteins of the matrix.

The Nuclear Matrix

The matrix is a fibrous skeleton that remains after the treatment of nuclei with various concentrations of salt to remove most of the chromatin and nucleoplasmic proteins (15–17). The composition of this matrix is 78% protein, 20% RNA, 1% phospholipid, and 1.5% DNA. The nucleic acids are tightly bound and can only be removed by extensive nuclease treatment that eventually reduces the amounts of RNA and DNA to 1.2% and 0.1%, respectively. Half of the protein content of the matrix is made up from just three proteins with $M_r = $ 60,000–70,000. The other half consists of a number of proteins with M_rs of 40,000–50,000, although one protein has an M_r of 200,000. No histones are present among the matrix proteins. Collectively the matrix proteins are predominantly acidic, but they also contain approximately 30% of hydrophobic residues.

The matrix appears to be generated by polymerization of these proteins to give a 20-Å-diameter fibril, which can further polymerize to generate fibers with diameters of 40–300 Å. The matrix polymers are in close contact with the inner nuclear membrane. In fact, the three major matrix proteins make up about 25% of the inner-membrane protein, being concentrated around the nuclear pore complexes.

There is now considerable evidence that the matrix plays an important role in DNA replication and perhaps also in the transcription process. All three of the major matrix proteins are DNA-binding proteins. This binding appears to be non-sequence-specific, but the proteins do prefer to bind to A:T-rich sequences. This is an interesting observation and suggests that the A:T-rich spacer sequences that punctuate the DNA may be the sites of attachment of the DNA to the matrix. There are three levels of interaction between DNA and the matrix. Seventy-five percent is non-matrix-associated, 24% is loosely associated, and 1% is very tightly associated. Experimentally, the 75% of DNA that is nonassociated can be removed by magnesium chloride precipitation. The other 25% of the DNA is magnesium chloride-soluble by virtue of the presence of RNA, suggesting that this is euchromatin that is actively being transcribed and has become loosely associated with the matrix during transcription. This loosely associated DNA can be removed by salt, indicating that it binds to the matrix via ionic bonds. The 1% of the DNA that is tightly, perhaps covalently, bound to the matrix (18) is referred to as the residual matrix DNA. In liver cells stimulated to proliferate by partial hepatectomy, examination of DNA replication by ^{3}H-thymidine labeling reveals that radioactivity first appears in the residual matrix DNA, then in the loosely associated DNA, and finally in bulk DNA, suggesting that before replication the DNA must first become attached to the matrix (17). Whether attachment to the matrix is a prerequisite for transcription remains to be established.

In common with many nuclear proteins, the proteins of the matrix can be covalently modified by phosphorylation (2), and the introduction of the negatively charged phosphate group may influence the interaction between the matrix and DNA. In regenerating liver, phosphorylation of the matrix proteins occurs during the first 12 hours, suggesting that it may play a role in the increase in RNA transcription that takes place before the increase in DNA replication. Heterogeneous nuclear RNA is associated with the matrix, and it appears to act as a structural support for the nucleic acid during processing and transport to the nuclear pores. As the properties of the matrix proteins become better understood, it is likely that they will provide great insight into the mechanisms of replication and transcription.

Other Structural Proteins of the Nucleus

The microtubule subunit proteins α- and β-tubulin have been identified as nuclear proteins (19), as have the contractile proteins actin, myosin, troponin C, and tropomyosin (20). In fact, the 200,000-molecular-weight matrix-associated protein is probably myosin. The function of these proteins is not yet known, but it is clear that the nucleus has a highly ordered substructure capable of supporting and moving the various macromolecules and macromolecular structures in a well-organized way.

The Nuclear Membrane

The inner nuclear membrane, as already described, has extensive connections with the nuclear matrix. DNA is bound to the inner nuclear membrane, which indicates that it has a functional role in the structural organization of the genetic material. The outer membrane has been well studied, and its properties are described in recent reviews by Wunderlich et al. (21) and Zbarsky (22). It has been found to contain 30–40 enzymes including cytochromes, cytochrome C reductase, and other oxido–reductases sufficient to give it a capacity for energy production. The outer membrane also contains the nucleoside triphosphatases that are necessary to provide energy for the transport of molecules into and out of the nucleus. For example, the various ribonucleoprotein particles are transported through the pore complexes into the cytoplasm, and there is extensive transport of proteins, including the histones, that are synthesized on cytoplasmic polysomes and then translocated into the nucleus. Some of these proteins may be synthesized on outer-membrane polysomes and inserted directly through the nuclear membrane.

THE STRUCTURAL ORGANIZATION OF DNA

The early work on the structure of chromatin was confined to studies on condensed heterochromatin since this was the most easily discerned microscopically. These studies revealed an extreme complexity of organization that was difficult to resolve in terms of a simple model of interaction between DNA and histones. There was little progress until the more sophisticated techniques of electron microscopy, together with advances in sample preparation, were applied to the study of both condensed and dispersed chromatin. These studies revealed that DNA exists at several levels of structural organization, from an almost completely relaxed form to the highly compacted form of the metaphase chromosome. There are several excellent reviews on this work (1, 2, 23, 24). Intermediate between these two extremes are two levels of structural complexity that are very relevant to the mechanism of transcription. These are seen in the electron microscope as a fiber of about 10 nm in diameter and a fiber of 20–30 nm in diameter as shown in Fig. 3-4. The appearance of the 10-nm fiber is remarkably similar to that seen in bacterial cells.

A major breakthrough in the study of DNA organization came from biochemical experiments carried out by Hewish and Burgoyne (25). They showed that when DNA is digested with nucleases, fragments of DNA are released that are either 200 base pairs long or multiples of 200 base pairs. When digestion is

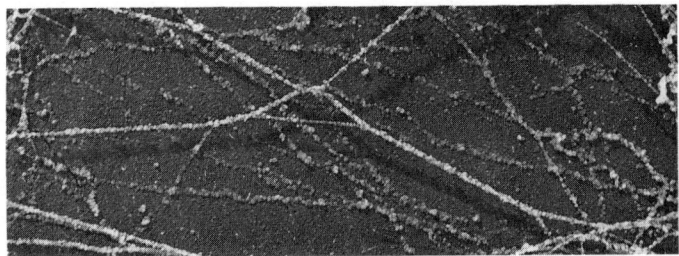

Fig. 3-4 Electron micrograph of the 10-nm and 20–30-nm fibers of dispersed chromatin. The DNA was obtained from newt erythrocytes and shows two long strands of 20–30 nm generated by condensing thinner 10-nm strands (×48,240). [The photograph was kindly supplied by Professor Hans Ris and is reproduced from Fig. 13 of Ref. 2, with permission.]

continued for a longer time period, these fragments are reduced uniformally to one of about 140 base pairs. This implied a subunit structure for DNA that may be the basis of all the higher-order structures of DNA seen in the electron microscope. The 200-base-pair and 140-base-pair fragments are easily isolated by either density-gradient centrifugation or chromatography. When studied under the electron microscope they were found to correspond in size to a single "bead" of the 10-nm "beads on a string" fiber. Biochemical studies revealed that each "bead" consisted of the fragment of DNA complexed with histones. These particles have become known as nucleosomes and are presently the objects of intensive research. The structure and composition of nucleosomes have been the subject of a number of recent reviews (Refs. 26–29, and see also Refs. 1 and 14 and Chapter 4), and some of these properties will be discussed below.

The Histone Core of the Nucleosome

The molecular weight of the 140-base-pair particle is about 200,000, with the DNA accounting for about half of it, the other half being protein. Analysis of these proteins revealed the presence of four histones: H2A, H2B, H3, and H4. Since the combined molecular weight of all four is only 54,000, it appears that there are two molecules of each histone associated with the 140-base-pair fragment of DNA. It has also been established that histone H1 is released when the 200-base-pair nucleosome is digested to the 140-base-pair particle. Thus the basic composition of the nucleosome is 200 base pairs of DNA, one molecule of histone H1, and two molecules each of H2A, H2B, H3, and H4 (30, 31).

All histones are small, very basic proteins with many lysine and arginine residues. The relative proportions of these two amino acids are used to characterize each histone as shown in Table 3-2. These charged amino acid residues are not evenly distributed throughout the molecule, but tend to be concentrated at the N- and C-terminals of the protein, producing a marked charge asymmetry as illustrated diagrammatically in the table. The central part of each histone

Table 3-2 Properties of the Histones

Histone	Characteristics	Former designation	Lys/Arg ratio	M_r	No. of residues	Charge Asymmetry	Approximate Structure
H1	lysine rich	f1	22	21,000	215		
H2A	slightly lysine rich	f2a2	1.2	14,000	129	NH$_2$ —[very basic]—[hydrophobic]—[basic]— COOH	NH$_2$ —●— COOH
H2B	slightly lysine rich	f2b	2.5	14,000	125		
H3	arginine rich	f3	0.7–0.8	15,000	135		
H4	arginine rich	f2a1	0.7–0.8	11,000	102		
H5	a lysine rich histone which replaces or partially replaces H1 in some cells.						

is relatively hydrophobic, thus the overall shape of the molecule is a globular central portion with positively charged "fingers" being free on the surface. The amino acid sequence of the histones has been remarkably well conserved during evolution. Figure 3-5 shows that histones H3 and H4 have undergone only one or two amino acid substitutions during millions of years of evolution (the amino acid sequences of the histones are given in Ref. 28). The other histones have shown more divergence, although certain critical sequences are extremely highly conserved. Such extreme conservation is testimony to the critical role histones play in the organization of the nucleosome.

Chemical methods, although indirect, have been applied to try to elucidate the organization of the histones in the core of the nucleosome after removal of the DNA. Chemical cross-linking studies have revealed that the core is indeed an octamer with two each of the histones H2A, H2B, H3, and H4 interacting by hydrophobic bonding. Using agents that span different, fixed distances it appears that histones H3 and H4 interact to form a tightly bound $H3_2$–$H4_2$ tetramer to give the particle its stability. H2A and H2B interact as H2A–H2B dimers and are bound more loosely to the tetramer. If these dimers are removed the $H3_2$–$H4_2$ tetramer is still capable of forming nucleosomelike particles with DNA. A clearer picture of the organization of the histones has emerged from an electron microscopy image-reconstruction study carried out by Klug et al. (32). As shown in Fig. 3-6a the $H3_2$–$H4_2$ tetramer forms a spool around which the DNA is wrapped. The two H3 molecules are in contact while the H4 molecules are not, in agreement with the cross-linking studies. Two dimers of H2A–H2B associate with the tetramer as shown in Fig. 3-6b (the DNA strand is omitted for clarity) and are also involved in DNA binding (see below). Also included in Fig. 3-6b is the probable binding site of histone H1 (to be considered in more

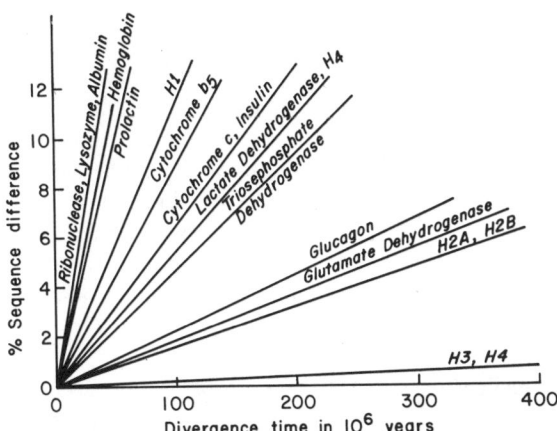

Fig. 3-5 A graphic representation of the extreme conservation of the sequences of the histones throughout evolution. [Reproduced from Ref. 30, with permission of Dr. Isenberg.]

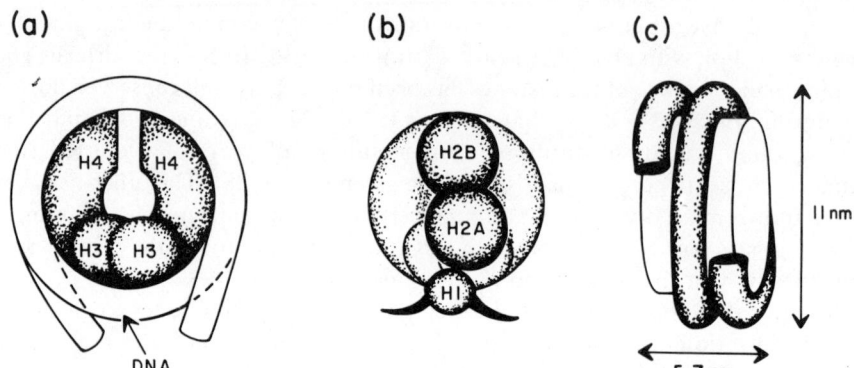

Fig. 3-6 Diagrammatic representation of the structure of the nucleosome based on the work of Klug et al. (32) and other data described in the text. (*a*) The DNA helix is wrapped around the $(H3)_2$–$(H4)_2$ tetramer. (*b*) The nucleosome histone core showing a H2A–H2B dimer superimposed on the $(H3_2$–$(H4)_2$ tetramer and the likely position of histone H1. (*c*) 1.75 turns of DNA wound on to the histone core.

detail later). This structure is actually based on model-building since the exact organization of the histone molecules has not yet been examined by a direct method. However, this model does satisfy all the cross-linking data (histone–histone and histone–DNA).

Organization of DNA on the Histone Core

The histone core is approximately 7×3.5 nm in size, and neutron scattering and X-ray diffraction studies both place the DNA on the surface of the nucleosome core (Figs. 3-6*a*, 3-6*c*). A DNA *supercoil* is wrapped around the core with 1.75 turns per nucleosome, equivalent to 146 base pairs per nucleosome. Adjacent nucleosomes are linked together by a segment of spacer DNA of about 60 base pairs in length, although it may vary from as few as 10 base pairs to as many as 100 base pairs, depending on the source of DNA. Figure 3-6*c* shows the path of DNA around the histone core (represented as a disk and turned 90° with respect to Figs. 3-6*a* and 3-6*c* for simplicity). The central turn of DNA binds to the $H3_2$-$H4_2$ tetramer, while the outer turns have more contact with the H2A-H2B dimers. It is believed that the positively charged histone "fingers" bind DNA to the core. Ionic bonds between the positively charged amino acid residues and the negatively charged DNA phosphates give the structure its stability.

Polynucleosomes and the Arrangement of Histone H1

There is now little doubt that the 10-nm "beads on a string" fiber seen in the electron microscope is a linear array of nucleosomes linked by spacer DNA. This is shown with particular clarity in the electron micrograph of Fig. 3-7, in

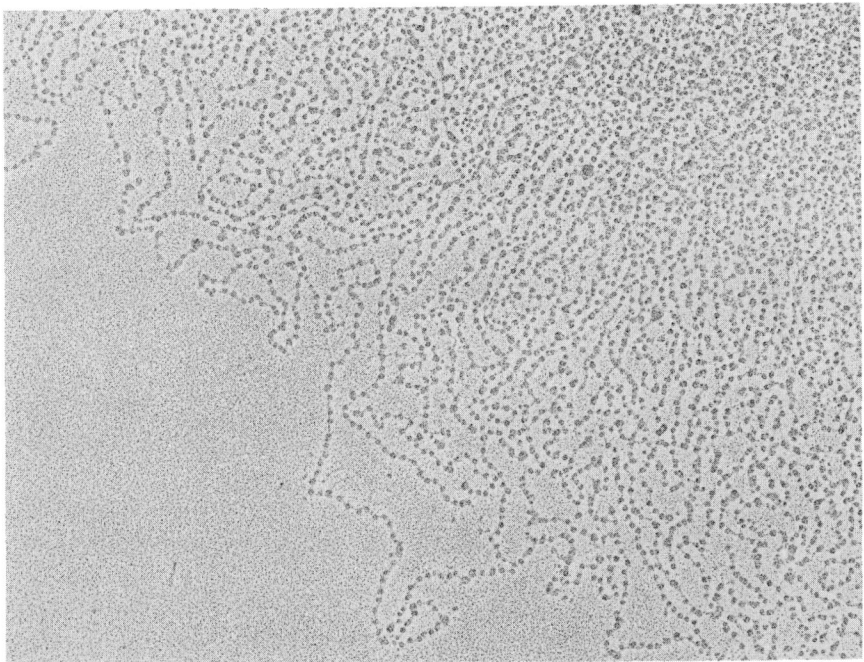

Fig. 3-7 Nucleosomal DNA. In this preparation of chromatin from *Drosophila melanogaster* the nucleosomal arrangement of the 10-nm fiber can be clearly seen (×44,075). [This micrograph was kindly provided by Drs. Steven L. McKnight and Oscar L. Miller, Jr. Reprinted from Fig. 2 of *The Cell Nucleus*, Vol. 7, pp. 97–127, Academic Press, with permission.]

which the DNA is prepared in a very relaxed state. There is evidence, however, that, rather than being uniform, the nucleosomes are arranged with a dinucleosome repeat. For example, the 10-nm fiber is often seen in a zigzag pattern in the electron microscope with the nucleosome "beads" lying on alternate sides of the "string" (28, 29). Furthermore, DNase-I studies carried out under conditions where the nuclease can only approach the fiber from one side digests DNA to dinucleosomes rather than mononucleosomes (33). These observations suggest that the 10-nm fiber is not made up simply of nucleosomes stacked end to end with DNA being maintained in a continuous supercoil. Mathematical analysis of the arrangement of DNA on the nucleosome also rules out such a simple model (29, 34).

From a study of the organization of spacer DNA, Lohr and Van Holde (35) concluded that there was indeed a dinucleosome repeat and proposed an alternative model for the winding of DNA on the nucleosomes. In this model the DNA is wound differently on alternate nucleosomes, as shown in Fig. 3-8. Such a model satisfies the mathematical criteria for nucleosomal DNA and explains the zigzag pattern and the nuclease digestion studies. When organized in this

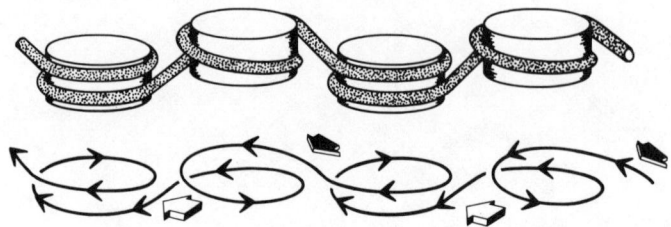

Fig. 3-8 A diagrammatic representation of the proposed winding pattern of DNA on the histone core (drawn as a disk), showing the origin of the dinucleosome repeat. The path of DNA is traced by the line drawing underneath the nucleosomes. The arrows show how this polymer could be cleaved into dinucleosomes (see text). [Adapted from Lohr and Van Holde (35) and Worcel et al. (34).]

way the fiber may exist in various states, being very compact as in Fig. 3-4 or very relaxed as in Fig. 3-7. Each state may depend upon the way in which the sample is prepared, and does not necessarily exist *in vivo*.

Some early work (reviewed in Ref. 29) suggested that nucleosomes were associated with particular DNA sequences, and the phenomenon became known as *nucleosome phasing*. However, evidence for a random arrangement was also presented (29). More recent studies (see, e.g., Ref. 36) suggest that the length of spacer DNA between nucleosomes may vary, and the nucleosomes appear to be arranged in phase with respect to particular DNA sequences. Such a nonrandom distribution of nucleosomes suggests that they are not only units of structure but may also be units of function (see Chapter 4).

Histone H1 is not a core histone, and, unlike the other histones, only one molecule per nucleosome is present. H1 is released from chromatin during digestion with nucleases, leading to speculation that it is bound onto the spacer DNA rather than being associated with the core. The precise location of H1 has still not been established, but there are indications from cross-linking and electron microscopy studies that it is located at the point where spacer DNA enters and exits the nucleosome (29, 37), as shown in Figs. 3-6*b* and 3-9. Allan et al. (37) and Boulikas et al. (38) have proposed that histone H1, which has a typical globular, hydrophobic core (2.8 nm in diameter) and polar NH_2- and COOH-terminal regions, is located in between the DNA strands at the point of entry and exit from the nucleosome. Since the pitch of the helix is also 2.8 nm, the core of H1 could conveniently fit between the strands. Cross-linking studies have shown that H1 does, indeed, make contact with core histones H3 and H2A (see Fig. 3-6*b*). The binding of histone H1 to this region of polynucleosomes is now considered to be of major importance in the formation of higher orders of chromatin structure.

Higher Orders of Chromatin Structure

The bulk of DNA in the nucleus exists as a fiber 20–30 nm in diameter, usually referred to as the *native fiber*. It is present in dispersed euchromatin and is

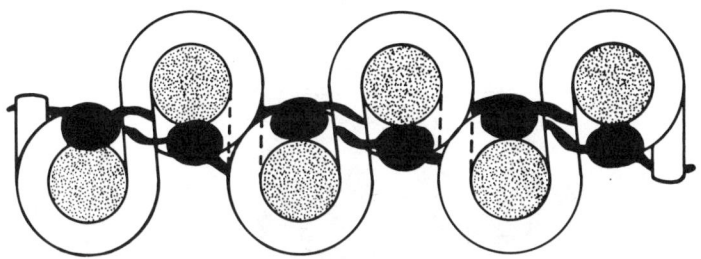

Fig. 3-9 A possible arrangement of histone H1 molecules along a string of nucleosomes. Based on a number of studies (see text), H1 makes contact with the core and with the spacer DNA strands. The binding of histone H1 compacts the nucleosome chain, increasing the DNA content per nucleosome to two full turns and generates a higher-order structure. This diagram represents an intermediate stage of this compaction and folding process.

folded to generate condensed heterochromatin. This fiber appears to be relaxed to form the 10-nm "beads on a string" fiber during gene expression, but there is still considerable uncertainty over the way in which the nucleosomes of the 10-nm fiber are packed into the native fiber and how the transition takes place. Some models favor a 25–30-nm-diameter solenoid structure generated by winding nucleosomes into a helix with six or seven nucleosomes per helical turn. Other models favor a "superbead" structure in which six to eight nucleosomes are "aggregated" together to form each superbead. The first model would generate a very uniform fiber, whereas the latter model would generate a very irregular one. Electron micrographs have been obtained that support each model. However, the conditions of sample preparation play a large role in determining the end result. For example, chromatin isolated at physiological ionic strengths and *in the presence of divalent cations* (Mg^{2+} or Ca^{2+}) produces somewhat irregular native fibers. Removal of the cations or lowering the ionic strength relaxes this fiber to the 10-nm form. Depending on the exact salt concentration, however, either solenoid or superbead structures can be obtained.

Since electron microscopy cannot yield an unequivocal answer to the structure of this fiber, other more indirect techniques have been used. Physical studies, such as X-ray diffraction and circular dichroism, support the solenoid model, whereas biochemical studies, such as nuclease digestion, support the superbead model. All of this work has been described in detail in the reviews of Ris and Korenberg (2), McGhee and Felsenfeld (29), and Hozier (39).

The nuclease digestion studies are interesting in that they show the 20–30-nm fiber to be cut initially into discrete units with each containing six to eight nucleosomes. These results are interpreted in favor of the superbead model with cuts between the beads. The solenoid-type models must explain how their apparently uniform structures could be cleaved every six to eight nucleosomes.

A more recent model that takes into account the dinucleosome repeat (Fig. 3-8) involves the stacking of every second nucleosome to generate a ribbon as

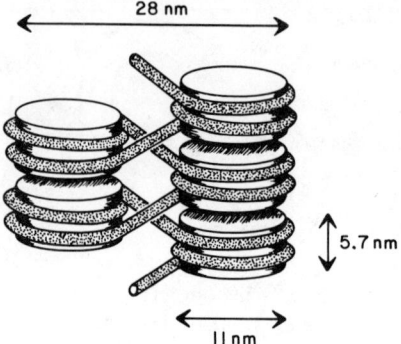

Fig. 3-10 A diagrammatic representation of the stacking of nucleosomes of the 10-nm fiber to generate an ~28-nm fiber. This diagram is based on a model proposed by Worcel et al. (34), taking into account the proposed winding of DNA on the histone cores as depicted in Figs. 3-8 and 3-9. In this structure histone H1 would bind along the central axis of the ribbon. [Adapted from Ref. 34, with permission.]

shown in Fig. 3-10. This ribbon has dimensions of the native fiber, and it is proposed (34) that the irregular appearance of this fiber is caused by local variations in either the length or the orientation of the spacer DNAs. Some spacers may be particularly sensitive to nuclease attack, yielding superbeadlike particles upon digestion. This model also satisfies the mathematical requirements for the folding of DNA into higher-order structures.

Although there is controversy over which model is correct or whether the native fiber is capable of adopting each of these structures in a particular microenvironment, it is unanimously agreed that histone H1 is involved in maintaining the integrity of the native fiber (see the reviews cited above). The 20–30-nm fiber is reduced to a 10-nm fiber when H1 histone is selectively removed from chromatin (40). Moreover, polynucleosomes can be reversibly interconverted between 10-nm and 20–30-nm fibers by manipulating their interaction with histone H1 (39).

All of the models for the native fiber place the H1 histone in the center of the solenoid or superbead such that H1–H1 interactions maintain the structure. For example, in the model described in Fig. 3-10 the H1 molecules would be postulated to bind along the central axis of the ribbon (cf. Fig. 3-9), making contact with spacer DNA and with each other.

A role for histone H1 in the packing of the nucleosome into the native fiber is supported by chemical cross-linking studies (41, 42). Exposure of the native fiber to certain cross-linking agents yields a H1-homopolymer with as many as 12 H1 molecules cross-linked together. The lengths of the cross-linking agents indicate that H1 molecules are less than 1 nm apart and may even be in contact with each other. The fact that only polymers with 12 H1 molecules, or multiples of 12, are found indicates some structural feature that alters the distance between (or replaces) H1 molecules after every twelfth nucleosome. It is likely that this change confers nuclease sensitivity since superbeads with 12 nucleosomes have been isolated from liver nuclei digested with an endogenous nuclease (43). Further studies on the position of the H1 in this fiber should contribute greatly to our knowledge of higher orders of chromatin structure. At the present time it appears that H1 folds the linker DNA into a higher-order configuration (32, 37) such that there are about two turns of DNA per nucleosome rather than 1.75, bringing the nucleosomes closer together. The regulation of this process is of considerable importance in the mechanism of gene expression.

The packing ratio (i.e., the amount by which the DNA molecule is forshortened) of DNA in the native fiber is 28:1, whereas in the metaphase chromosome it is several thousand to one. Therefore the native fiber must be folded considerably to generate such a compact form. The structural basis of these still higher levels of chromatin structure is poorly understood and will not be discussed further in this section (see Refs. 2 and 23 for more details).

Other Structural Proteins Associated with Nucleosomes

At the present time comparatively little is known about other proteins that may be associated with the nucleosomes, with the exception of one group of proteins, the high-mobility group (HMG) proteins. Two other nucleosome-associated proteins, ubiquitin and protein A24, have also been characterized. The HMG proteins are so named because of their very high mobility in gel electrophoresis systems. They were initially found as contaminants of histone preparations, and since they were usually run off the edge of the gels during electrophoresis of histones they were ignored until quite recently. During the past few years they have become recognized as chromatin-associated proteins of some importance, although their exact function has not been elucidated. An extensive review of their properties has recently been published (44). There are four major proteins of this group—HMG1, HMG2, HMG14, and HMG17—and a number of other minor components have recently been identified (45).

HMG1 and HMG2 have molecular weights of approximately 26,000, and they appear to be very similar proteins; for example, there is only one difference in the sequence of the first 20 amino acids. Both molecules are very highly charged with a marked asymmetry in charge distribution. Their N-terminal regions are rich in lysine residues and are very basic, whereas the C-terminal regions are rich in aspartic acid and glutamic acid residues, making them very acidic. The proteins carry an overall negative charge. In addition to sequence similarity near the N-terminus they both have a sequence near the middle of the molecule that is

$$\ldots \text{LYS-LYS-LYS-(ASP/GLU)}_{41}\text{-LYS-LYS} \ldots$$

in which there is a sequence of 41 residues consisting of only aspartic and glutamic acids. In HMG1 the ratio of ASP/GLU is 3:1, and in HMG2 it is 5:1. In addition there are hydrophobic areas that give both molecules an organized secondary structure in solution.

Unusual amino acid sequences are also found in HMG14 and HMG17. Both of these proteins of $M_r = 10,000$ are also highly charged molecules. HMG17 has a sequence of 6 prolines, 3 lysines, and a glutamic acid residue in the center of the molecule, a very unusual combination of amino acids. HMG17 has been found to share sequence homology with histone H1 and with HMG protein H6 (this is not a histone, but a HMG protein from trout testis). H6 also has some sequence similarity to HMG14.

There are approximately 10^6 molecules of each HMG protein per nucleus. This is too high a number to be regulatory (cf. the 10^4–10^5 polymerase molecules per nucleus) and too low a number for there to be one molecule per nucleosome (there are approximately 10^7 nucleosomes per nucleus). Nuclease digestion studies (46–48) suggest that the HMG proteins are present on the sequences of DNA that are actively transcribed. However, they also seem to be present in satellite DNA (49), ruling out the possibility that they are associated exclusively with active sequences.

HMG14 and HMG17 appear to bind to two sites on each nucleosome. The sites are located near the points of entry and departure of the DNA, and either HMG14 or HMG17 may bind (50, 51). There is no marked conformational change in the nucleosome upon binding, but the nucleosomes are rendered more sensitive to nucleases such as DNase-I that attack core-bound DNA. HMG1 and HMG2, on the other hand, do not bind to the nucleosome but may replace histone H1 on some DNA spacers (52). This also leads to an increased sensitivity to nucleases, particularly those that cut internucleosomal DNA (micrococcal nuclease, for example). HMG1 and HMG2 do not alter DNA binding at the nucleosome level, but apparently they alter the higher-order structure of DNA. Cross-linking experiments have shown that HMG1 molecules must, when present, be very close together since cross-linked trimers can be isolated (53). Thus it appears likely that all four HMG proteins bind to chromatin fibers at specific intervals, causing a change in the 25–30-nm fiber such that it can be transcribed more easily. It has recently been shown (54, 55) that HMG1 and HMG2 prefer to bind to single-stranded DNA. Such proteins are *helix destabilizing* proteins since they melt the DNA in order to expose the single strand, and this may be the mechanism by which they alter DNA structure or facilitate transcription.

Ubiquitin, sometimes referred to as HMG20, is a small protein of $M_r = 8000$ that is associated with chromatin of nearly all cells. Its function is not known but appears to be structural since it is sometimes covalently attached to histone 2A. This ubiquitin–H2A complex is also known as protein A24 (56). There is approximately one ubiquitin molecule for every five nucleosomes, and since it is transiently lost from chromatin during mitosis (57) it almost certainly is involved in maintaining the configuration of interphase DNA. Further work on ubiquitin and on the other HMG proteins should reveal interesting insight into the higher-order structure of DNA and, more importantly, how structure is related to function.

REFERENCES

1. B. Lewin (1975). *Gene Expression*, Vol. 2. John Wiley & Sons, London.
2. H. Ris and J. Korenberg (1979). Chromosome structure and levels of chromosome organisation. In D. M. Prescott and L. Goldstein (Eds.). *Cell Biology: A Comprehensive Treatise*, Vol. 2. Academic Press, New York, pp. 268–361.
3. C. J. Bostock and A. T. Sumner (1978). *The Eukaryotic Chromosome*. Elsevier/North-Holland Biomedical Press, Amsterdam.
4. R. J. Britten and D. E. Kohne (1968). Repeated sequences in DNA. *Science* **161**, 529–540.
5. E. H. Davidson (1976). *Gene Activity in Early Development*. Academic Press, New York, pp. 187–243.
6. J. Bonner, J. M. Sala-Trepat, W. R. Pearson, and J.-R. Wu (1978). Mammalian chromatin: structure, expression and sequence organisation. In H. Busch (Ed.). *The Cell Nucleus*, Vol. 6. Academic Press, New York, pp. 369–407.
7. E. M. Bradbury, N. Maclean, and H. R. Matthews (1981). *DNA, Chromatin and Chromosomes*. Blackwells, Oxford.
8. C. J. Bostock (1980). The organisation of DNA sequences in chromosomes. In L. Goldstein and D. M. Prescott (Eds.). *Cell Biology: A Comprehensive Treatise*, Vol. 3. Academic Press, New York, pp. 1–59.
9. J. Moreau, L. Matyash-Smirniaguina, and K. Scherrer (1981). Systematic punctuation of eukaryotic DNA by A+T-rich sequences. *Proc. Natl. Acad. Sci. USA* **78**, 1341–1345.
10. W. F. Doolittle and C. Sapienza (1980). Selfish genes, the phenotype paradigm and genome evolution. *Nature* **284**, 601–603.
11. L. E. Orgel and F. H. C. Crick (1980). Selfish DNA: the ultimate parasite. *Nature* **284**, 604–607.
12. I. L. Cameron and J. R. Jeter, Jr. (Eds.) (1974). *Acidic Proteins of the Nucleus*. Academic Press, New York.
13. D. W. Fitzsimmons and G. E. W. Wolstenholme (Eds.) (1975). *The Structure and Function of Chromatin*, CIBA Foundation Symposium No 28. Elsevier, New York.
14. J. O. Thomas (1978). Chromatin structure. In B. F. C. Clark (Ed.). *Biochemistry of Nucleic Acids II*, Vol. 17, International Review of Biochemistry. University Park Press, Baltimore, pp. 181–232.
15. D. E. Comings (1978). Compartmentalization of nuclear and chromatin proteins. In H. Busch (Ed.). *The Cell Nucleus*, Vol. 4. Academic Press, New York, pp. 345–371.
16. R. Berezney (1979). Dynamic properties of the nuclear matrix. In H. Busch (Ed.). *The Cell Nucleus*, Vol. 7. Academic Press, New York, pp. 413–456.
17. J. H. Shaper, D. M. Pardoll, S. H. Kaufmann, E. R. Barrack, B. Vogelstein, and D. S. Coffey (1979). The relationship of the nuclear matrix to cellular structure and function. *Adv. Eng. Reg.* **17**, 213–248.
18. W. Krauth and D. Werner (1979). Analysis of the most tightly bound proteins in eukaryotic DNA. *Biochim. Biophys. Acta* **564**, 390–401.
19. W. M. LeStourgeon, R. Totten, and A. Forer (1974). The nuclear acidic proteins in cell proliferation and differentiation. In I. L. Cameron and J. R. Jeter, Jr. *Acidic Proteins of the Nucleus*. Academic Press, New York, pp. 159–190.
20. W. M. LeStourgeon (1978). The occurrence of contractile proteins in nuclei and their possible functions. In H. Busch (Ed.). *The Cell Nucleus*, Vol. 7. Academic Press, New York, pp. 305–326.
21. F. Wunderlich, R. Berezney, and H. Kleinig (1976). The nuclear envelope: An interdisciplinary analysis of its morphology, composition and functions. In D. Chapman and D. F. H. Wallach (Eds.). *Biological Membranes*. Academic Press, London, pp. 241–333.

22. I. B. Zbarsky (1978). An enzyme profile of the nuclear envelope. *Int. Rev. Cytol.* **54**, 295-360.
23. G. P. Georgiev, S. A. Nedospar, and V. V. Bakayaev (1978). Supranucleosomal levels of chromatin organisation. In H. Busch (Ed.). *The Cell Nucleus*, Vol. 6. Academic Press, New York, pp. 35-74.
24. U. Scheer, H. Spring, and M. F. Trendelenburg (1979). Organisation of transcriptionally active chromatin in lampbrush chromosome loops. In H. Busch (Ed.). *The Cell Nucleus*, Vol. 7. Academic Press, New York, pp. 4-47.
25. D. R. Hewish and L. A. Burgoyne (1973). The digestion of chromatin DNA at regularly spaced sites by a nuclear deoxyribonuclease. *Biochem. Biophys. Res. Commun.* **52**, 504-510.
26. R. A. Garrett (1979). The structure of chromatin. In R. E. Offord (Ed.). *Chemistry of Macromolecules, IIB*, Vol. 25. International Review of Biochemistry. University Park Press, Baltimore, pp. 179-203.
27. D. M. J. Lilley and J. F. Pardon (1979). Structure and function of chromatin. *Ann. Rev. Genet.* **13**, 197-233.
28. R. L. Rill (1979). Nucleosomes: composition and substructure. In J. H. Taylor (Ed.). *Molecular Genetics, Part III*. Academic Press, New York, pp. 247-313.
29. J. D. McGhee and G. Felsenfeld (1980). Nucleosome structure. *Ann. Rev. Biochem.* **49**, 1115-1156.
30. I. Isenberg (1977). Physical studies of the inner histones (H2a, H2b, H3, H4). In B. Kaminar (Ed.). *Search and Discovery: A Tribute to Albert Szent Györgi*. Academic Press, New York, pp. 195-215.
31. I. Isenberg (1978). Protein-protein interactions of histones. In H. Busch (Ed.). *The Cell Nucleus*, Vol. 4. Academic Press, New York, pp. 135-154.
32. A. Klug, D. Rhodes, J. Smith, J. T. Finch, and J. O. Thomas (1980). A low resolution structure for the histone core of the nucleosome. *Nature* **287**, 509-516.
33. L. A. Burgoyne and J. D. Skinner (1981). Chromatin superstructure: the next level of structure above the nucleosome has an alternating character. A two nucleosome based series is generated by probes armed with DNase-I acting on isolated nuclei. *Biochem. Biophys. Res. Commun.* **99**, 893-899.
34. A. Worcel, S. Strogatz, and D. Riley (1981). Structure of chromatin and the linking number of DNA. *Proc. Natl. Acad. Sci. USA* **78**, 1461-1465.
35. D. Lohr and K. E. Van Holde (1979). Organization of spacer DNA in chromatin. *Proc. Natl. Acad. Sci. USA* **76**, 6326-6330.
36. E. R. Shelton, P. M. Wassarman, and M. L. DePamphilis (1980). Structure, spacing, and phasing of nucleosomes on isolated forms of mature simian virus 40 chromosomes. *J. Biol. Chem.* **255**, 771-782.
37. J. Allan, P. G. Hartmann, C. Crane-Robertson, and F. X. Aviles (1980). The structure of histone H1 and its location in chromatin. *Nature* **288**, 675-681.
38. T. Boulikas, J. M. Wiseman, and W. T. Garrard (1980). Points of contact between histone H1 and the histone octamer. *Proc. Natl. Acad. Sci. USA* **77**, 127-131.
39. J. C. Hozier (1979). Nucleosomes and higher levels of chromosomal organisation. In J. H. Taylor (Ed.). *Molecular Genetics, Part III*. Academic Press, New York, pp. 315-385.
40. F. Thoma and Th. Koller (1977). Influence of histone H1 on chromatin structure. *Cell* **12**, 101-107.
41. A. V. Itkes, B. O. Glotov, L. G. Nikolaev, S. R. Preen, and E. S. Severin (1980). Repeating oligonucleosomal units. A new element in chromatin structure. *Nuc. Acid Res.* **8**, 507-527.
42. J. O. Thomas and J. A. Khabaza (1980). Cross-linking of histone H1 in chromatin. *Eur. J. Biochem.* **112**, 501-511.

REFERENCES

43 W. H. Strätling and R. Klingholz (1981). Supranucleosomal structure of chromatin: Digestion by calcium/magnesium endonuclease proceeds via a discrete size class of particles with elevated stability. *Biochemistry* **20**, 1386–1392.

44 G. H. Goodwin, J. M. Walker, and E. W. Johns (1978). The high mobility group (HMG) nonhistone chromosomal proteins. In H. Busch (Ed.). *The Cell Nucleus*, Vol. 6. Academic Press, New York, pp. 181–219.

45 G. H. Goodwin, E. Brown, J. M. Walker, and E. W. Johns (1980). The isolation of three new high mobility group nuclear proteins. *Biochim. Biophys. Acta* **623**, 329–338.

46 V. V. Bakayev, V. V. Schmatchenko, and G. P. Georgiev (1979). Subnucleosome particles containing high mobility group proteins HMG-E and HMG-G originate from transcriptionally active chromatin. *Nuc. Acid Res.* **7**, 1525–1540.

47 S. Weisbrod, M. Groudine, and H. Weintraub (1980). Interaction of HMG14 and 17 with actively transcribed genes. *Cell* **19**, 289–301.

48 E. I. Georgieva, I. G. Pashev, and R. G. Tsanev (1981). Distribution of high mobility group and other acid-soluble proteins in fractionated chromatin. *Biochim. Biophys. Acta* **652**, 240–244.

49 C. G. P. Mathew, G. H. Goodwin, T. Igo-Kemenes, and E. W. Johns (1981). The protein composition of rat satellite chromatin. *FEBS Lett.*, 25–29.

50 J. K. W. Mardian, A. E. Paton, G. J. Bunick, and D. E. Olins (1980). Nucleosome cores have two specific binding sites for nonhistone chromosomal proteins HMG14 and HMG17. *Science* **209**, 1534–1536.

51 S. C. Albright, J. M. Wiseman, R. A. Lange, and W. T. Garrard (1980). Subunit structures of different electrophoretic forms of nucleosomes. *J. Biol. Chem.* **255**, 3673–3684.

52 J. B. Jackson and R. L. Rill (1981). Circular dichroism, thermal denaturation, and deoxyribonuclease I digestion studies of nucleosomes highly enriched in high mobility group proteins HMG1 and HMG2. *Biochemistry* **20**, 1042–1046.

53 A. V. Itkes, B. O. Glotov, L. G. Mikolaev, and E. S. Severin (1980). Clusters of nonhistone chromosomal protein HMG1 molecules in intact chromatin. *FEBS Lett.* **118**, 63–66.

54 J. B. Jackson, J. M. Pollock, Jr., and R. L. Rill (1979). Chromatin fractionation procedure that yields nucleosomes containing near-stoichiometric amounts of high mobility group nonhistone chromosomal proteins. *Biochemistry* **18**, 3739–3748.

55 K. Javaherian and M. Sadeghi (1979). Nonhistone proteins HMG1 and HMG2 unwind DNA double helix. *Nuc. Acid Res.* **6**, 3569–3580.

56 I. L. Goldknopf and H. Busch (1978). Modification of nuclear proteins: the ubiquitin-histone H2A conjugate. In H. Busch (Ed.). *The Cell Nucleus*, Vol. 6. Academic Press, New York, pp. 149–180.

57 S.-I. Matsui, B. K. Seon, and A. A. Sandberg (1979). Disappearance of a structural chromatin protein A24 in mitosis: implications for molecular basis of chromatin condensation. *Proc. Natl. Acad. Sci. USA* **76**, 6386–6390.

58 W. W. Franke, U. Scheer, H. Spring, M. F. Trendelenburg, and H. Zentgraf (1979). Organisation of nuclear chromatin. In H. Busch (Ed.). *The Cell Nucleus*, Vol. 7. Academic Press, New York, pp. 49–95.

CHAPTER FOUR

Regulation of Gene Expression at the Transcriptional Level in Eukaryotes

Genes that are actively being transcribed, or are available for transcriptional activation, reside in the dispersed euchromatin, whereas those not to be expressed probably reside in highly compacted regions of heterochromatin. Dispersed chromatin consists of two fibers—the 10-nm fiber and the 20–30-nm native fiber. The relative amount of the two fibers *in vivo* is unknown, but since as much as half of the total nuclear DNA is dispersed, much of it must be in the more compact 20–30-nm fiber. Intuitively, one expects a gene in the process of transcription to exist either as a fully relaxed Watson–Crick double helix or at most as a 10-nm "beads on a string" fiber. It is difficult to imagine how a RNA polymerase molecule could initiate transcription on DNA folded into 20–30-nm fibers. However, this does not exclude the possibility that potentially active genes, not in the process of transcription, reside in this higher-level structure. Furthermore, there are DNA sequences interspersed among active genes that are not transcribed, for example, the spacer DNA between rDNA transcription units (Fig. 3-2). In fact, most of the single-copy DNA is interspersed with these repetitive DNA sequences. This raises the question of the structural organization of active and inactive DNA such that RNA polymerase can bind to and transcribe active genes. The first part of this chapter concentrates on this question because it is of fundamental importance for an understanding of the mechanism of transcription in eukaryotic cells and the regulation of this process.

THE NATURE OF ACTIVE GENES

The Nuclease Sensitivity of Active Gene Sequences

When nuclei are incubated *in vitro* with nucleases such as micrococcal (staphylococcal) nuclease, DNase-I, or DNase-II, a portion of the chromatin is released as discrete fragments consisting of nucleosomes and polynucleosomes. Examination of the released DNA with complementary DNA (cDNA) probes derived from cytoplasmic messenger RNA reveals that it is enriched in active gene sequences [see the review of Allfrey (1) for references]. For example, the digestion of hen oviduct nuclei with any of these nucleases results in the preferential solubilization of the ovalbumin gene. However, if liver nuclei are digested the ovalbumin gene is not solubilized, because this gene is not active in liver cells. Similar experiments have established that in chicken reticulocytes globin sequences are preferentially solubilized, while ovalbumin sequences are not. Thus, these nucleases are able to recognize and cleave DNA in an active configuration in preference to the inactive DNA. (This is only preferential and not absolute since prolonged digestion leads to the eventual solubilization of inactive DNA). This observation has permitted a biochemical examination of the DNA of active genes.

The mechanism of action of these nucleases is not known, but it is believed that they recognize a structural feature of chromatin that is characteristic of *active* chromatin. Since the mechanism of action of individual nucleases is different, active chromatin must have several structural differences from inactive chromatin. In general, micrococcal nuclease cleaves DNA between the nucleosomes whereas DNase-I cleaves both internucleosomal DNA and nucleosomal DNA. If digestion is continued long enough, DNase-I cuts DNA into 10 base-pair fragments, indicating that the winding of DNA onto the nucleosome core gives a structural change every 10 base pairs (2). When the digestion conditions are sufficiently mild, micrococcal nuclease cleaves active genes first, followed by inactive gene sequences and finally bulk chromatin (3, 4). The active gene sequences are released essentially intact, which indicates that the nuclease recognizes a structural change in chromatin that lies very close to the beginning and end of transcription units. If these released gene sequences are then challenged in a second digestion with DNase-I they are found to be 20–50 times more sensitive than bulk chromatin, showing that the DNA within the gene is in an altered configuration (5).

Nuclease Sensitivity and Nucleosome Phasing

In Chapter 3 it was concluded that, despite evidence that appeared to favor a random distribution of nucleosomes along DNA (6), nonrandom phasing of nucleosomes probably does occur. This idea is supported by recent work on the nucleosomal arrangement of a number of genes including tRNA genes (7), 5S rRNA genes (8), histone genes (3), and "heat-shock" protein genes (4). Thus, rather than being simply units of structure, the nucleosomes also contribute to the function of chromatin.

The organization of the nucleosomes along the 5000 base-pair repeat unit of the amplified histone genes of *Drosophila* is a particularly good example of the nonrandom distribution of nucleosomes (3). All five histone genes are located close to each other on the *Drosophila* genome, as shown in Fig. 4-1. Transcription of these genes is divergent, with each arginine-rich histone (H3 and H4) and each lysine-rich histone (H2A and H2B) being transcribed in opposite

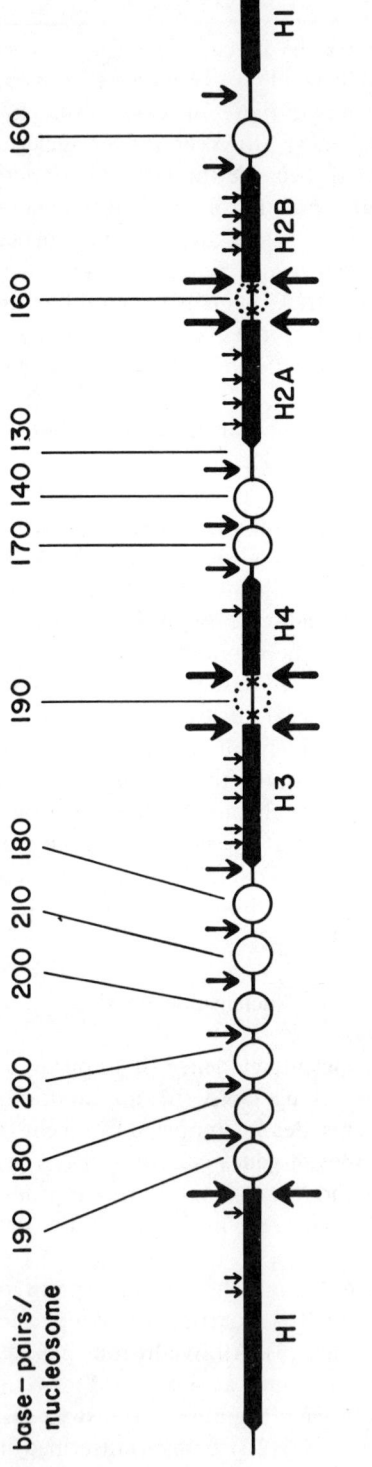

Fig. 4-1 The structural organization of the histone genes of *Drosophila*. A repeating unit is shown illustrating the direction of transcription of each gene. The circles represent the position of nucleosomes along the spacer DNA, and the number of base pairs of DNA associated with each is given. The vertical arrows above the genes represent sites of micrococcal nuclease cleavage. The arrows below the genes represent the DNase-I-sensitive sites. The size of the arrows gives an indication of the relative sensitivity of the sites. The × symbols near H2A, H2B, H3, and H4 are the sites of the —TATA— box (see text). [Reproduced from Fig. 5 of Ref. 3 with the permission of Dr. A. Worcel. Copyright 1981 The Massachusetts Institute of Technology.]

directions. The sequences between the genes are not transcribed. This unit is repeated as many as 100 times along the *Drosophila* genome. Fine-structure mapping of the micrococcal nuclease and DNase-I cleavage sites produced the nucleosome arrangement shown in Fig. 4-1. The intergene spacer DNA is not cleaved into uniform fragments. Instead, fragments ranging in size from 130 base pairs to more than 210 base pairs are found, indicating that each nucleosome has a characteristic amount of associated DNA.

All five genes have a site very close to the start of transcription that is exceptionally sensitive to both micrococcal nuclease and DNase-I. A similar nuclease sensitivity is found at the 5′ end of the preproinsulin-II gene in tissues in which it is expressed, but not in tissues such as liver or kidney that do not synthesize insulin (9). The 5′ end of the *Drosophila* "heat-shock" protein gene is similarly hypersensitive (see the references in Ref. 9).

In contrast, the coding sequences of the genes are much less sensitive to micrococcal nuclease and almost totally insensitive to DNase-I. The sites on the genes that are cleaved by micrococcal nuclease are atypical in that the intervals do not correspond to the usual internucleosome repeat. This suggests that the arrangement of DNA in the active regions is very different from that of the adjacent inactive regions. It is possible that either chromatin in the act of transcription is not organized into nucleosomes in the conventional way, or that they are continually undergoing changes during transcription. The lack of DNase-I sensitivity is in contrast to that described earlier for the ovalbumin and other active genes. Genes as small as the histone genes may be atypical in that proteins associated with transcription such as RNA polymerase and perhaps regulatory proteins occupy and protect a much larger fraction of the total gene sequence than in longer genes. The longer genes may have segments that are similarly arranged, together with segments that are more conventionally organized into nucleosomes. Thus, nucleosome phasing may be "induced" by DNA-binding proteins rather than being caused by the association of histones with specific DNA sequences.

Structural Correlates of Nuclease Sensitivity

Active genes are nuclease-sensitive because of structural changes in the arrangement of DNA in the 10-nm fiber. At the present time these changes are not understood, although a few clues have been found. For example, in the last chapter we discussed that the HMG proteins are associated preferentially, although not exclusively, with active gene sequences. Moreover, the binding of HMG proteins 14 and 17 to nucleosome cores rendered the DNA more sensitive to DNase-I. In addition the presence of HMG proteins 1 and 2 on spacer DNA rendered it more sensitive to micrococcal nuclease. It is likely that the presence of these proteins and possibly other nonhistone proteins produces marked changes in the phasing of nucleosomes and on the amount of DNA associated with each nucleosome. Moreover, it has also been shown that the histones themselves can be modified. For example, the histone H4 released from "active" nucleosomes is heavily acetylated. Histone H1 may also be covalently modified, causing it to interact differently with histone cores or with spacer DNA in such a way that nucleosome phasing is altered (10). Histone modifications and the role they play in the regulation of transcription are discussed in a later section.

The net effect of these changes is the alteration of the *structure* of the gene in the vicinity of the promoter, and this change can be recognized by RNA polymerase. The enzyme binds, and transcription commences. More than 85% of RNA polymerase-II is released from chromatin during brief digestions that solubilize only 8–10% of the total DNA (5).

The Binding of RNA Polymerase to Active Genes

For the more than 100 years that chromatin has been known to consist of a mixture of histone and DNA it has been generally assumed that the histones are responsible for repressing the part of the DNA that is transcriptionally inactive. Transcriptionally active DNA was presumed to be free of histones. The discovery of the nucleosome as the unit of chromatin structure, together with the observation that "active" DNA is organized into nucleosomes, has brought about a fundamental change in our view of gene expression. We are now faced with the task of explaining how the sense strand of DNA can be transcribed when it is wound around the histone octamer. Additionally, we must address the problem of the nature of the interaction between regulatory molecules and nucleosome-associated DNA.

The winding of DNA onto the histone octamer does not necessarily prevent its interaction with other proteins. As we have already discussed, nucleases can bind to and cleave the DNA, particularly the DNA associated with active genes. However, mammalian RNA polymerases are much bigger molecules; indeed, they approach the size of nucleosomes. The question is whether polymerase can form an initiation complex with DNA still bound to the histone core or whether the DNA must first be dissociated. Furthermore, how can polymerase translocate along DNA that is wound into nucleosomes? This problem has been approached by a number of people (11, 12) who have used *E. coli* RNA polymerase and tested its ability to transcribe DNA organized into nucleosomes. These studies show that RNA polymerase *can* bind to nucleosomes and transcribe the 150–200 base pairs of associated DNA. Moreover, a stable initiation complex can be formed without the DNA having to be completely liberated from the histone core (although in these experiments the strength of the histone–DNA interaction was first weakened by the presence of a high salt concentration). Nucleosomes treated with cross-linking agents to strengthen the histone–DNA interaction could not be transcribed.

Translocation is believed to involve some kind of displacement of the histones, however transiently, but the mechanism is not known. It should be emphasized that this work was carried out with the bacterial enzyme that is known to bind to, and initiate transcription at, nonspecific sites on mammalian

DNA. Thus it does not necessarily follow that mammalian RNA polymerases, known to be much more fastidious in their binding requirements, can also transcribe nucleosome-organized DNA.

In a related study Chao et al. (13, 14) used a novel approach to demonstrate that regulatory proteins can recognize and bind to specific DNA sequences on nucleosome-organized DNA. They took a fragment of DNA containing the *lac*-operon operator sequence and used histone octamers to form nucleosomes, then challenged the structure with *lac* repressor. Surprisingly, they found that the protein could locate and bind very tightly to the *lac* operator. Moreover, binding still occurred when the DNA was cross-linked to the histone core but at a lower affinity, suggesting that a conformational change in the nucleosome is required for high-affinity binding.

Since *lac* repressor protein is known to recognize and bind to one "edge" of the DNA helix as outlined in Chapter 2, the DNA must be wound onto the core in such a way that this "edge" of the operator sequence is exposed, providing more evidence that DNA does not associate randomly with nucleosomes but adopts particular configurations such that certain sequences are exposed on the outside of the structure (see also Ref. 2).

This work demonstrates that the winding of DNA onto nucleosome cores does not prevent either its transcription or the recognition of regulatory sequences by protein molecules. However, the histones may have to be modified to give proteins access to the DNA, and this is the basis of transcriptional activation. It is likely that without such modifications the DNA is too tightly bound to the histone octamer to permit interaction with other proteins.

THE TRANSCRIPTION PROCESS

Although still poorly understood, it is known that portions of the genome that are available for transcription differ structurally from the bulk of the chromatin that is not transcribed. The modifications appear to involve both the presence of nonhistone proteins and changes in the histones of the active nucleosomes. In this section we consider two more aspects of the transcription process, the nature of mammalian RNA polymerases and the sequences that they may recognize on the DNA template.

Multiplicity of Mammalian RNA Polymerases

There are three different RNA polymerase enzymes in the nuclei of eukaryotic cells, and a number of recent reviews summarize their properties (15–18). The generally accepted terminology for the enzymes is polymerases I, II, and III, although they are sometimes referred to as A, B, and C, respectively (see Ref. 18). All three enzymes are large multisubunit molecules with molecular weights of several hundred thousand. The structural and functional properties of the enzymes are summarized in Table 4-1.

Table 4-1 The RNA polymerases of Eukaryotic Cells[a]

Polymerase	Location	Stable structures	Subunits ($M_r \times 10^{-3}$)									Total M_r	Product
			large subunits			small subunits							
Type I	Nucleolus	I_A	195	130		—	49	40			19	384,000 ⎫	45 S rRNA precursor
		I_B	195	130		65	49	40			19	450,000 ⎭	
Type II	Nucleoplasm	II_A	220	140				41	29	19.5	19	⎫	HnRNA
		II_B	214	140				41	29	19.5	19	~520,000	
		II_C	180	140				41	29	19.5	19	⎭	
Type III	Nucleoplasm	III_A	155	138	89	70	53	49	41	32	29	19 ⎫ ~700,000	4S, 5S RNA
		III_B	155	138	89	70	53	49	41	32	29	19 ⎭	

[a] Each polymerase can exist in a number of stable configurations with the probable number of subunits indicated (M_r = molecular weight ratio).

In general, all three enzymes, each of which can exist in a number of stable forms, consist of two large subunits in a 1:1 ratio together with a number of smaller subunits. The smaller subunits are usually in a 1:1 ratio, although the exact number of subunits and their ratios vary from species to species. Some of the smaller subunits appear to be common to more than one enzyme. For example, the $M_r = 19{,}000$ protein is found in all three polymerases, whereas the $M_r = 49{,}000$ protein is present on type-I and -III polymerases only. In addition to the three nuclear enzymes a different enzyme is present in mitochondria and chloroplasts. This enzyme is a monomer with $M_r = 60{,}000$ and is capable of carrying out all the functions of a DNA-dependent RNA polymerase. This suggests that the eight or nine additional subunits of polymerases I–III could be involved in the *regulation* of enzyme activity.

The type-I enzyme is restricted to the nucleolus and transcribes only rDNA to generate the 45S rRNA precursor. The enzyme has been purified to homogeneity from rat liver (18) and is capable of transcribing 45S RNA from isolated nucleoli. The enzyme exists in two states; I_A and I_B. Type I_B is the active form of the enzyme that results from the presence of an additional subunit and is found bound to nucleolar chromatin. The type-III enzyme also exists in two states, is nucleoplasmic, and is responsible for the transcription of the genes for transfer RNA and other small nuclear RNA species. The type-II enzyme exists in three states, but it is not yet known how these forms are interrelated. All three forms of this enzyme are nucleoplasmic and are responsible for the synthesis of heterogeneous nuclear RNA, the nuclear RNA that is processed to give messenger RNA, and also some small nuclear RNA species.

Studies on these enzymes have revealed some anomalous properties, one of which is particularly disturbing. The enzymes have considerable difficulty initiating transcription on intact double-stranded DNA. Where activity has been detected it can usually be attributed to the presence of single-strand sites that are produced by nicks or breaks in the DNA. It is possible that the lack of activity on double-stranded DNA is due to the loss, during isolation, of a "sigma-like" factor that dictates the binding of the enzyme. Alternatively, the enzyme may only initiate transcription on single-stranded DNA *in vivo*, requiring the presence of a helix-destabilizing protein in the vicinity of the transcription-start site.

Nucleotide Sequences Important for the Transcription of Structural Genes

In bacterial cells the nucleotide sequence immediately upstream from the start of transcription of structural genes defines the site of RNA polymerase binding and the sites of binding of a number of regulatory molecules. Consequently, studies on eukaryotic genes have concentrated on the sequences preceding the start of transcription. More than 60 structural genes have now been sequenced [see the review by Breathnach and Chambon (19) and also Ref. 20], and several common features are apparent as shown in the consensus sequence illustrated in Fig. 4-2. Approximately 24 nucleotides (i.e., about 2 turns of double helix)

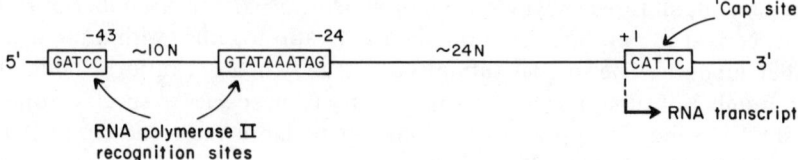

Fig. 4-2 A consensus sequence of the 5' end of structural genes. Transcription starts at +1 and goes toward the 3' end. Each box represents a highly conserved sequence that may be a site of RNA polymerase interaction.

upstream from the start of transcription of many genes is a sequence very similar to the Pribnow box sequences of bacterial cells (cf. Fig. 2-7 in Chapter 2). This is usually referred to as the Hogness box, or "—TATA—" box. A second, shorter segment of sequence homology is found about 10 nucleotides (1 helical turn) even further upstream. These two sequences are thought to be important signals for RNA polymerase. (Note that in bacterial cells transcription is initiated about 5 nucleotides downstream from the Pribnow box, but in eukaryotic cells transcription commences about 24 nucleotides downstream from the —TATA— box.) In addition to these sequences it has been shown that other sequences, *even further upstream*, are necessary for the correct initiation of transcription. Typically these sequences are approximately 80 and 150 base pairs upstream. Immediately following the start of transcription there is a pentanucleotide sequence that is thought to be the recognition signal, in the transcript, for the enzyme that adds the 5' cap (see Chapter 5).

The relative contributions of all these sequences to polymerase binding has been studied by a number of groups (21–23) using the recently developed *in vitro* transcription systems. In these experiments gene fragments with substantial amounts of 5' flanking sequence were cloned and then "mutated" *in vitro* by either deletion of segments of sequence or, in some cases, the introduction of point mutations. These altered fragments were then transcribed by purified RNA polymerase II in the *in vitro* system. Single base changes in the —TATA— box reduce the *rate* of initiation of transcription substantially, but even if the entire sequence is deleted some transcripts are still initiated. However, rather than starting precisely at the "cap" site, transcripts are initiated at different loci. Thus the —TATA— box sequence ensures an accurate *starting* location for the initiation of transcription. Since transcripts can be initiated in the absence of the —TATA— box, other sequences must be involved in the initial binding of RNA polymerase II to the gene. Deletion experiments have established that two sequences, 80 and 150 base-pairs upstream, are absolutely essential for polymerase binding. When these sequences are deleted no transcripts are initiated even if the —TATA— box is left intact.

Although RNA polymerase II is a large molecule (~10 nm in diameter), it is unlikely to be able to span 150 base-pairs of DNA double helix. However, if the DNA is organized into nucleosomes with 80 base pairs per turn, then the sequences at −150 and −80 become precisely aligned with the transcription start site (see Fig. 4-3).

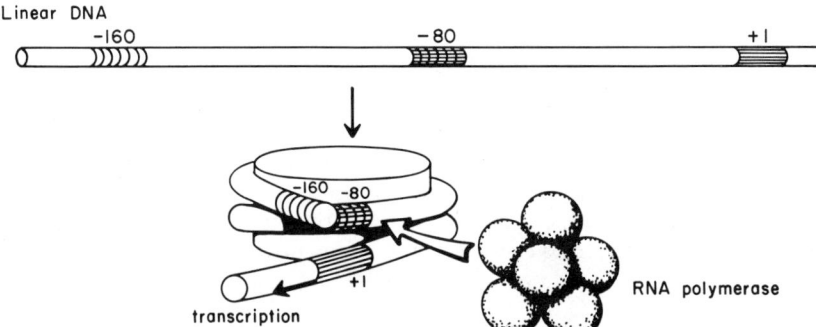

Fig. 4-3 The interaction of RNA polymerase with DNA. Binding Sites at +1, −80, and −160 are brought into precise alignment when there are two turns of DNA per nucleosome.

The nucleotide sequence as far as 500 base-pairs upstream from the start of the sea urchin histone H2A gene is essential for its transcription (24). Thus the entire spacer between the H3 gene and the H2A gene, although not transcribed, plays an important role in determining efficient and accurate transcription of the H2A gene. This observation, together with the earlier discussion on nucleosome phasing on DNA spacers (see Fig. 4-1, but note that the arrangement of the histone genes on the *Drosophila* repeat units is different from that of the sea urchin), suggests a model for the binding of RNA polymerase II to active genes.

In this model the DNA immediately upstream from the actual start of transcription is organized into nucleosomes, with the nucleosomes being phased in a very precise way such that a particular conformation, recognizable by RNA polymerase, is created. This conformation exposes specific nucleotide sequences that facilitate polymerase binding and its location of the correct site for the initiation of transcription. The nucleotide sequence around the —TATA— box is usually A:T rich, which favors either DNA breathing or polymerase-induced melting so that the coding strand is exposed.

The introduction of any protein to, or the modification of proteins already existing on, these critical nucleosomes may alter the efficiency at which transcription is initiated and regulate the expression of the gene. The roles played by histone modification, HMG protein binding, and the binding of other regulatory molecules will be discussed in a later section.

The genes of bacterial cells have a specific sequence that determines the end of transcription to prevent the polymerase enzyme from translocating onto the next structural gene or operon. The 3' sequence of eukaryotic genes has also been examined for the presence of terminator sequences. Two sequences have been found that are common to many genes (25) as illustrated in Fig. 4-4. One of these sequences immediately precedes the last transcribed nucleotide in most genes, although in some it lies a few base-pairs downstream. This sequence is quite different from that of bacterial cell terminators. It is A:T rich, but it does not have the numbers of consecutive thymidine residues seen in bacterial cells

```
····T AATAAA AACA TGT T TAAGC AAACAC TTT TCACTTGT A····   ovalbumin
····A AATAAT CA AG AAAGAAT····                           histone H2B
····T AATAAA GGAAATTTA TTTTCATTGC AAT AGTGTGT TGGAATTT···· rabbit β-globin
····A AATAAA GCATTT TTTTCACTGC ATTCTAGT TGTGGTTTGT····    SV-40 early
····C AATAAA CAAGT TAACAACAAC AATTGCATT CATTTTATG····    SV-40 late
····CCAAAACGGCTC TTTTCAGAGC CACCAAATAATCAAGAAA····       histone HI
····CTC AATAAA CTAAATCAGCAAC ACTCC T TTGTCTT····         conalbumin
```

Fig. 4-4 The nucleotide sequence near the 3' end of a number of eukaryotic genes. The two highly conserved sequences are boxed. Transcription stops immediately after the cytidine residue in bold type. [These sequences are reproduced from Fig. 8 of Ref. 25, with permission.]

and is not preceded by a G:C-rich region. Moreover, the 3' sequences do not have the dyad symmetry that can give rise to the base-paired "terminator" structures in the RNA transcripts that are an essential feature of the mechanism of termination in bacterial cells. It is likely that a different mechanism of transcription termination operates in eukaryotic cells. For example, it is possible that the common sequence is recognized by one of the polymerase subunits, causing termination of transcription. It is also possible that this sequence phases one or two nucleosomes into an arrangement that destabilizes polymerase by steric hindrance.

The sequence —AATAAA— is also present at the 3' end of many genes. With the exception of histone H2B (see Fig. 4-4), this sequence is not present on the histone genes, and since the histones are not polyadenylated it has been suggested that the transcribed sequence —AAUAAA— in the RNA transcript acts as a signal for binding of poly-A polymerase, the enzyme responsible for adding the poly-A tail to many RNA transcripts. It has also been proposed that poly-A polymerase may bind to the RNA transcript during transcription and cause termination by interfering with RNA polymerase.

Other RNA Polymerases Use Different "Promoter" Sequences

Eukaryotic cells differ from bacterial cells by using three different RNA polymerase molecules in gene transcription. Two of these, RNA polymerases I and III, transcribe RNA molecules that are not translated into protein. These are the ribosomal RNA components and the small nuclear RNAs, respectively. The genes coding for these RNAs are amplified and are present in hundreds or in some cases thousands of copies per nucleus, and the regulation of transcription of these genes is currently being studied by a number of groups (26-32). Although many of these studies are still in their infancy, a number of important features have already been described.

It appears that none of these genes are preceded by the —TATA— box sequence found near the start of structural genes. This includes the gene for small nuclear RNA-UI, which is transcribed by RNA polymerase II. Examples of the sequences near the start of transcription of the ribosomal genes of a number of species are shown in Fig. 4-5. The transcription-start sites for *Xenopus* and *Drosophila* have been accurately determined, whereas that for the

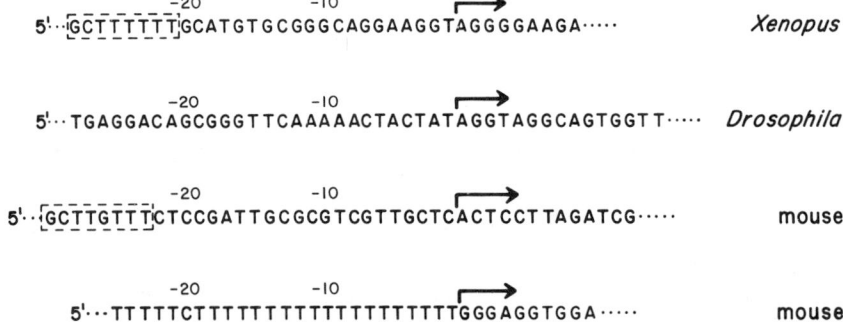

Fig. 4-5 The nucleotide sequence near the start of transcription of the *ribosomal* RNA genes of a number of species. The start of transcription by RNA polymerase I is shown by the arrows, and the nucleotides to the right are transcribed. The only conserved sequences are boxed.

mouse is still controversial. The difficulty in establishing the transcription-start site stems from the problem of isolating intact 45S rRNA precursors so that the DNA and RNA sequences can be aligned to establish the exact start site.

There is little sequence homology between the species at any of these sites. If the upper sequence is correct for the mouse it shares some homology with *Xenopus* DNA approximately 20 base pairs upstream from the start of transcription. The lower sequence suggests that the start of transcription immediately follows an extremely long run of A:T base pairs. Since the sequence of the spacer DNA upstream from the ribosomal genes in the other species also tends to be A:T rich, RNA polymerase I may bind to single-stranded DNA exposed by local DNA breathing at these loci. In this case a specific, *sequence-defined* RNA polymerase I binding site may not be important.

There is also a lack of sequence homology between the 5' *transcribed* portions of the rRNA genes (Fig. 4-5). This may seem surprising until it is recalled (see Fig. 3-2) that the first 2000 or so nucleotides of the precursor transcript are discarded during the processing of the rRNA precursor. Therefore, the actual start site of transcription may not be so important, since the cell may rely on its ability to determine accurately the site of RNA processing at the beginning of the 18S rRNA. Thus, although transcribed, there has been no evolutionary pressure to conserve the leader sequence, whereas the sequences of DNA within the gene that actually code for the 18S, 28S, and 5.8S rRNAs are remarkably well-conserved (see an example in Fig. 6-1 in Chapter 6).

The nucleotide sequence near the start of transcription of a number of small RNA genes has been established, as shown in Fig. 4-6. The tRNA and 5S rRNA genes are transcribed by RNA polymerase III, whereas the U1Sn RNA gene is transcribed by RNA polymerase II. Once again there is little sequence homology except for the short sequence —^A_CAAAG— that occurs in all three genes. The significance of this short sequence is not clear, because it is too short to be a major binding site for polymerase.

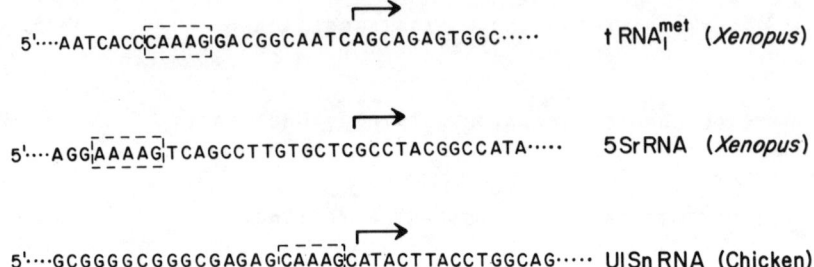

Fig. 4-6 The nucleotide sequences immediately before the start of transcription of a number of *small nuclear* RNA genes transcribed by either RNA polymerase III (tRNA$_1^{met}$, 5S rRNA) or RNA polymerase II (U1Sn RNA). The start of transcription is shown by the arrows, and the bases to the right are transcribed. The conserved sequences are boxed.

In the cases of the tRNA$_1^{met}$ gene and the 5S rRNA gene, *in vitro* deletion studies have shown that the 22–26 nucleotides immediately preceding the start of transcription are sufficient to ensure accurate and complete transcription of the gene (26, 28, 29). Surprisingly, virtually all of this 5' flanking region can be removed and yet transcription still takes place, but the *efficiency* of transcription is decreased, and there is some loss of accuracy at the start of transcription. The transcription of the 5S rRNA gene in the frog *Xenopus laevis* has been studied in great detail by Brown's group (28, 29, 33) and to date is the only eukaryotic gene shown directly to be controlled at the level of transcription.

In the oocyte of *Xenopus* and other related frogs there are approximately 100,000 copies of the 5S rRNA gene per nucleus. Each gene, of about 120 base pairs, is organized into a repeating unit of about 700 base pairs. The first half of the repeating unit consists of A:T-rich repetitive DNA sequences, whereas the second half is predominantly G:C rich (see Fig. 4-6) and includes the 5S gene. Only the gene is transcribed. The gene has been isolated, purified, and sequenced together with much of the spacer DNA. The intact gene is faithfully transcribed by RNA polymerase III *in vivo* and *in vitro*. It is a small gene and is very amenable to genetic manipulation *in vitro*. Figure 4-7 gives a diagrammatic representation of the experimental approach used to assess the role of DNA sequence in the regulation of the gene and illustrates the usefulness of the so-called *in vitro*, or surrogate, genetic technique.

The central portion of the figure shows the 5S gene together with about 80 base-pairs of DNA at each end. The horizontal lines above or below the gene show the amount of DNA deleted from either the 5' or 3' ends of the DNA in each of a series of "mutants" created *in vitro*. The deleted DNA is replaced by an equivalent length of foreign DNA, and each mutant is then tested for its ability to be transcribed (+ or −). Surprisingly, near normal transcription occurs with many of these altered fragments. Only four are not transcribed, and since all four delete a portion of the center of the gene it is apparent that a regulatory site exists in the *center* of the gene, not at the 5' end.

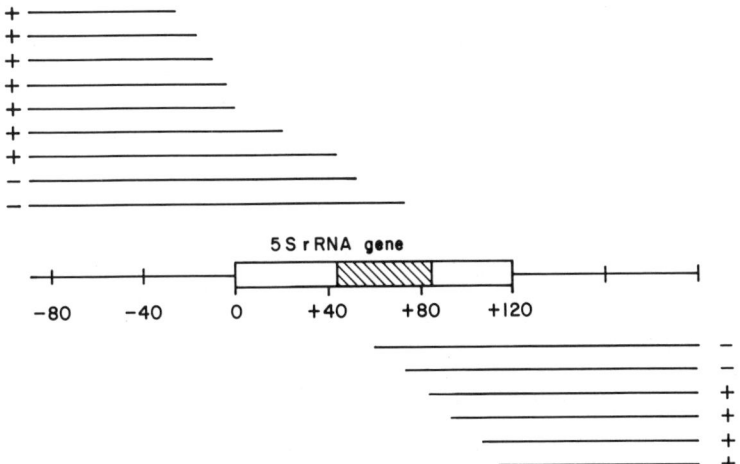

Fig. 4-7 Protocol for the study of transcription of the 5S rRNA gene in *Xenopus* oocytes. Each horizontal line represents the amount of DNA deleted (and substituted with foreign DNA) from either the 5' or 3' end of the gene in a series of mutants. Each mutant was tested for transcription in an *in vitro* system. The + symbol denotes mutant was transcribed; − denotes no transcription. [Reproduced from Fig. 9 of Ref. 20, with permission. Copyright 1980 Massachusetts Institute of Technology.]

This internal regulatory sequence stretches from +55 to +80 base-pairs (Fig. 4-7), and when it is intact transcription starts approximately 50 base-pairs upstream, independent of the sequence. However, *accurate and highly efficient* transcription only occurs when the 26 base-pairs immediately before the start of transcription are intact. RNA polymerase III does not bind to either the gene or the internal regulatory sequence when tested *in vitro*. However, a protein of M_r = 37,000–40,000 has been found that binds tightly to the regulatory sequence (33, 34). In the presence of this protein, RNA polymerase III rapidly and accurately transcribes the gene. The protein binds to the DNA sequence between +45 and +97 base-pairs.

The overall structure of the gene and its regulatory proteins is depicted in Fig. 4-8. The binding site of the regulatory protein contains a sequence —AGCA GGT— that has been found to occur in a number of tRNA and other small RNA genes (cf. the sequence immediately following the start of transcription of the tRNA gene in Fig. 4-6), but in general there is little sequence homology between these various genes. The termination signal is —GCTT— and deletion studies have shown that this is all that is necessary to ensure accurate transcription termination.

The regulatory significance of the interaction between the protein and the gene became apparent when it was recognized that this protein was the same molecule as the protein known to form a ribonucleoprotein complex with the 5S rRNA molecule in the cytoplasm of the oocyte. During the early phase of oocyte development there is a massive synthesis of 5S rRNA, *but the major*

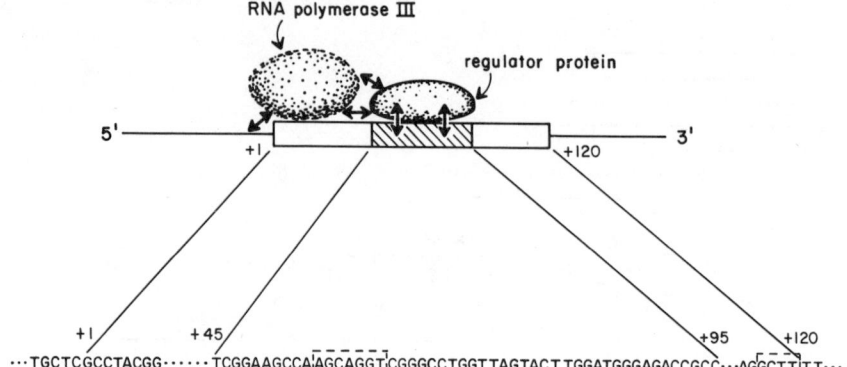

Fig. 4-8 The structure of the 5S rRNA gene of *Xenopus laevis* and its interaction with regulator protein and RNA polymerase III as described in the text. The nucleotide sequences of the 5' end, the central regulatory region, and the 3' end of the gene are shown. Conserved sequences in the regulatory region and termination sequence are boxed.

ribosomal RNA gene (18S, 28S, and 5.8S rRNA) remains silent. The 5S rRNA is synthesized and stored in the cytoplasm as an inactive RNP complex with the protein. Presumably, synthesis of this protein commences first during development, and it binds to the 5S rRNA genes. RNA polymerase III recognizes it, and the genes are transcribed as long as there is excess protein present. When all the protein is complexed with RNA in the cytoplasm there is none left for activation of the gene, and transcription stops. Thus the transcription of the genes is controlled by the concentration of free protein, and 5S rRNA prevents further transcription when its concentration reaches a level that complexes all the protein.

This is a novel mechanism, and it remains to be seen whether it operates only at this stage of oocyte development or whether it is more generally applicable. The *Xenopus* oocyte is somewhat anomalous in that the 5S rRNA gene is present in so many copies, whereas in other cells there are only 100–200 such genes, and a similar regulatory mechanism may not be applicable.

In summary, promoterlike sequences of DNA near the 5' end of genes are not found in eukaryotic cells. All three polymerases appear to interact with the genome by different mechanisms, with DNA sequence and DNA conformation playing important roles. The initial interaction between polymerase and DNA appears to rely more on a particular conformation (at the nucleosome level) than on sequence or, in the case of RNA polymerase III in *Xenopus*, the presence of a regulatory protein. Once the polymerase locates the correct binding site it uses a specific DNA sequence, for example, the —TATA— box, to determine the exact site at which transcription is to begin. Termination of transcription has not been well studied, but here again structure rather than sequence is likely to play a major role. Since genes are usually widely separated there is less necessity to have an accurate termination site, particularly if the primary transcript is to be polyadenylated or processed in other ways at its 3' end.

Gene expression is regulated at the transcriptional level by modulating the rate at which RNA polymerase binds to, and transcribes, the gene. This is achieved by either activating (or inhibiting) the enzyme or by altering the availability of the template. With the exception of the 5S rRNA gene we have little knowledge of how these changes are brought about, despite a great deal of effort.

THE REGULATION OF TRANSCRIPTION

Activators of RNA Polymerases

Recent progress in the development of *in vitro* transcription assays (35, 36) has demonstrated beyond any doubt that proteins other than the RNA polymerases are required for accurate and highly efficient transcription. Both RNA polymerase II and RNA polymerase III require a specific set of protein modulators. A number of individual proteins have been identified as transcriptional activators and have been shown to activate polymerase by either binding to the enzyme or by binding to chromatin. In general, these proteins remain poorly characterized, mainly because of the difficulty in assessing their specific function in crude systems. Some of the proteins that bind to the enzyme may be subunits that are dissociated from the multimeric enzyme at certain times and subsequently reassociate and activate it. RNA polymerase II has also been shown to be activated by phosphorylation of one or more of its subunits by a cyclic nucleotide-independent protein kinase (37).

Considerable progress could be made in this field with the development of cleaner assay systems involving polymerase and individually cloned genes so that the various steps of transcription initiation, DNA melting, translocation, and termination can be studied individually. It is particularly important to determine if any of these modulators can direct the transcription of a specific subset of genes as opposed to generalized increases in the rate of transcription of all genes.

Histone Modifications and Changes in Template Availability

Histones are stable molecules in the sense that they have no known turnover rate. However, they are *conformationally unstable* because of a high degree of covalent modification of amino acid residues. Thus the histones have been shown to undergo phosphorylation, acetylation, methylation, and ADP-ribosylation. All of these forms of modification decrease the net positive charge on the proteins or create considerable steric hindrance. Since the stability of nucleosome-organized chromatin is derived from the ionic bonds formed between positively charged amino acid residues of the histones and the negatively charged phosphates of the DNA backbone, any alteration in the ability to form these bonds is likely to have dramatic effects on DNA conformation and template availability.

The types of covalent modification undergone by the histones are reviewed extensively by Allfrey (1) and are summarized in Table 4-2. A great deal of attention has been paid to phosphorylation and acetylation because changes in the extent of these modifications can be positively correlated with changes in template availability.

All the histones can be phosphorylated, but most of the work has been carried out on the phosphorylation of histone H1. As many as eight molecules of phosphate can be added to each molecule of H1, and peptide mapping techniques have shown that the sites of phosphorylation lie in the head and tail regions of the molecule (cf. Table 3-2 and Fig. 3-9 in Chapter 3). Different protein kinases preferentially phosphorylate specific amino acid residues. For example, the cAMP-dependent protein kinase phosphorylates the serine residue at position 37. This site is also phosphorylated by a cGMP-dependent kinase that also phosphorylates sites in the COOH-terminal region of the molecule (39). Cyclic nucleotide-independent kinases also phosphorylate histone H1, forming P—O bonds with either serine or threonine or P—N bonds with arginine or lysine (40, 41). The importance of the P—N bond is only just being realized; its acid lability makes it particularly difficult to study because most histone extraction and separation techniques utilize acid. Strictly speaking, the formation of these bonds should be referred to as *phosphoramidation* because the whole phosphate group is not transferred (see Fig. 7-1; covalent modifications of all kinds are discussed in Chapter 7).

In most eukaryotic cells histone H1 exists as a family of variants that have significant differences in amino acid composition. For example, one known variant does not have a serine residue at position 37 and cannot be phosphorylated by cAMP-dependent protein kinase. The distribution of these variants in chromatin has not yet been established, but their existence may serve to establish subsets of nucleosomes that have specific requirements for covalent modification.

Table 4-2 The Covalent Modifications of Histones

Reaction	Amino acid substituted	Net charge change	Predominant histone
O-phosphorylation	serine, threonine	adds 2 −ve	all histones
N-phosphoramidation	arginine, lysine	adds 2 −ve	H1, H4
acetylation	serine, lysine	removes 1 +ve	core histones
N-methylation	histidine, arginine, lysine	increases hydrophobicity	H3, H4
ADP-ribosylation	glutamic acid	adds 1.5 −ve/residue	H1, H2B

Table 4-3 The Extent of Phosphorylation of Histone H1 Variants H1A and H1B During the Cell Cycle[a]

	NUMBER OF PHOSPHATES / HISTONE		
	G1	S	M
Histone H1A	0-1	1-2	4-6
Histone H1B	0-3	1-4	4-8

[a] Data obtained from Refs. 41 and 42.

There are a number of correlations between histone H1 phosphorylation and changes in either nucleosome conformation or template availability (see Ref. 1). In view of the important role of this histone in the formation (and disruption) of higher orders of chromatin structure that can directly influence template availability, it is likely that H1 phosphorylation is an essential prerequisite to the expression of many genes.

Probably the best examples of the effects of the modification of histone H1 on chromatin structure are the changes observed during DNA replication and mitosis. During DNA replication all the DNA must be completely relaxed in order to be copied, whereas in mitosis all the DNA becomes highly compacted during chromosome segregation. Ajiro et al. (41, 42) have measured the extent of phosphorylation of two histone H1 variants (H1A and H1B) during the cell cycle of HeLa cells; their results are summarized in Table 4-3. In the late G1 phase of the cell cycle the level of histone H1 phosphorylation is very low. However, during the S phase of the cell cycle (DNA replication) there is an increase in the extent of phosphorylation by about 1 phosphate molecule per histone, and during mitosis there is a marked increase in the extent of phosphorylation. Moreover, there are differences in the location of the phosphate residues. They are located almost exclusively in the C-terminal region of the molecule during replication but are located in both N- and C-terminal regions during mitosis. During replication two different C-terminal residues were found to be phosphorylated; one site is phosphorylated immediately before replication, and the second site is phosphorylated immediately afterward. Virtually all of the sites are rapidly dephosphorylated immediately after mitosis. Thus there is a clear correlation between H1 phosphorylation and chromatin condensation.

The core histones have also been shown to be phosphorylated, and a number of specific kinases have been identified. For example, Whitlock et al. (43) and Shoemaker and Chalkley (44) have identified kinases that are chromatin-associated and specifically phosphorylate histone H3. Histone H3 modification is of considerable importance because it will likely have a major effect on DNA conformation, this histone being located at the point of entry and exit of DNA from the nucleosome (Fig. 3-6 in Chapter 3) and possibly influencing the number of base pairs of DNA associated with the core and perhaps even facilitating dissociation of DNA. Histone H2A can also be phosphorylated, and there appears to be a direct correlation between the number of H2A molecules phosphorylated and the amount of constitutive heterochromatin (45), suggesting that this modification locks DNA into an inactive, higher-order structure.

In summary, histone phosphorylation appears to play a major role in determining the amount of DNA in the various levels of chromatin structure and in increasing the availability of the template for transcription. The nucleus contains histone-specific kinases that are activated by cAMP, cGMP, calmodulin, and other regulatory second messenger molecules (46) which indicates that this form of covalent modification is a major mechanism for the regulation of gene transcription by hormones and other inducers. Phosphoprotein phosphatases that remove the phosphate groups are an essential part of this mechanism, but, to date, they have been poorly studied. The importance of these latter enzymes is emphasized in the recent publication (47) that demonstrates calcium–calmodulin-dependent dephosphorylation of histones H1, H2A, and H2B by a specific phosphatase.

All four core histones are modified by acetylation, but histone H1 is not. This observation suggests immediately that this form of modification is used to alter histone–nucleic acid interaction at the level of the nucleosome. Acetylation occurs mainly in the N-terminal regions of the core histones. All four are acetylated on the N-terminal residue immediately after synthesis in the cytoplasm. This is a stable alteration in contrast to the acetylation of the molecules that occurs in the nucleus, which has a rapid turnover. Intranuclear acetylation occurs at multiple sites, for example, histone H3 can be acetylated at residues 4, 9, 14, 18, and 23, reducing the net positive charge of this part of the molecule by at least 50%. Most of these sites are rapidly deacetylated ($t_{1/2} = 3$ min), although some groups appear to have a half-life of several hours (1).

Physicochemical studies on nucleosomes before and after acetylation (48) show clearly that DNA becomes more loosely associated with the histone core. Hyperacetylation increases the DNase-I sensitivity of the nucleosomes and also increases the number of bacterial RNA polymerase initiation sites, indicating a marked increase in template accessibility. Allfrey (1) describes many examples of the correlation between increases in gene expression and increases in core histone acetylation [although it has recently been reported that butyrate-induced hyperacetylation inhibits the induction of ovalbumin and transferrin by steroids in chick oviduct (49)].

Histone acetylase and deacetylase enzyme activities have been demonstrated in the nuclei of many cells. The reaction catalyzed is shown in Fig. 7-5 in Chapter 7. However, at this stage little is known about the regulation of the activity of these enzymes. It is not clear whether they are responsible for the initial reactions leading to an increase in template availability or whether they are activated only when the gene is actually being transcribed. Acetylation and phosphorylation appear to be closely linked. For example, hyperacetylation of histone H3 makes the molecule more sensitive to phosphorylation. Moreover, phosphorylation and acetylation occur simultaneously during DNA replication.

Core histones H3 and H4 are modified by the substitution of one to three lysine residues with a methyl group. The reaction is shown in Fig. 7-5 in Chapter 7. Methylation takes place in the nucleus immediately after replication, but before mitosis. It is a stable alteration, and it is believed to increase the hydrophobicity of the histones, enhancing the stability of the histone octamer of the nucleosome.

Histones H1 and, to a lesser extent, H2B are modified by the addition of ADP-ribosyl groups to glutamic acid residues by the reaction shown in Fig. 7-4 in Chapter 7 catalyzed by poly(ADP-ribose)polymerase. This enzyme is present in the nucleus of all eukaryotic cells. The ADP-ribosyl groups can be polymerized and it is believed, but not yet definitively established, that this polymer serves as an intranuclear cross-linker. For example, H1 molecules involved in the formation of higher orders of chromatin structure may be cross-linked to increase the stability of the structure. This and other aspects of poly(ADP-ribosylation) are discussed in a recent monograph (50).

In summary, the nuclei of eukaryotic cells possess enzyme activities that can modify the interaction between histones and DNA such that the double helix can become more accessible to the transcriptional apparatus. Figure 4-9 gives a diagrammatic summary of some of these reactions. Thus the histones are in a dynamic state, serving either to repress genes or activate them. Some modifications such as ADP-ribosylation or histone H2A phosphorylation may move DNA into transcriptionally inactive regions of chromatin, whereas other modi-

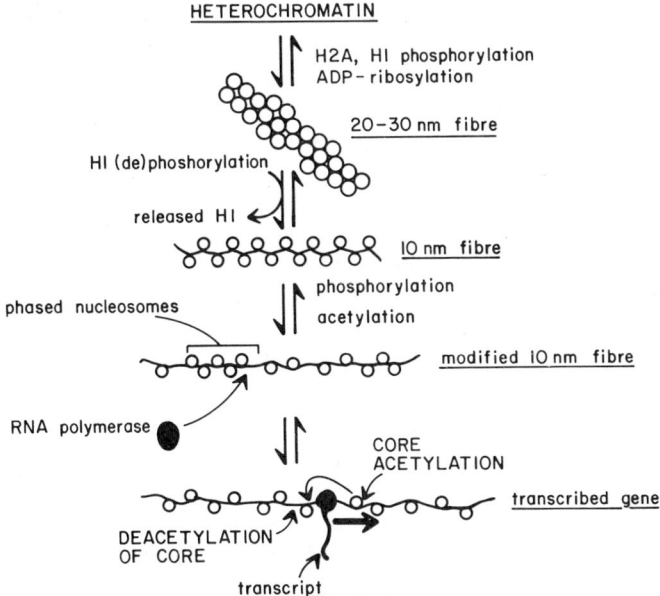

Fig. 4-9 The role of covalent modifications in the interconversion of chromatin structures in the initiation of transcription. The various steps are described in the text.

fications such as histone H1 phosphorylation (at appropriate sites, for other sites promote chromosome condensation) and core histone acetylation make DNA available for transcription.

These reactions are by no means completely understood, and two important questions remain to be answered. The first of these is how does a cell transmit to its daughter cells the extent and nature of the covalent modifications of its histones such that a faithful copy of chromatin structure is passed along? Up until this point replication was seen solely as a duplication of the nucleotide sequence of the DNA. It now appears that the G-2 period of the cell cycle is devoted to modifying chromatin so that following chromatin condensation at mitosis the chromatin subsequently disperses as an exact *structural* copy of the parental chromatin.

The second question relates to the specificity of gene activation. If all the histones are susceptible to the same kinds of covalent modification, how does the cell use these reactions to activate individual genes or subsets of genes? There is no doubt that much of the chromatin can be packaged in such a way that the histones are not accessible for modification, and even among the dispersed euchromatin certain histone modifications (or the presence of histone variants) may preclude the formation of others. However, it is unlikely that there are enough permutations and combinations of modification that each subset of genes is uniquely represented in chromatin structure. Thus other factors must be involved in determining the response of the genome to the various inducers that cause specific changes in gene expression. The only example we have is the regulatory protein that binds to, and activates, the 5S rRNA gene. In the next section we discuss the evidence for the involvement of the nonhistone proteins in the regulation of transcription.

The Role of Nonhistone Proteins

There is little doubt that, as a group, the NHPs play a major role in determining the phenotypic expression of the genome. This has been classically demonstrated by taking chromatin from the cells of tissue A, removing the NHP, and incubating the DNA with NHP from the cells of tissue B. Under these conditions the DNA of tissue A expresses genes characteristic of tissue B that were normally repressed in tissue A. Unfortunately, attempts to identify individual NHP molecules that are responsible for the activation of specific structural genes have been uniformly unsuccessful.

Mild digestion of chromatin by micrococcal nuclease releases into solution a number of chromatin-associated NHPs (see Ref. 51 for a typical example), indicating that these proteins are intimately associated with actively transcribed genes. However, the conclusion drawn by many people that they must have been associated with the nuclease-sensitive, internucleosomal spacer DNA is erroneous because nuclease digestion releases not only nucleosomes but also HnRNP particles, together with components of the nuclear matrix, since transcription appears to take place in close association with the matrix. (This does not rule out an involvement in transcription for these NHP.) Moreover, when mononucleosomes are isolated from the digests, as many as 20 NHPs remain attached to these particles (not all mononucleosomes contain all 20 proteins, they are probably heterogeneous). Thus there is evidence for a localization of NHP in the vicinity of active genes.

A characteristic of these NHPs is that they are virtually all phosphoproteins. Indeed, the average number of phosphate groups per protein molecule in the nucleus is 20 to 30 times higher than in the cytoplasm, which suggests that this form of covalent modification may play a major role in the regulation of gene expression at the transcriptional level. Studies with the radioisotope ^{32}P show that many NHPs undergo cycles of phosphorylation and dephosphorylation that appear to correlate with changes in gene expression [see Table IV of the review by Kleinsmith (52)]. As already mentioned, a number of protein kinases are present in the nucleus, two of which appear to be specific for NHP rather than histones (53, 54). Each kinase ($M_r = 37,000$ and $M_r = 140,000$) is cyclic nucleotide-independent and phosphorylates a characteristic subset of NHP, including subunits of RNA polymerase II (55).

The striking effect that phosphorylation can have upon the activity of chromatin-associated NHP is demonstrated by the properties of a protein isolated and purified from calf thymus nuclei (56). This protein ($M_r = 55,000$) binds tightly to the histone core when it is phosphorylated. Upon dephosphorylation the protein becomes more tightly associated with the DNA of the nucleosome, rendering the DNA 100 times more transcribable by *E. coli* RNA polymerase. However, since the *in vivo* location of this protein has not yet been found, it is difficult to assess its true function.

The fact that many of the NHPs and their kinases are released from the chromatin, even under very mild digestion conditions, makes it difficult to establish their location with reference to particular aspects of chromatin structure. It appears that if these molecules are involved in the regulation of transcription they exert their effects while being in only loose association with the transcription complex. This transcription complex consists of nucleosomal DNA, RNA polymerase, and the primary transcript, together with all its associated proteins. The complex is in a loose association with the nuclear matrix, and there is no doubt that it will be a formidable task to assess how all these components interact.

Attempts are being made to develop *in vitro* model systems for specific gene transcription. Despite considerable technical problems there are three model systems currently being studied in detail: estrogen-stimulated chick oviduct chromatin (57), histone mRNA synthesis by chromatin isolated from proliferating HeLa cells (58), and globin gene transcription by mouse fetal liver (59). These systems are active *in vitro*; for example, chromatin from the livers of fetal mice synthesizes globin mRNA, but chromatin from adult liver does not. Similarly ovalbumin mRNA is synthesized in greater quantities by chromatin obtained from estrogen-stimulated oviduct cells than from untreated cells.

The histone genes are only transcribed in the S phase of the cell cycle.

Chromatin isolated at this time is active *in vitro*. However, if S-phase chromatin is stripped of all NHP and a NHP fraction from cells in the G-1 phase of the cell cycle is used, the histone genes are not expressed. Moreover, if the S-phase NHPs are added to G-1 DNA, the histone genes are expressed. Thus specific proteins must be present in the S-phase NHP fraction that positively activate the histone genes (58). These proteins have been fractionated, and 96% of the stimulatory activity is present among a small group of proteins of $M_r = 40,000$-60,000. Further studies should identify the protein(s) specifically responsible for the activation of these genes. Further work along these lines is required for other eukaryotic genes. However, these systems all use *E. coli* RNA polymerase rather than mammalian RNA polymerase II, and the results obtained must be treated with caution until proteins considered to be regulatory are shown to be active in a totally mammalian *in vitro* system.

There is one group of nonhistone proteins that is definitely involved in the activation of specific genes (or groups of genes), and these are currently being studied in some detail. These are the receptor for thyroid hormone and the acceptor proteins for steroid hormone–receptor complexes. The discussion here is limited to the interactions of these molecules at the nuclear level; a discussion of the role of these proteins in the physiological actions of the hormones can be found in Chapter 11.

The thyroid hormone triiodothyronine interacts with target cells by directly penetrating the nucleus and binding to a specific, chromatin-associated receptor protein. The receptor is a protein of $M_r = 50,000$, and, since 50–60% is released under conditions where only 5% of the DNA is digested, it appears to be bound to a transcriptionally active subset of nucleosomes (60). The receptor is not released as a single protein but appears complexed with a number of other NHPs, indicating that the receptor alone is insufficient to transmit all the information from the hormone to chromatin. The nature of this complex and its location within chromatin have not been further studied. However, it has been shown (61) that acetylation of core histones causes the release of the hormone–receptor complex, which suggests that the response to the hormone depends on prior modifications to the DNA template.

Steroid hormones are small hydrophobic molecules that easily penetrate the plasma membrane of target cells and interact with a specific receptor protein in the cytoplasm. A steroid–receptor complex (SRC) is formed, and it is rapidly translocated to the nucleus, where it can elicit some highly specific changes in gene transcription. For example, in unstimulated chicken oviduct cells there are about 10 molecules of ovalbumin mRNA per cell, and the concentration of ovalbumin protein is very low. Upon stimulation by estrogen the number of mRNA molecules increases rapidly to about 50,000 copies per cell (50% of the total mRNA population), whereas there is no change in the rates of transcription of approximately 14,000 other genes.

The SRC interacts with a protein-acceptor molecule on chromatin prior to the change in gene expression. Each steroid hormone (estrogens, androgens, glucocorticoids, progesterone) has its own cytoplasmic receptor and its own nuclear acceptor. The nuclear acceptors tend to be small molecules, $M_r = 14,000$

in the case of the androgen–receptor and 12,000–16,000 in the case of the progesterone–receptor, whereas the estrogen–receptor in the uterus is somewhat larger ($M_r = 70,000-80,000$). These molecules are released from chromatin by mild nuclease digestion, which indicates that they are associated with transcriptionally active DNA.

The interaction of the progesterone–receptor complex with DNA has been studied in the most detail (see Chapter 11). The SRC consists of two dissimilar subunits A and B (Fig. 12-4). Subunit B binds to the progesterone acceptor located among the NHPs but does not bind to DNA. Subunit A, on the other hand, binds to DNA, but not to NHP. It has been proposed that the dimer enters the nucleus and interacts with the acceptor, which is bound to chromatin near the start of transcription of progesterone-activated genes. The interaction between the B subunit and the acceptor brings the DNA-binding A subunit into contact with DNA at a point where it can effect transcriptional activation. The nature of the DNA-binding reaction has been studied recently in an *in vitro* system using purified A subunit and fragments of DNA derived from clones of the ovalbumin gene, one of the genes induced by progesterone (62). This test system overcomes many of the problems of attempting to assess specific DNA binding in whole chromatin, where a large number of low-affinity sites may mask high-affinity binding, and is directly analogous to the systems used for bacterial cells in which specific, high-affinity ($K_a = 10^{13} M^{-1}$) binding of regulatory molecules can be demonstrated.

DNA from 1200 base pairs upstream to 7000 base pairs beyond the 3' end of the ovalbumin gene was tested in an effort to detect sequence-specific binding. However, the A subunit showed no specific binding, and although the binding of the protein to DNA was of high affinity ($K_a = 10^{10} M^{-1}$), it was lower than that normally required for sequence-specific binding. The A subunit exhibited a preference for single-stranded DNA and appears to be a helix-destabilizing protein.

In summary, a role for specific NHP in the activation of transcription of individual structural genes has not yet been established. Several proteins have been described that interact with chromatin at various structural levels, but mechanisms of action are still unknown. The data obtained from studies on the steroid-acceptor proteins suggest that several NHPs are involved in the mechanism of activation of any one gene. This notion is supported by the observation (see Table 11-4) that nonhistone proteins that mask and prevent transcriptional activation by steroid-acceptor molecules are present in certain tissues. Thus the transcription of inducible genes may be regulated by a "committee" of proteins, each receiving different signals from other parts of chromatin, the cytoplasm, and even outside of the cell.

General Comments on the Regulation of Transcription in Eukaryotes

The vast majority of genes in somatic cells of multicellular organisms are expressed constitutively, and most tissues have little capacity for adaptation. During the process of differentiation the phenotype of the cells of such tissues is determined by alterations in the types of protein associated with the DNA and by patterns of covalent modification of both the DNA and the histone and nonhistone components of chromatin. Constitutively expressed genes are probably transcribed with a frequency determined by the ability of RNA polymerase to interact with particular structural alterations of the arrangement of nucleosomes along the 10-nm fiber immediately upstream from the gene. These structural alterations probably involve nucleosome phasing and helix destabilization induced by NHP.

The cells of a number of tissues retain the capacity to alter the frequency of transcription of certain genes. In fact, for tissues such as liver this is a major physiological function. Most of these changes are in response to humoral and neural signals originating from outside the tissue, and, rather than calling for the induction of only one protein, these signals tend to elicit programs of change that involve the transcription of a number of genes. The nature of these signals and the programs of change they bring about are discussed extensively in Section 4. Consequently, a search for the mechanism of activation of individual sequences may obscure more generalized regulatory mechanisms.

Information is transmitted to the nucleus in one of three ways (see Chapter 11 and Fig. 11-1). Thyroid hormones enter the nucleus directly and interact with chromatin-associated receptors, steroid hormones interact with a receptor in the cytoplasm before entering the nucleus, whereas polypeptide and other hormones induce the synthesis, or release, of a second messenger that enters the nucleus and alters the activity of an enzyme capable of covalently modifying one or more chromatin constituents. This modification is usually phosphorylation, although acetylation may also be used. The net result is the interaction of the activated thyroid receptor, the activated steroid-hormone–receptor:acceptor complex, or the covalently modified (or modifying) protein with nucleosome-organized DNA such that RNA polymerase transcribes the gene at a higher rate (or the affinity of the template is decreased where there is repression). The mechanisms of these interactions are the very basis of transcriptional control, and although they remain poorly understood some general statements can be drawn from current knowledge.

It is unlikely that genes are activated and repressed by the sequence-dependent interaction of inducers and repressors as they are in bacterial cells, for several reasons. The first is the fact that the concentration of each single-copy structural gene is far less than that of its bacterial counterpart. Second, if 12 nucleotides is the minimum length of a unique sequence on the bacterial cell genome, a far longer sequence will be required to represent a unique sequence on the eukaryotic genome, together with a large increase in the number of "like" sequences to which a putative sequence-specific regulatory protein would bind. Moreover, such a sequence may be too long for most proteins to span, particularly when most NHP proteins that do bind to DNA are quite small. All of this makes it unlikely that a regulatory protein could ever locate a single unique binding site on the eukaryotic genome in a time frame compatible with the observed rates of gene expression. This conclusion is borne out by the only example we have of the specific interaction between a gene and a protein—the 5S rRNA gene. Rather than being represented as a single copy, this gene is amplified 100,000-fold so that its concentration is greatly increased. Moreover, the regulatory protein, rather than being present in a few molecules per cell, is present at a high concentration to ensure a high rate of occupancy of its binding site. The expression of the gene is regulated by the gene product competing for the inducer until its concentration is reduced such that it can no longer interact with the gene. This is a novel mechanism but clearly impractical for the regulation of single-copy genes.

There is a considerable amount of evidence to indicate that the molecules that interact with the genome to effect a change in transcription do so by inducing a conformational change in chromatin. This may be a change in fiber size (20–30 nm ↔ 10 nm) or a change in the arrangement of nucleosomes along the 10-nm fiber. It is highly likely that this involves helix destabilization exposing single-stranded DNA for which RNA polymerases have a high affinity. The observation that RNA polymerases have little or no affinity for intact, double-stranded DNA is highly significant, for this would prevent the nonspecific sequestering of the small number of polymerase molecules by the vast amount of double-stranded DNA on the genome.

Protein–protein interactions are a key feature of transcriptional control. In the nucleus proteins may recognize other proteins more easily, more quickly, and much more specifically than proteins could ever interact with DNA. The specificity of transcription is determined by the placement of particular proteins at strategic sites in chromatin during the process of differentiation. These proteins remain in place near the sites of activation of genes or groups of genes and are able to destabilize the DNA (or restrict its availability) when appropriately stimulated. Several proteins may be involved in these interactions, including kinases, phosphatases, acetylases, and deacetylases, in order to effect the changes in chromatin structure immediately before and during transcription. The regulatory properties of these enzymes is a key part of the regulation of transcription.

REFERENCES

1. V. G. Allfrey (1980). Molecular aspects of the regulation of eukaryotic transcription: Nucleosomal proteins and their postsynthetic modifications in the control of DNA conformation and template function. In L. Goldstein and D. M. Prescott (Eds.). *Cell Biology: A Comprehensive Treatise*, Vol. 3. Academic Press, New York, pp. 347–437.

2. A. V. Belyavsky, S. G. Bavykin, E. G. Goguadze, and A. D. Mirzabekov (1980). Primary organization of nucleosomes containing all five histones and DNA 175 and 165 base-pairs long. *J. Mol. Biol.* **179**, 519–536.

3. B. Samal, A. Worcel, C. Louis, and P. Schedl (1981). Chromatin structure of the histone genes of *D. melanogaster. Cell* **23**, 401–409.

4. A. Levy and M. Noll (1981). Chromatin fine structure of active and repressed genes. *Nature* **289**, 198–203.

5. G. J. Dimitradis and J. R. Tata (1980). Subnuclear fractionation by mild micrococcal-nuclease treatment of nuclei of different transcriptional activities causes a partition of expressed and non-expressed genes. *Biochem. J.* **187**, 467–477.
6. J. D. McGhee and G. Felsenfeld (1980). Nucleosome structure. *Ann. Rev. Biochem.* **49**, 1115–1156.
7. B. Wittig and S. Wittig (1979). A phase relationship associates tRNA structural gene sequences with nucleosome cores. *Cell* **18**, 1173–1183.
8. C. Louis, P. Schedl, B. Samal, and W. Worcel (1980). Chromatin structure of the 5S RNA gene of *D. melanogaster*. *Cell* **22**, 387–392.
9. C. Wu and W. Gilbert (1981). Tissue-specific exposure of chromatin structure at the 5' terminus of the rat preproinsulin II gene. *Proc. Natl. Acad. Sci. USA* **78**, 1577–1580.
10. J. R. Davie and E. P. M. Candido (1980). DNase-I sensitive chromatin is enriched in the acetylated species of histone H4. *FEBS Lett.* **110**, 164–168.
11. H. G. Hodo III, C. G. Sashasrabbuddhe, M. F. Plishker, and G. F. Saunders (1980). Use of RNA polymerase as an enzymatic probe of nucleosome structure. *Nuc. Acid Res.* **8**, 3851–3864. 3864.
12. B. Wasylyk and P. Chambon (1980). Studies on the mechanism of transcription of nucleosomal complexes. *Eur. J. Biochem.* **103**, 219–226.
13. M. V. Chao, J. D. Gralla, and H. G. Martinson (1980). *lac* Operator nucleosomes. 1. Repressor binds specifically to operator within the nucleosome core. *Biochemistry* **19**, 3254–3260.
14. M. V. Chao, H. G. Martinson, and J. D. Gralla (1980). *lac* Operator nucleosomes. 2. *lac* Nucleosomes can change conformation to strengthen binding by *lac* repressor. *Biochemistry* **19**, 3260–3269.
15. P. Chambon (1975). Eukaryotic Nuclear RNA polymerases. *Ann. Rev. Biochem.* **44**, 613–638.
16. R. E. Roeder (1976). Eukaryotic Nuclear RNA polymerases. In R. Losick and M. J. Chamberlin (Eds.). *RNA polymerase*. Cold Spring Harbor Laboratory, pp. 285–329.
17. M. Muramatsu, T. Matsui, T. Onishi, and Y. Mishima (1979). Nucleolar RNA polymerase and transcription of nucleolar chromatin. In H. Busch (Ed.). *The Cell Nucleus*, Vol. 7. Academic Press, New York, pp. 123–161.
18. T. J. C. Beebee and P. H. W. Butterworth (1977). Eukaryotic deoxyribonucleic acid-dependent ribonucleic acid polymerases: A critical assessment of current ideas concerning their multiplicity, specificity and function and their role in the regulation of gene expression. In P. B. Garland and A. P. Mathias (Eds.). *Biochemistry of the Cell Nucleus, Biochemical Society Symposium*, Vol. 42. The Biochemical Society, London, pp. 75–98.
19. R. Breathnach and P. Chambon (1981). Organization and expression of eucaryotic split genes coding for proteins. *Ann. Rev. Biochem.* **50**, 349–383.
20. M. Busslinger, R. Portmann, J. C. Irminger, and M. L. Birnstiel (1980). Ubiquitous and gene-specific regulatory 5' sequences in a sea urchin histone DNA clone coding for histone protein variants. *Nuc. Acid Res.* **8**, 957–977.
21. S. Y. Tsai, M.-J. Tsai, and B. W. O'Malley (1981). Specific 5' flanking sequences are required for faithful initiation of *in vitro* transcription of the ovalbumin gene. *Proc. Natl. Acad. Sci. USA* **78**, 879–883.
22. B. Wasylyk and P. Chambon (1981). A T to A base substitution and small deletions in the conalbumin TATA box drastically decrease specific *in vitro* transcription. *Nuc. Acid Res.* **9**, 1813–1824.
23. C. Benoist and P. Chambon (1981). *In vivo* sequence requirements of the SV40 early promoter region. *Nature* **290**, 304–315.
24. R. Grosschedle and M. L. Birnstiel (1980). Spacer DNA sequences upstream of the TATAAATA sequence are essential for promotion of H2A histone gene transcription *in vivo*. *Proc. Natl. Acad. Sci. USA* **77**, 7102–7106.

REFERENCES

25 C. Benoist, K. O'Hare, R. Breathnach, and P. Chambon (1980). Ovalbumin gene-sequence of putative control regions. *Nuc. Acid Res.* **8**, 127–142.

26 J. L. Telford, A. Kressmann, R. A. Kosbi, R. Grosschedl, F. Müller, S. G. Clarkson, and M. L. Birnstiel (1979). Delimitation of a promoter for RNA polymerase III by means of a functional test. *Proc. Natl. Acad. Sci. USA* **76**, 2590–2594.

27 B. Sollner-Webb and R. H. Reeder (1979). The nucleotide sequence of the initiation and termination sites for ribosomal RNA transcription in *X. laevis*. *Cell* **18**, 485–499.

28 S. Sakonju, D. F. Bogenhagen, and D. D. Brown (1980). A control region in the center of the 5S RNA gene directs specific initiation of transcription: I. The 5' border of the region. *Cell* **19**, 13–25.

29 D. F. Bogenhagen, S. Sakonju, and D. D. Brown (1980). A control region in the center of the 5S RNA gene directs specific initiation of transcription: II. The 3' border of the region. *Cell* **19**, 27–35.

30 D. R. Roop, P. Kristo, W. E. Stumph, M. J. Tsai, and B. W. O'Malley (1981). Structure and expression of a chicken gene coding for U1 RNA. *Cell* **23**, 671–680.

31 R. Bach, I. Grummt, and B. Allet (1981). The nucleotide sequence of the initiation region of the ribosomal transcription unit from mouse. *Nuc. Acid Res.* **9**, 1559–1569.

32 Y. Urano, R. Kominami, Y. Mishima, and M. Muramatsu (1980). The nucleotide sequence of the putative transcription initiation site of a cloned ribosomal RNA gene of the mouse. *Nuc. Acid. Res.* **8**, 6043–6058.

33 H. R. B. Pelham and D. D. Brown (1980). A specific transcription factor that can bind either the 5S RNA gene or 5S RNA. *Proc. Natl. Acad. Sci. USA* **77**, 4170–4174.

34 B. M. Honda and R. G. Roeder (1980). Association of a 5S gene transcription factor with 5S RNA and altered levels of the factor during cell differentiation. *Cell* **22**, 119–126.

35 J. Segall, T. Matsui, and R. G. Roeder (1980). Multiple factors are required for the accurate transcription of purified genes by RNA polymerase III. *J. Biol. Chem.* **255**, 11,986–11,991.

36 T. Matsui, J. Segall, P. A. Weil, and R. G. Roeder (1980). Multiple factors required for accurate initiation of transcription by purified RNA polymerase II. *J. Biol. Chem.* **255**, 11,992–11,996.

37 M. E. Dahmus (1976). Stimulation of ascites tumor RNA polymerase II by protein kinase. *Biochemistry* **15**, 1821–1829.

38 C. E. Zeilig, T. A. Langan, and D. B. Glass (1981). Sites in histone H1 selectively phosphorylated by guanosine 3':5'-monophosphate-dependent protein kinase. *J. Biol. Chem.* **256**, 994–1001.

39 C. C. Chen, B. B. Bruegger, C. W. Kern, Y. C. Lin, R. M. Halpern, and R. A. Smith (1977). Phosphorylation of nuclear proteins in rat regenerating liver. *Biochemistry* **16**, 4852–4855.

40 M. Sikorska and J. F. Whitfield (1982). Isolation and purification of a new 105 K protein kinase from rat liver nuclei. *Biochim. Biophys. Acta* **703**, 171–179.

41 K. Ajiro, T. W. Borun, and L. H. Cohen (1981). Phosphorylation states of different histone 1 subtypes and their relationship to chromatin functions during the HeLa S-3 cell cycle. *Biochemistry* **20**, 1445–1454.

42 K. Ajiro, T. W. Borun, S. D. Shulman, G. M. McFadden, and L. H. Cohen (1981). Comparison of the structures of human histones 1A and 1B and their intramolecular phosphorylation sites during the HeLa S-2 cell cycle. *Biochemistry* **20**, 1454–1464.

43 J. P. Whitlock, Jr., R. Augustine, and H. Schulman (1980). Calcium-dependent phosphorylation of histone H3 in butyrate-treated HeLa cells. *Nature* **287**, 74–76.

44 C. B. Shoemaker and R. Chalkley (1980). H3-Specific nucleohistone kinase of bovine thymus chromatin. *J. Biol. Chem.* **255**, 11,048–11,055.

45 M. S. Halleck and L. R. Gurley (1980). Histone H2A subfractions and their phosphorylation in cultured *Peromyscus* cells. *Exp. Cell Res.* **125**, 377–388.

46. M. Sikorska, J. P. MacManus, P. R. Walker, and J. F. Whitfield (1980). The protein kinases of rat liver nuclei. *Biochem. Biophys. Res. Commun.* **93**, 1196–1203.
47. D. J. Wolff, J. M. Ross, P. N. Thompson, M. A. Brostrom, and C. O. Brostrom (1981). Interaction of calmodulin with histones. *J. Biol. Chem.* **256**, 1846–1860.
48. J. Bode, K. Henco, and E. Wingender (1980). Modulation of the nucleosome structure by histone acetylation. *Eur. J. Biochem.* **110**, 143–152.
49. G. S. McKnight, L. Hager, and R. D. Palmiter (1980). Butyrate and related inhibitors of histone deacetylation block the induction of egg white genes by steroid hormones. *Cell* **22**, 469–477.
50. M. E. Smulson and T. Sugimura (Eds.) (1980). *Novel ADP-Ribosylations of Regulatory Enzymes and Proteins*. Elsevier/North-Holland, New York.
51. V. Zongza and A. P. Mathias (1981). The variation with age of nuclear phosphoproteins released during micrococeal-nuclease digestion and nucleosomal phosphoproteins in three cell types from rat liver. *Biochem. J.* **194**, 963–974.
52. L. J. Kleinsmith (1978). Phosphorylation of nonhistone proteins. In H. Busch (Ed.). *The Cell Nucleus*, Vol. 6. Academic Press, New York, pp. 221–261.
53. H. Stahl and R. Knippers (1980). Chromatin-associated protein kinases specific for acidic proteins. *Biochim. Biophys. Acta* **614**, 71–80.
54. M. E. Dahmus (1981). Purification and properties of calf thymus casein kinases I and II. *J. Biol. Chem.* **256**, 3319–3325.
55. M. E. Dahmus (1981). Phosphorylation of eukaryotic DNA-dependent RNA polymerase. *J. Biol. Chem.* **256**, 3332–3339.
56. J. D. Saffer and J. E. Coleman (1980). Reversible phosphorylation of a nucleosome binding protein that stimulates transcription of nucleosome deoxyribonucleic acid. *Biochemistry* **19**, 5874–5883.
57. M.-J. Tsai, S. Y. Tsai, and B. W. O'Malley (1979). *In vitro* chromatin transcription. In H. Busch (Ed.). *The Cell Nucleus*, Vol. 7. Academic Press, New York, pp. 163–199.
58. G. S. Stein, S. Hochhanser, and J. L. Stein (1979). Histone genes: Their structure and control. In H. Busch (Ed.). *The Cell Nucleus*, Vol. 7. Academic Press, New York, pp. 259–307.
59. R. S. Gilmour (1978). Structure and control of the globin gene. In H. Busch (Ed.). *The Cell Nucleus*, Vol. 6. Academic Press, New York, pp. 329–367.
60. D. B. Jump and J. H. Oppenheimer (1980). Thyroid hormone receptor-containing fragment released from chromatin by deoxyribonuclease I and micrococcal nuclease. *Science* **209**, 811–813.
61. H. H. Samuels, F. Stanley, J. Casanova, and T. C. Chas (1980). Thyroid hormone nuclear receptor levels are influenced by the acetylation of chromatin-associated proteins. *J. Biol. Chem.* **255**, 2499–2508.
62. M. R. Hughes, J. G. Compton, W. T. Schrader, and B. W. O'Malley (1981). Interaction of the chick oviduct progesterone receptor with deoxyribonucleic acid. *Biochemistry* **20**, 2481–2491.

SECTION TWO

Regulation of Enzyme Synthesis by Post-Transcriptional Mechanisms

The second stage of enzyme synthesis is the translation of the messenger RNA transcript of the gene into a polypeptide. In bacterial cells this is relatively straightforward because transcription and translation are tightly coupled. In eukaryotic cells the situation is very different. First, the primary transcript of the gene is not a functional, information-carrying molecule. Second, the processes of transcription and translation take place in different parts of the cell, and messenger RNA molecules must be translocated from one location to the other. In Chapter 5 these differences are discussed, starting with the remarkable observations of split genes and the nature of the primary RNA transcript and progressing to the processing of this transcript into a mature messenger RNA molecule and its translocation to the cytoplasm. The regulatory potential of these additional steps in enzyme synthesis is emphasized.

The mechanisms of protein synthesis in bacterial and eukaryotic cells are compared and contrasted in Chapter 6. In bacterial cells there is little evidence of regulation of enzyme synthesis at the translational level, although the translational machinery may mediate transcriptional control under some circumstances. In eukaryotes there is considerable confusion over the extent and significance of translational control in the regulation of enzyme synthesis. There is no doubt that the translational process is a highly coordinated series of reactions and that alterations to any of the catalytic proteins change the rate of the overall process. Moreover, in certain highly specialized cells (reticulocytes) and under certain pathological conditions (viral infection), translation is modulated by covalent modification of initiation factor eIF-2. However, this is a nonspecific effect on the rate of translation of all proteins and, by its very nature, cannot modulate the rate of translation of individual mRNAs.

Now that the concentration and kinetics of turnover of individual mRNAs can be measured it is becoming clear that mRNAs have an intrinsic translatability. Furthermore, this can be modulated by changing the availability for translation of specific mRNAs by factors acting at the mRNP level. As more is learned about the transport of mRNP particles and their fate in the cytoplasm, a major new level of regulation is emerging. Since this involves individual mRNA molecules, gene expression can be regulated in a specific way.

The final stage of gene expression for many proteins and enzymes is the incorporation of the newly synthesized molecule into a specific subcellular organelle. In addition, other proteins and enzymes undergo some form of covalent modification to make them physiologically functional molecules. The mechanisms involved and their regulatory significance are described in Chapter 7. This chapter also illustrates, briefly, the role that reversible covalent modification of proteins plays in the regulation of enzyme activity, a major level of metabolic control in both prokaryotes and eukaryotes that will not be covered in this book.

CHAPTER FIVE

Regulation by Post-Transcriptional Modification of RNA Transcripts

Messenger RNA is the intermediate carrier of information in the process of gene expression in all cells. In bacterial cells mRNA is the immediate product of RNA polymerase and is a direct copy of a contiguous structural gene sequence. The information content of the mRNA molecule is rapidly turned into a protein by ribosomes that commence translation of the RNA transcript even before transcription is complete (see Fig. 2-8). Each mRNA molecule is translated by several ribosomes, and then the message is degraded. It is not unusual for transcription, translation, and degradation of the mRNA molecule to be occurring simultaneously, since the average half-life of bacterial cell mRNAs is 1–2 min.

In eukaryotic cells the situation is very different. For most genes mRNA is not the immediate product of transcription of the gene by RNA polymerase II. Moreover, there are considerable differences in the stability of the messenger since eukaryotic cell mRNAs have average half-lives of 20–24 h. In addition, the processes of transcription and translation occur in separate subcellular compartments, the nucleus and cytoplasm, respectively. Thus in eukaryotic cells there are several extra facets to the metabolism of mRNA that have great regulatory potential.

This chapter deals with the nature of the primary RNA transcript in eukaryotes, its processing to a mature mRNA molecule, and its transport to the cytoplasm. Relatively few RNA processing events occur in bacterial cells, but where they have been shown to occur novel reaction mechanisms are often involved. These are discussed in this chapter.

INTERVENING SEQUENCES AND THE NATURE OF THE PRIMARY RNA TRANSCRIPT

Surely the least predictable and most spectacular discovery in recent years was the finding that the sequence of DNA that codes for many proteins is not a continuous sequence on the eukaryotic genome. This phenomenon was discovered following the work of Berget et al. (1) and Chow et al. (2) on animal-cell viruses and has now been found to occur in virtually all mammalian structural genes so far examined, with the exception of the histone genes. Thus noncoding sequences intervene between the coding sequences, giving rise to split genes. The origins and significance of this remarkable aspect of eukaryotic gene structure are far from understood, but its existence has several implications for the mechanism by which the genome expresses its information.

Split Genes

The sequence organization of a number of mammalian structural genes is given in Table 5-1. These genes are split into two to eight coding pieces as shown diagrammatically in Fig. 5-1 for the ovalbumin, β-globin, and α-amylase genes. Quite clearly the amount of noncoding sequence in all these genes is far greater than the amount of coding sequence. (The word *gene* is used here to describe the sequence of DNA stretching from the 5' end of the first coding sequence to

Table 5-1 Eukaryotic Split Genes[a]

Gene	Source	No. of coding pieces	No. of intervening sequence	Base-pairs coding	Total gene	Ref.
ovalbumin	chicken	8	7	1,872	7,600	3,4
ovomucoid	"	8	7	820	5,600	3
transferrin	"	7	6	2,400	10,000	6
β-globin	rabbit	3	2	591	1,316	5
β-globin	human	3	2	618	1,718	5
conalbumin	chicken	7	6	900	4,600	4
prolactin	rat	3	2	1,000	7,000	7
α-actin	chicken	4	3	2,000	2,500	8
α-amylase	rat	8	7	1,600	8,800	9
lysozyme	chicken	4	3	1,000	6,000	10
DHF reductase	mouse	6*	5*	1,600	42,000	11
Insulin I	rat	2	1	450	570	12
Insulin II	rat	3	2	620	1,060	12
*minimum number			TOTAL	15,471	98,764	
			AVERAGE	1,190	7,597	

[a] The number of coding and intervening sequences is given for each gene, together with the number of base pairs in the coding sequences and the total gene.

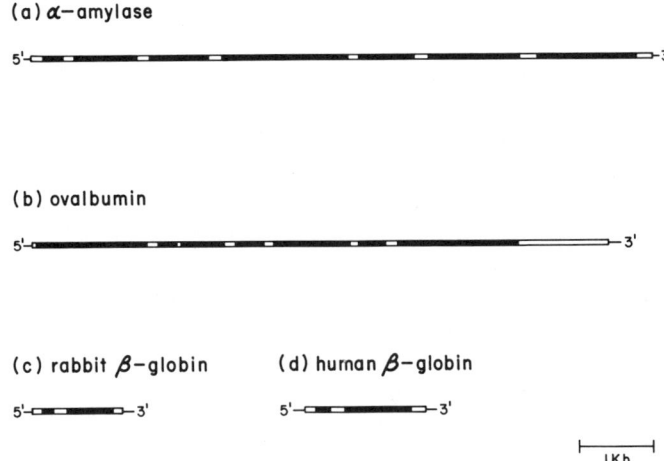

Fig. 5-1 The arrangement of coding (white) and intervening (black) sequences for the α-amylase, ovalbumin, and β-globin genes. All the genes are drawn to the same scale.

the 3' end of the last coding sequence, but we recognize that not all of the sequence has coding information.) Moreover, the coding sequences are usually fragments of DNA smaller than the amount associated with a single nucleosome.

Hybridization studies have shown that much of the DNA of the intervening sequences is single-copy DNA, even though it has no coding information. Moreover, when coding and intervening sequence regions of the same gene in different species are compared the coding sequences are very highly conserved, whereas the intervening sequences show little conservation.

The four genes in Fig. 5-1 are all drawn to the same scale to illustrate the considerable differences in length. Furthermore, these genes do not represent the extremes—the dihydrofolate reductase gene spans *42,000* base pairs of DNA, whereas the procollagen gene contains *52* intervening sequences.

For a complete understanding of gene expression it is necessary to know whether all of the gene, including the intervening sequences, is transcribed or whether RNA polymerase II skips the intervening sequences and only transcribes the coding regions. This problem is relatively easy to approach since, if the whole gene is transcribed, one should be able to find primary transcripts among the heterogeneous nuclear RNA (HnRNA) that are equal in size to the gene. If only the coding sequences are transcribed there should be no sequences in HnRNA larger than mRNA. It has now been established that the whole gene is, indeed, transcribed, and primary transcripts of several thousand nucleotides can be detected when HnRNA is probed with cDNA complementary to cytoplasmic mRNA (see, e.g., Ref. 13 for typical studies of the ovalbumin gene primary transcript). This has been confirmed by constructing probes to intervening sequence DNA and showing that they do exist in primary transcript RNA.

Observations that the primary transcripts contain both coding and noncoding sequences explain, in part, some of the more confusing aspects of nuclear RNA metabolism. We have known for some time that much of the HnRNA in the nucleus is degraded very rapidly. Moreover, there are many more single-copy sequences present in HnRNA than are ever found in the cytoplasm. If the primary RNA transcript were cleaved to eliminate the intervening sequences, which are then degraded, it would explain these high rates of turnover and loss of sequences. This is now known to occur. It has become known as RNA splicing and is just one of a number of modifications to the RNA transcript that take place in the nucleus; these will be described in the section on RNA processing.

However, first we reexamine the issue of the apparent vast excess of single-copy DNA in light of the fact that approximately six times more single-copy DNA is transcribed per "gene" than was previously thought (Table 5-1).

Sequence Complexity of HnRNA

Several recent studies [reviewed by Williamson (14), see also Ref. 15] indicate that in tissues such as liver and brain as much as 30–50% of the single-copy DNA can be hybridized to HnRNA, indicating that this is the amount transcribed. In other cells, such as thymus, only about 9% of the single-copy DNA is transcribed. Thus enough single-copy DNA to code for 90,000 (thymus) to 500,000 (liver and brain) different polypeptides is actually transcribed (assuming each 1000–2000 nucleotide transcript codes for one protein). However, as described above, we now know that about six times more DNA is transcribed than is actually needed. Implementing this correction the amount of DNA transcribed could code for 15,000–85,000 polypeptides. The lower figure (15,000) is within a reasonable range, but the higher figure is somewhat more than most estimates of total numbers of polysomal mRNA sequences (20,000–30,000) in liver and brain. This implies the presence of transcribed sequences in the HnRNA of these tissues that never appear in the cytoplasm. If this is the case, it is of regulatory importance because there must be a selection process that determines the sequences to be saved for processing and exporting to the cytoplasm.

Evidence that there are sequences of RNA confined to the nucleus has been obtained from studies on regenerating rat liver (16). When these cells are stimulated to proliferate, no *new* mRNAs can be detected in the cytoplasm, whereas approximately 5000–10,000 *new* sequences appear in the nucleus. These HnRNA sequences are polyadenylated, but are not transported to the cytoplasm. It is possible that these sequences are regulatory and induce changes in the *relative abundance* of mRNAs appearing in the cytoplasm. Thus, the mRNAs required to direct regeneration are always present in the cytoplasm, but are maintained at too low a concentration to be effective until they are increased following proliferative activation.

Although large primary transcripts are difficult to work with, more details are needed on changes in sequence complexity if we are to secure a complete understanding of gene expression. It is clear that at least half of the single-copy DNA is actually transcribed by RNA polymerase II, although only a small proportion of it actually has mRNA coding information. The cell has evolved sophisticated techniques for handling this mass of HnRNA and ensuring its efficient processing and transport to the cytoplasm.

The Structural Organization of HnRNP Particles

The long primary transcripts generated by RNA polymerase II are wound into a series of HnRNP particles linked together by spacer RNA as shown in Fig. 5-2. These particles have been studied in the electron microscope and by biochemical techniques, and a number of reviews of this work have been published (17-21).

These particles are believed to correspond to the perichromatin granules that are associated with the dispersed chromatin (see Fig. 3-3). The particles appear to be formed during transcription, and in rat hepatocytes the particles are fairly evenly distributed along the transcript (22), whereas in some tissues there is evidence for a nonrandom phasing of the particles along the transcript (23).

Polymers of HnRNP particles are difficult to isolate because of their size and their acute sensitivity to RNases during extraction. Usually the polymers become degraded to unit particles of diameter 20-30 nm and sedimentation coefficients of 30-40S. Approximately 500-1000 nucleotides of RNA are associated with each particle; current models envision a protein core around which the

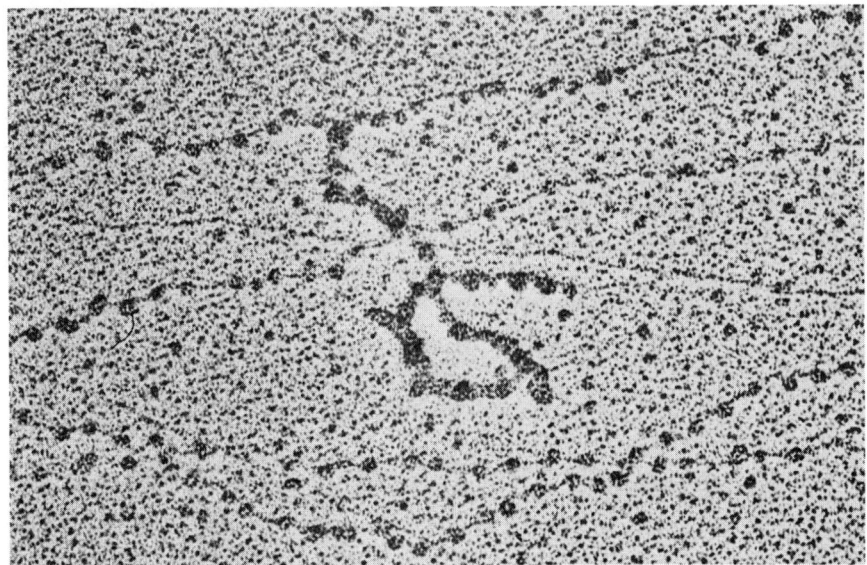

Fig. 5-2 Electron micrograph of HnRNP particles in rat liver nuclei (×130,900). Note the relative sizes of the RNA particles and nucleosomes. [The photograph was kindly supplied by Dr. F. Puvion-Dutilleul and is reprinted from Fig. 5 of Ref. 22, with permission.]

RNA transcript is wrapped in a similar way to DNA on the histone core. However, since the RNA is a single strand that has considerably more flexibility than double-helical DNA and since it is likely to have precise secondary and tertiary structures, the nature of the winding is probably somewhat different.

Considerably more protein is associated with HnRNA in these particles than with DNA on the nucleosomes (protein:nucleic acid ratio of 4–7:1 compared to 1:1). The actual number of proteins associated with HnRNP monomer particles is still a matter of controversy. More recent studies with 2D gel electrophoresis (24, 25) suggest that as many as 50–100, predominantly basic, proteins are present. However, it is very difficult to extract these particles from nuclei in such a way that nonspecific association of proteins can be ruled out. Moreover, since these particles are intimately involved with a transcriptional complex that is made up of many proteins in loose association with each other, artifactual rearrangements could easily occur. The proteins range in molecular weight from 20,000 to 150,000, and since the total mass of protein associated with each monomer is 800,000–1,000,000 daltons all the particles cannot contain all the proteins.

A small number of proteins (M_r = 30,000–40,000) account for a large proportion of the mass of protein, and since they are found in HnRNPs from all species, they may be the structural "core" proteins that interact with the transcript to generate the particle. Three other proteins with M_r = 60,000–70,000 are also present on the particles. They are very acidic and appear identical to the three major matrix proteins. It is now thought that transcription, RNA processing, and particle transport all take place in close association with the nuclear matrix. Indeed, under mildly disruptive conditions 80% of the HnRNA remains associated with the matrix (26).

Other proteins of the particles are likely to be functional proteins involved in the processing of the transcripts into mature mRNP and are probably only associated with the particles transiently. In view of the amount of RNA that is removed the particles must be considered to be dynamic structures changing in composition and size. Many of the associated proteins can be modified by phosphorylation, and a cyclic nucleotide-independent protein kinase is present in these particles (27). Other proteins are known to be methylated, but so far no relationship between this modification and function has been established.

RNA PROCESSING IN EUKARYOTIC CELLS

Modification of the 5' End—Capping

The earliest detectable covalent modification to the primary transcript is the addition of a guanosine residue, via a 5'—5' sugar triphosphate bond, to the 5' nucleotide of the transcript (see Fig. 5-3). This process has been reviewed by Shatkin (28) and appears to occur soon after the initiation of transcription. Indeed, since methyl group donors are known to stimulate transcription, cap-

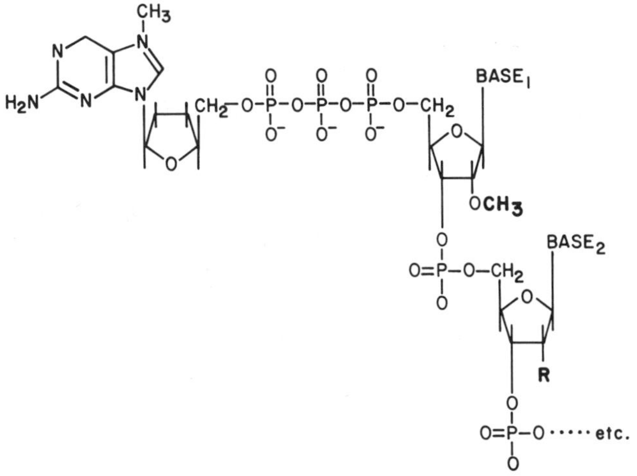

Fig. 5-3 The 5' cap structure of HnRNA and mRNA. Base$_1$ is the 5' nucleotide of the transcript. In Cap-I structures R = OH (on the penultimate nucleotide); in Cap-II structures R = OCH$_3$.

ping may be an important event in the initiation of structural gene transcription by RNA polymerase II. Capping does not occur on rRNA and tRNA molecules, although some small nuclear RNAs (perhaps those transcribed by RNA polymerase II) have a modified cap structure.

The addition of the cap is catalyzed by two enzymes: guanylyl transferase, which adds the guanosine residue, and an N7 methylase, which adds a methyl group to the guanine ring. A third enzyme methylates the 2'-o-position of the ribose moiety of the 5' residue to complete the Cap-I structure as shown in Fig. 5-3.

The cap protects the 5' end of the RNA transcript from ribonucleases since the 2'-o-methyl group on the sugar residue precludes the formation of the 2',3'-cyclic nucleotide, which is an essential intermediate in the exonucleolytic cleavage of RNA. Pulse-labeling studies (29) have shown that the cap structure is conserved during processing and appears in mature cytoplasmic mRNA. Functions other than protection of the RNA transcript from degradation have been suggested, but it is not clear to what extent they occur *in vivo*. For example, it has been proposed that the presence of the cap ensures rapid translation of the mRNA. A protein eIF 4E (M_r = 24,000) associated with initiation factor eIF 4B (see Chapter 6 for a discussion on initiation factors) recognizes and binds to the cap structure and promotes the formation of a stable translation-initiation complex.

Some cytoplasmic mRNAs have a modified cap structure, known as Cap II, generated by a cytoplasmic enzyme that methylates the 2'-o-position of the ribose of the penultimate nucleotide at the 5' end (Fig. 5-3). Cap-II structures are never found in HnRNA molecules and analysis of the cytoplasmic distribution of Cap-I and Cap-II mRNAs reveals a threefold concentration of Cap-II mRNAs on the polysomes, suggesting that these molecules are preferentially translated. However, the possibility that Cap-II formation takes place only on the polysomes has not been ruled out.

This capping process could have regulatory significance by selecting for particular HnRNA or mRNA sequences and protecting them from degradation. However, since the capping sequence —CATTC— appears to be present on all primary RNA transcripts and all appear to be capped, it is not clear how selectivity could be obtained (without the presence of additional proteins bound to the RNA transcript to prevent capping). Furthermore, there is little evidence for the selective removal of caps from cytoplasmic mRNAs. Thus capping appears to protect the 5' end of all transcripts against degradation, and its presence may also assist in the formation of the translational initiation complex by delimiting the 5' end of the molecule.

Methylation of Internal Adenyl Residues of HnRNA and mRNA

In addition to methylation of the residues associated with the cap structure there is also some methylation of the N6 position of adenyl residues along the primary transcripts [reviewed by Rottman (30)] in cells of the higher eukaryotes. In comparison with tRNA and rRNA the extent of methylation of HnRNA is quite low, and only a few nucleotides per transcript are modified. However, although rare, the reaction is extremely site-specific. Analysis of the distribution of radioactivity in cells pulse-labeled with a radioactive methyl donor shows that the methylated residue is present in one of two sequences: either —AA^{m}C— or —GA^{m}C—.

Messenger RNA has fewer methylated nucleotides than HnRNA, which indicates that either some methyl groups are removed by demethylation during processing or some of the methylated residues are among those removed during splicing. The function of methylation at any of these internal sites is not known, but, clearly, it could help to define binding sites for splicing enzymes or for regulatory proteins. It may also place constraints upon the secondary and tertiary structures that the transcript or messenger RNA may adopt.

Modification of the 3' End—Polyadenylation

A high proportion of mature mRNA molecules in the cytoplasm have a long sequence of adenyl residues attached to the 3' end, usually referred to as the poly-A tail. Polyadenylation takes place in the nucleus, and recent work on primary transcripts shows that it is a very early event. Thus RNA transcripts, several thousand nucleotides long and still containing intervening sequences, are often already polyadenylated. The length of the poly-A tails of HnRNA and mRNA species varies considerably but is usually 200–250 nucleotides long. It is not synthesized by transcription; it is a post-transcriptional event formed by the stepwise addition of adenylyl residues to the 3' end of the transcript by the enzyme poly-A polymerase.

Although a considerable amount of effort has been devoted to the determination of the role of the poly-A tail, there is no consensus on its function. Several recent reviews (30–33) have discussed some of the more confusing aspects of the study of this phenomenon. Generally, five possible functions have been considered:

1. It is involved in transcription termination.
2. It stabilizes the 3' end of the transcript.
3. It is involved in the selection of sequences that are to be transported to the cytoplasm.
4. It plays a role in the transport mechanism.
5. It is involved in mRNA stability and translatability.

One of the major problems in trying to assess the poly-A tail's function is the fact that as many as 30% of the mRNA sequences present in the cytoplasm is not polyadenylated. Moreover, sequence complexity studies suggest that there are sequences in the nucleus that are polyadenylated that never appear in the cytoplasm.

The sequence —AAUAAA— is found near the 3' end of virtually all primary transcripts (see Fig. 4-4 in Chapter 4) and is believed to be the signal that is recognized by poly-A polymerase, or an associated protein. Direct evidence for this has been obtained by studying deletion mutants of the SV40 viral genome in the region of transcription termination of the "late protein" mRNAs (34). If the —AAUAAA— sequence is deleted the transcripts are not polyadenylated. Polyadenylation usually commences 11–30 nucleotides past the signal sequence. The base composition of these 11–30 nucleotides has no effect on polyadenylation. However, alteration in the sequence to the *5' side* of the signal sequence results in an inaccurate start site for polyadenylation, indicating that some additional sequence, or sequence-induced structure, is involved.

The temporal relationship between transcription, termination, and polyadenylation has not yet been established. There are two possibilities—either RNA polymerase II stops at the last residue to be transcribed and releases the chain, which is then polyadenylated, or it proceeds past this point, and the signal sequence or another sequence is recognized by poly-A polymerase or a nuclease. The binding of the enzyme(s) may either destabilize the transcription complex, causing termination, or cause the nuclease to cleave and liberate the transcript (Fig. 5-4). This is a particularly attractive mechanism since, as we have discussed in Chapter 4, eukaryotic genes do not have strong terminator sequences. The true 3' end of the transcript would be generated by the endonuclease, which may recognize the —UUUUCAUAGC— sequence found near the 3' end of many genes (see Fig. 4-4 in Chapter 4). However, if polyadenylation is merely part of the termination process, this does not explain why it needs to be 200 nucleotides long or how nonpolyadenylated transcripts are terminated.

The poly-A tail is conserved during processing and appears in the mature mRNA (except in those polyadenylated sequences that are completely de-

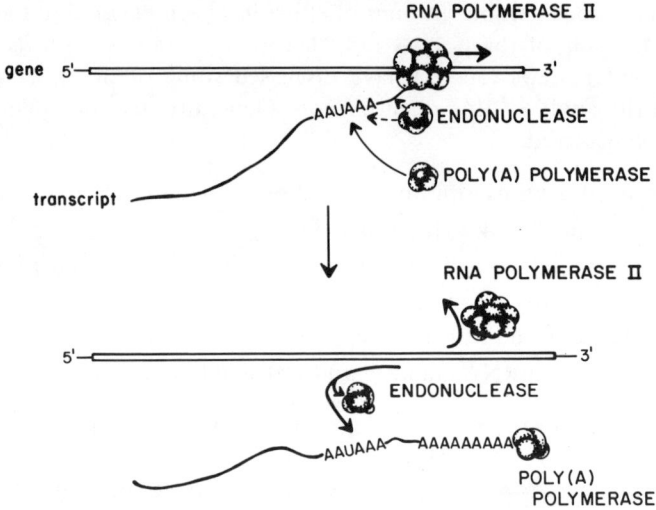

Fig. 5-4 A possible role for poly-A polymerase or endonuclease in transcription termination. The enzymes bind to the transcript as soon as their binding sites are transcribed and release it from the transcription complex. RNA polymerase II subsequently dissociates from the gene.

graded). During the time the transcript is in the nucleus the poly-A tail is packaged as a specific particle attached to the other HnRNP particles of that transcript. This particle can be isolated; it sediments at 15S in sucrose gradients and has a discrete class of associated proteins. Since it is not in intimate contact with other parts of the RNA transcript, it is unlikely to contribute a great deal to the overall stability of the transcript at this stage.

The mechanism by which HnRNP particles are transported from their site of formation to the nuclear pores is not understood, although they appear to be bound to the matrix. It is possible that the poly-A tail is involved in this process. When mature mRNA molecules are translocated out of the nucleus, virtually all of the HnRNP proteins remain in the nucleus, and the naked RNA molecule associates with a new set of proteins in the cytoplasm. The only exception is a protein bound to the poly-A tail, which remains bound during the transport process. The refolding of the molecule during transport to the cytoplasm may bring the poly-A tail into contact with internal oligo (U) sequences interspersed throughout the mRNA providing the basis for a stabilized mRNP particle.

Although the poly-A tail does protect the 3' end of the molecule from exonucleolytic attack there has been no clear correlation established between length of poly-A tail and longevity of the mRNA molecule in the cytoplasm. Moreover, there is no correlation between presence or absence of a poly-A tail and translatability of the message. Thus the function of this sequence remains an enigma that is unlikely to be resolved until we have a clearer understanding of the structural organization of the HnRNA as it is processed into mRNA and transported to the cytoplasm to be translated.

Splicing and Joining of Coding Sequences to Generate a Contiguous mRNA Sequence

The most spectacular part of RNA processing is the removal of the long intervening sequences of noncoding RNA from the primary RNA transcript, reducing it in size until all the coding sequences are contiguous. Often this involves the removal of about 80% of the sequence, although in extreme cases it involves removal of more than 95% of the primary transcript and may involve more than 50 splicing events. The reaction mechanisms are not understood; indeed, none of the enzymes have yet been isolated. However, there has been considerable speculation on the nature of the reaction. But before becoming involved in this discussion it is worthwhile to consider some of the features of the process:

1 During each splicing event the endonucleolytic reactions must occur in close proximity for a ligase to efficiently join the coding sequences.
2 Accuracy is essential so that the coding sequences remain in the correct reading frame.
3 For transcripts with more than one intervening sequence the processing events must take place in an orderly fashion to permit the coding sequences to be joined in the proper order.

Attention has been focused on the way in which the ends of an intervening sequence are brought together and, assuming that the splicing mechanism is quite general (i.e., each splicing reaction does not require a unique protein), how can it bring together sequences that are 4000 base pairs apart as well as those only 200 base pairs apart? Two possibilities have been considered. One model requires the primary transcript to take up a thermodynamically stable secondary structure in which the intervening sequences are looped out such that the ends to be spliced are brought together. The second model uses specific small RNA molecules to ensure that the sequences to be spliced and ligated are brought into close proximity.

Trapnell et al. (35) have used computer simulations of secondary structures for a number of primary transcripts to demonstrate how highly organized, base-paired, stem-loop structures could bring the sequences to be spliced into very close proximity. An example is illustrated in Fig. 5-5 for the small intervening sequence of the β-globin gene (cf. Fig. 5-1c and 5-1d). The two ends of the intervening sequence (boxed) are brought together as part of a loop (Fig. 5-5a). The intervening sequence is cut out by an endonuclease, and the two coding sequences become continuous as the ligase joins the free ends (Fig. 5-5b). There is some sequence redundancy at the ends of the two coding sequences which makes it difficult to establish the exact bonds cut by the endonuclease. Figure 5-5c shows that codons for arginine and leucine could come from either the first coding sequence or the second, or a combination of both.

Although several other transcripts can adopt similar structures (35), it is not clear whether all transcripts could be spliced by a mechanism depending solely on intramolecular base-pairing. An alternative processing mechanism has been

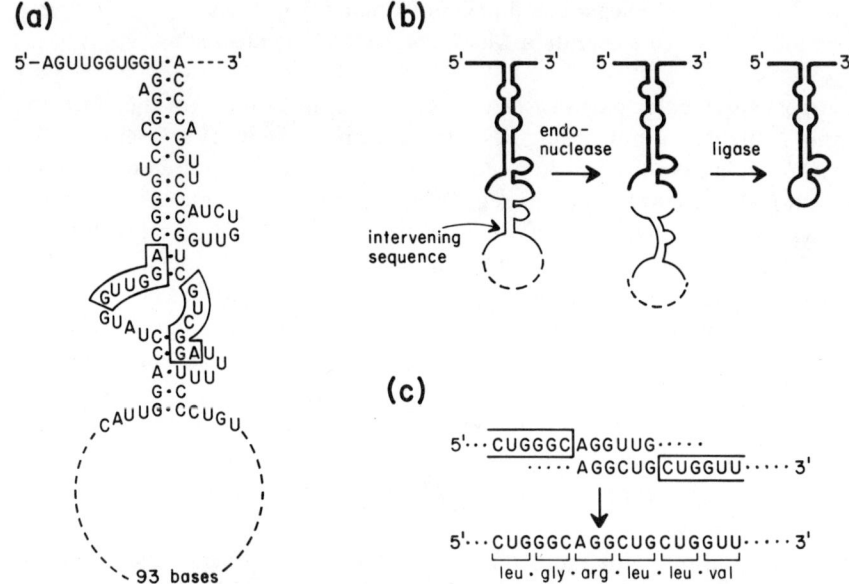

Fig. 5-5 A model for the processing of the small intervening sequence of the β-globin gene. (*a*) Base-paired, stem-loop structure of the primary transcript, which loops out the intervening sequence and brings the coding sequences (boxed) into close proximity. (*b*) The steps of processing; the coding sequences are drawn with the heavier lines. (*c*) Sequence redundancy near the splicing site; the codons for arg and leu could derive from either coding sequence. [Reproduced from Figs. 4 and 5 of Ref. 35, with permission.]

proposed, independently by a number of people (36–38), that uses a small RNA molecule, from a family of small nuclear RNAs (SnRNA), as a matrix upon which to align the ends of the intervening sequences so that splicing can occur. SnRNA molecules have been known to exist for some time and although several have been isolated and sequenced (see Ref. 39 for a recent review) their function has not been established. These molecules vary in size from 90 to 200 nucleotides and are found in the nuclei of all eukaryotic cells. All are capped with a modified Cap-II structure (the guanosine residue is trimethylated) and are extensively methylated at internal sites. Consequently, they are relatively stable molecules ($t_{1/2}$ = 24–48 h). They are known to coextract with HnRNP particles, making them likely candidates for their proposed role.

An example of this mechanism of splicing is given in Fig. 5-6. The sequence of the 5' end of SnRNA U1A is shown in Fig. 5-6a. In Fig. 5-6b the SnRNA is aligned with the two ends of intervening sequence G of the ovalbumin primary transcript. There is extensive complementarity in such a way that the intervening sequence is looped out and the two adjacent coding sequences are brought close together. When the appropriate bonds are cleaved the coding sequences (6 and 7) can be spliced together.

This mechanism depends upon intermolecular base-pairing and requires complementarity between the ends of intervening sequences and the SnRNA if it is to be a general mechanism. Figure 5-7 gives a partial listing of the nucleotide sequences found near a number of splice sites in various genes (a more complete list can be found in Refs. 38 and 39), showing that the high degree of sequence homology expected for this mechanism does exist. It is not clear if the sequence diversity found at these sites determines the efficiency at which splicing occurs or whether some sequences have a greater affinity for other molecules in the SnRNA family. Indeed, differences in efficiency may actually be an essential feature of the mechanism serving to establish the order in which intervening sequences are removed so that incorrect splicing cannot occur.

Studies on the processing of the ovalbumin primary transcript (40) established that, although intervening sequences are removed in a preferred sequence, it is not ordered in the sense that they are removed sequentially in the 5' to 3' direction. Indeed, the 3' intervening sequences are removed first (the preferred removal sequence is E,F>D,G>A,B,C, where A is the 5' intervening sequence and F is the 3' intervening sequence in Fig. 5-1b). This is interesting since, although the splicing reactions do not occur sequentially in the 5' to 3' direction, *the order in which the coding regions are joined together is always the same as they are represented in the gene.* These results also indicate that the primary transcript exists, transiently, as a full-length molecule that is presum-

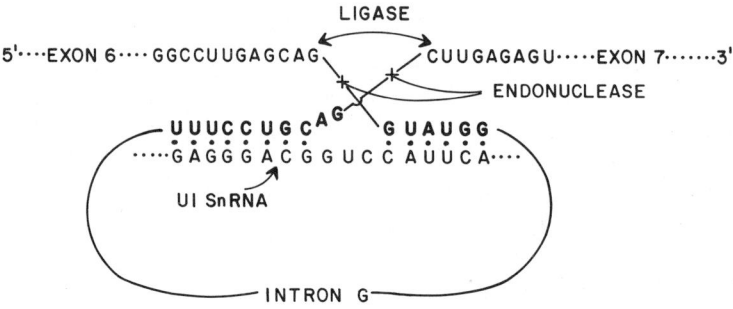

Fig. 5-6 The role of SnRNA U1 in processing. The 5' terminal sequence shown in (*a*) is used as a template (*b*) to align the splicing sites and loop out the intervening sequence (in this case intron G of the ovalbumin gene). The sequence of the joined coding sequences is shown in (*c*). [Reprinted by permission from *Nature* **283**, 220–224. Copyright © 1980 MacMillan Journals Ltd.]

	←Coding—	←— Intervening sequence —→	—Coding→
ovalbumin	UAAG UCAG GCCA AAUG UGAG GCAG	GUGA................UACAG GUAC................UUCAG GUAA................UACAG GUAA................UAAAG GUAU................UCCAG GUAU................UGCAG	GUUGU UGUGG GAAUA GAAUU CAAGA CUUGA
β-globin	GCAG CAGG	GUUG................UUUAG GUGA.................CACAG	GCUGC CUCCU
insulin	CAAG	GUAA................GGCAG	UGGCA
ovomucoid	CCAG UGAG GCAU UCCU AAGU GAGU UCGU	GUGA................CCCAG GUGA................UCGAG GUGU................UUCAG GUAA................CACAG GUUA................UUCAG GUGA................UGCAG GUAC................ UUCAG	AUGCU GUGGA AGAAU AUGAA AGAGC GUUGA GGAAA
α-amylase	CAUG* AAUG† GCAG	GUAU................UUUCAG GUGA................UUUCAG GUGU................UUGUAG	AAAU AAAU GUCU
consensus sequence	$\cdot^A_C$AG	GU^A_GA................U·CAG	G$^{UC}_{GG}$··

Fig. 5-7 Sequence homologies at the sites of splicing for the coding sequences of a number of eukaryotic genes. A consensus sequence, representing the frequency with which each nucleotide occurs in a given position is also shown (· represents no preferred nucleotide).

ably wound onto, or associated with, a number of HnRNA core particles. The nature of the association of the transcript with the particles is not known, but the particles must be phased to some extent so that the intervening sequences can be spooled out of the way and the coding sequences brought into proximity with the "active site" of the complex. It is possible that the SnRNA molecules (which exist as distinct SnRNP particles) first attach to binding sites near the ends of intervening sequences, and their presence determines the subsequent formation of the HnRNP structures.

Splicing and the Regulation of Gene Expression

Splicing is a critical stage of gene expression in eukaryotic cells because *it is the step at which a functional, information-carrying molecule is generated*. If a primary transcript is not processed the gene is not expressed. If the rate of processing is altered the concentration of mature mRNA will be altered. Moreover, if the number of splicing events is modulated *new* information-carrying mRNA molecules can be generated. The latter possibility makes this step as powerful as transcription itself in determining the phenotype of the cell.

At the present level of ignorance concerning the splicing mechanism and a lack of *in vitro* splicing model systems it is difficult to assess the extent to which splicing is regulated. The process could be regulated at a gross level by altering the availability of SnRNA or splicing enzymes. Moreover, the process may be selective if several SnRNA molecules are involved in splicing and each one is responsible for the processing of a specific subset of genes. Clearly those transcripts not requiring processing, for example, the histone mRNAs, have a distinct advantage.

Any molecule, protein or RNA, that binds to a primary transcript in the vicinity of the splicing junctions will affect its rate of processing and modulate the concentration of mature mRNA. The potential for regulation is almost without limit. However, at the present time it is impossible to distinguish between a regulatory event occurring at the transcriptional level (i.e., an increase in the rate at which polymerase transcribes the gene) from one occurring at the splicing level (i.e., an increase in efficiency of processing leading to an increase in the production of mature RNA), since both can only be assayed by measuring the increase in mRNA concentration.

We now have direct evidence that splicing *is* a regulatory process. It can combine sequences from a transcript in different ways to generate different mRNAs from *one* primary transcript. Several examples have been found among animal-cell viruses (see Refs. 12 and 41 for references), and there is at least one example in a eukaryotic genome. The primary transcript of the SV40 early genes can be spliced in two different ways as shown in Fig. 5-8a. Thus all three coding regions can be joined together to form a mature mRNA that codes for a small ($M_r = 15,000-20,000$) protein, called the t antigen (there is a translation-stop codon near the end of the second coding sequence). In an alternative joining pattern the second coding sequence is eliminated. Since the translation-stop codon is removed, the message is translated through to the end of coding sequence 3, producing a different protein, the T antigen ($M_r = 90,000-100,000$). Both proteins have the same N-terminal amino acid sequence (derived from coding region 1), but have different C-terminal regions (derived from coding regions 2 and 3, respectively).

A similar example of flexibility in splicing is shown in Fig. 5-8b for the processing of adenovirus primary transcripts. In this example *two* transcripts are produced from the same DNA sequence, starting at the same promoter. One transcript contains the three small sequences *a*, *b*, *c* and all three coding sequences 1, 2, 3; whereas the other transcript does not have the third coding sequence. During processing the intervening sequences between the small sequences *a*, *b*, *c* are removed, and each mature mRNA contains a nontranslated 5' sequence derived from *abc*. The transcript containing all three coding sequences is processed so that only the third coding sequence is retained. The shorter primary transcript is processed in one of two ways as shown in the figure, producing mature mRNAs containing either coding sequences 1 and 2

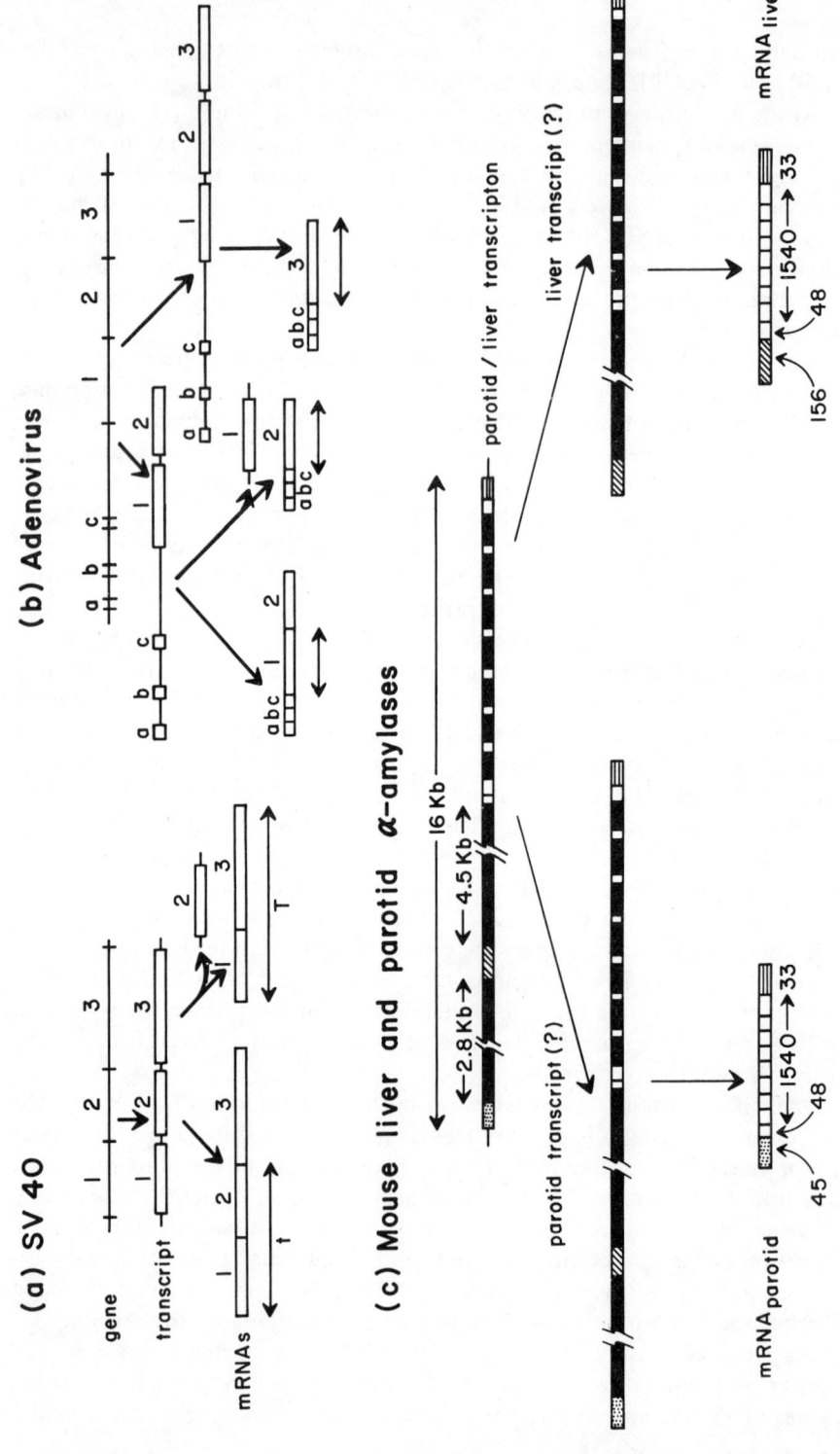

spliced together or only coding sequence 2 (the mRNA containing both coding sequences is only translated to the end of the first sequence before a translation-stop codon is encountered). Thus three different polypeptides are produced from the same set of DNA sequences, a considerable economy of gene expression. Such economy is of great advantage to viruses with very small genomes.

Economy of gene expression is certainly not a feature of eukaryotic cells; however, differential splicing has been found to occur in a way that suggests a role for splicing in differentiation and possibly adaptation. The enzyme α-amylase is sythesized in the cells of the parotid gland and also in the cells of the liver. The polypeptide is identical in each tissue, but the concentration of the enzyme in liver is only 1% that in the cells of the parotid. Surprisingly, the mRNAs coding for this identical polypeptide are *not* completely the same (42). The parotid gland mRNA has a nontranslated leader sequence of *93* nucleotides, a coding sequence of 1540 nucleotides followed by a 33-nucleotide nontranslated 3' tail (Fig. 5-8c). The liver mRNA has a *204*-nucleotide leader sequence of which the last nucleotides are identical to the last 48 nucleotides of the parotid leader. The coding and 3' tail sequences are also identical. Thus the mRNAs differ only in the sequence of their 5' leaders, a difference that could only arise by differential splicing in the two tissues.

The α-amylase gene and its 5' flanking sequence have recently been examined (43), revealing some unusual aspects of the structure of genes in eukaryotes. The 8 coding sequences of the structural gene are clustered together in a 8000-base-pair segment of DNA. However, the promoter and the sequence that codes for the 5' portion of the nontranslated leader of the liver mRNA lie *4500* base pairs upstream, and the corresponding sequence for the parotid cell mRNA lies *2800* base pairs even farther upstream (Fig. 5-8c). The sequence of DNA *from the "promoter" to the 3' end of the structural gene* will be referred to as a *transcripton.*

The nature of the primary transcripts in each tissue has not yet been established. In the parotid gland transcription starts at the "parotid" promoter and the 16,000-nucleotide transcripton transcribed. The transcript is processed by removing an intervening sequence *7300* nucleotides long (which includes the liver promoter) in order to splice the *45*-nucleotide 5' portion of the leader to the rest of the gene. If transcription starts at the "liver" promoter in parotid cells the transcript must be degraded, since only mRNA with the appropriate leader is found in the cytoplasm. It is likely, however, that either this promoter is blocked or initiation is much less efficient than at the "parotid" promoter. In liver, transcription presumably starts at the "liver" promoter, and the low efficiency

Fig. 5-8 The formation of multiple mRNAs from a single "gene" as a result of differential splicing. The examples shown are for the animal-cell viruses SV40 and adenovirus, and the α-amylase gene in liver and parotid gland cells. The processing events are described in the text [the horizontal arrow under each mRNA in (*a*) and (*b*) represents the coding sequence]. In (*c*) the coding sequences are in white, and intervening sequences are in black.

of this promoter may explain the lower concentration of α-amylase of mRNA in this tissue. However, the cells still perform the remarkable feat of cutting out a *4500*-nucleotide-long intervening sequence to join a *156*-nucleotide-long sequence to the remainder of the 5' noncoding region.

At the present time it is not known how the difference in nontranslated leader sequences affects the processing, transport, translatability, and stability of the two mRNAs. When the primary transcripts are identified we will be able to separate the differences in rates of transcription from differences in posttranscriptional events. Moreover, the α-amylase gene will represent the first example of a gene, expressed differently in two differentiated tissues, to be understood at the genetic level. In addition, we shall have some clues about the nature of switching mechanisms, at either the transcriptional or posttranscriptional (splicing) level, during differentiation. It is of interest to establish whether the same mechanism also operates for the expression of the ovalbumin gene, which is similar in many ways. For example, ovalbumin is synthesized at a much greater rate in oviduct cells than in liver cells, and it is known that the nontranslated leader of the mRNA is generated by splicing together two separate coding sequences. An examination of the liver mRNA should establish if it has the same leader sequence or not.

Finally, a new concept in gene expression is emerging, and a new term, *the transcripton*, is proposed to describe the functional unit of DNA that includes a transcription-start site (promoter) connected to a structural gene sequence.

Summary of the Processing of HnRNA → mRNA

A diagrammatic flowchart of the various stages of processing of the product of transcription into a mature mRNA molecule is presented in Fig. 5-9. A minimum of eight enzymes are involved in the process, and all catalyze covalent modifications of the RNA chain. At the present time none of these enzymes have been identified. Each enzyme appears to be directed toward the nucleotide(s) to be modified or excised by a specific sequence, although the secondary structure of the RNA molecules is also likely to be a major determinant in the reaction mechanism. Pulse-labeling experiments indicate that transcripts, for example, the ovalbumin gene primary transcript, are processed very quickly, and the mRNAs rapidly translocate to the cytoplasm. Furthermore, there is no evidence for the existence of pools of unprocessed HnRNA that could be processed under appropriate stimuli. Unfortunately, the overall efficiency of the system, and the error rate, cannot be assessed without being able to precisely quantitate the number of primary transcripts produced by RNA polymerase II.

As more and more genes and their primary transcripts are examined it becomes evident that splicing accounts for most of the rapid RNA turnover seen in the nucleus. However, there is evidence that this does not account for *all* the RNA turnover, and there may, indeed, be sequences with coding potential that are degraded in their entirety within the nucleus. When the sequence complexities of HnRNA and mRNA in *different* tissues are compared it is

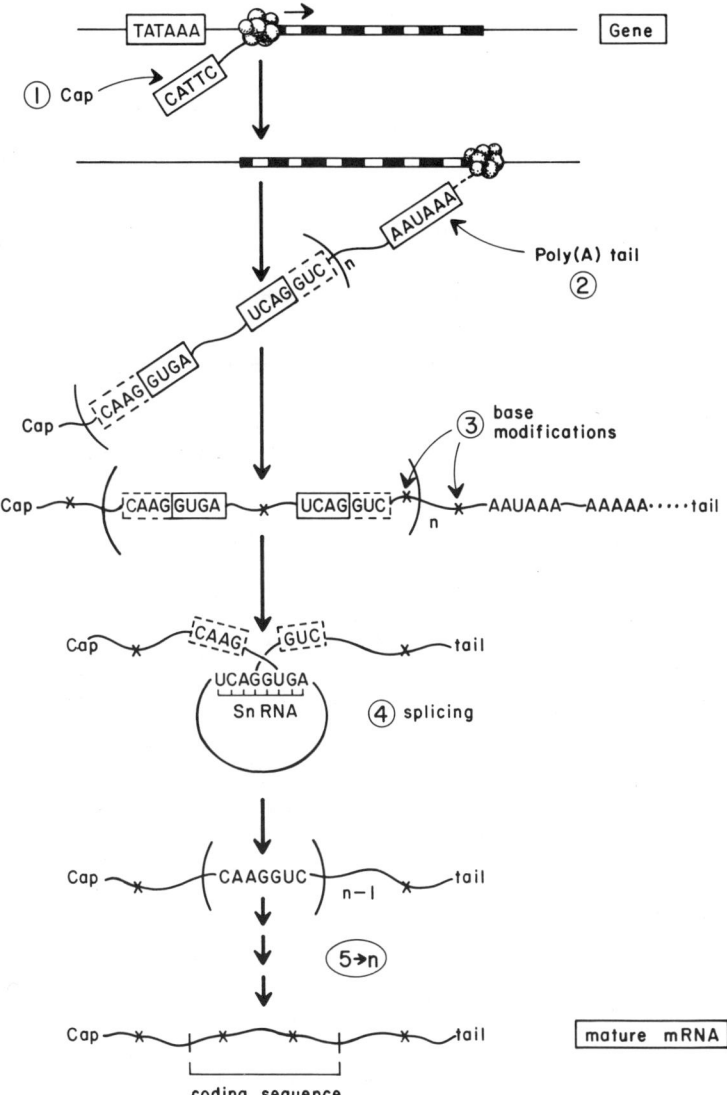

Fig. 5-9 Flowchart for RNA transcription and processing. The nucleotide sequences that act as control signals for the various processing enzymes are shown. n is the number of intervening sequences that must be removed.

found that there are greater similarities between the HnRNAs than the mRNAs, suggesting that nuclei of tissues such as liver and brain transcribe many of the same sequences, but a lot are degraded, and only those characteristic of each tissue's differentiated state appear in the cytoplasm. These observations have great implications for the way in which gene expression is regulated, particularly with regard to the expression of the differentiated state. Moreover, one of the dogmas of the process of differentiation, which states, "All cells contain the same set of DNA sequences, and each gene codes for only one polypeptide," must be modified. Although all cells may, indeed, have the same set of DNA sequences, the presence of multiple promoters coupled with differential splicing gives cells the capacity to generate new information-carrying products and to generate more than one polypeptide from the same DNA sequence.

Processing of Other RNA Species

Table 5-2 shows that HnRNA (and mRNA) are not the only kinds of RNA in the nucleus. Ribosomal RNA and transfer RNA are also synthesized in the nucleus, along with the other species of SnRNA already discussed. Both rRNA and tRNA molecules are synthesized as precursors that require RNA processing to generate biologically active molecules. The processing of these molecules will not be described in detail, and the reviews of Perry (44) and Busch (45) should be consulted for rRNA processing, together with the reviews of Altman (46) and Kang et al. (47) for the processing of tRNA.

In general, their processing is similar to HnRNA except that they are not polyadenylated and not always capped. The primary transcripts (see Fig. 3-2 for the structure of the rRNA gene and primary transcript) are trimmed at both the 5' and 3' ends, reactions that obviate the need for precise transcription-start and -stop sites. Both rRNA and tRNA transcripts contain intervening sequences which must be removed.

The enzymes involved in the processing of these molecules have also not yet been characterized. The only information that we have on the enzymatic reactions that are involved in RNA processing comes from studies on the small amount of RNA processing that takes place in bacterial cells.

Table 5-2 RNA Species Present in the Nucleus Showing Size[a] and Covalent Modifications

RNA species	Nucleotides	Size ($S_{20,w}$)	Cap	Methylated bases	Poly (A) tail
Hn RNA	~8,000	40S–100S	+	±	±
m RNA	~2,000	8S–30S	+	±	±
pre r RNA	14,000	45S	–	+	–
28S r RNA	5,000	28S	–	+	–
18S r RNA	2,000	18S	–	+	–
5.8S r RNA	120	5.8S	–	+	–
5S r RNA	158	5S	–	–	–
pre t RNA	100–120	5S	(+)	+	–
t RNA	~100	5S	(+)	+	–
Sn RNA	90–220	4–6S	+	+	–

[a] Length in nucleotides and sedimentation coefficient.

RNA PROCESSING IN BACTERIAL CELLS

There is no evidence for processing of the transcripts of bacterial *structural* genes. They are translated immediately and then degraded, reinforcing the strict transcriptional control of gene expression. However, RNA processing does occur in bacterial cells since rRNA and tRNA molecules are transcribed as larger primary transcripts. Moreover, the *bacteriophage* T7 processes a large primary transcript into five different mRNAs. These aspects of RNA processing have recently been reviewed (48, 49). The discussion here is confined to the novel reaction mechanisms of some of the enzymes involved.

Processing of Phage-T7 Early mRNAs

The transcription of phage-T7 DNA in infected cells is discontinuous, with a group of early mRNAs being transcribed first, followed by a second group, the late messengers. The early messengers are transcribed, not individually, but altogether as a primary transcript about 7500 nucleotides long. This primary transcript is then cleaved to generate five separate mRNAs by the host cell's RNase-III, an enzyme of $M_r = 50,000$.

There are several aspects of this reaction mechanism that are interesting. It occurs rapidly, with processing starting before transcription is complete. It appears that the enzyme does not locate the site to be cleaved by recognizing a

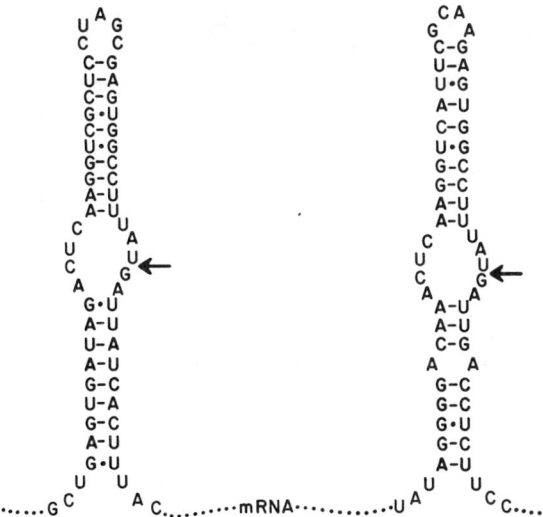

Fig. 5-10 The base-paired stem-loop structures generated in phage-T7 early mRNA primary transcripts. A stem-loop structure is formed between adjacent mRNAs and is recognized by RNase-III, which binds and cleaves the U—G bond in the central bubble. [Reproduced from F. W. Studier et al. (1979). *Miami Winter Symp.* **16**, 261–269, with the permission of Academic Press.]

specific nucleotide sequence, as the sequences on either side of the various splice sites show little sequence homology. However, all the sequences can generate the same secondary structure through intramolecular base-pairing of the transcript, as shown in Fig. 5-10 (49). All five cleavage sites have the same central "bubble" of unpaired bases, and the U—G bond is cleaved. The secondary structure of the transcript appears to be formed spontaneously and does not require any other proteins. The primary transcript is cleaved into five mRNAs, and a 29-nucleotide-long fragment is the only portion of the transcript that is not used.

Processing of Bacterial-Cell tRNA Molecules

The processing of *E. coli* $tRNA_1^{tyr}$ is the best studied system in that the sequences of the gene, primary transcript, and the mature tRNA are known. There are two copies of the gene on the *E. coli* genome, separated by a 200-base-pair nontranscribed spacer. The primary transcript of each gene contains 41 base pairs at the 5' end and 225 base pairs at the 3' end, which must be removed to generate mature $tRNA_1^{tyr}$.

Two enzymes are involved in processing the primary transcript. RNase P catalyzes the removal of the 5' leader sequence. This is an interesting enzyme, since it is found in association with a 300-nucleotide RNA molecule as a ribonucleoprotein complex (80% RNA:20% protein). It is probable, but not yet established, that the RNA moiety is essential for the functioning of the enzyme. Thus the mechanism of action of this enzyme is particularly relevant to current ideas about the involvement of SnRNA molecules in eukaryotic RNA processing. RNase P processes the primary transcripts of all tRNAs, and since there is little sequence homology around any of the splice sites it probably recognizes a common structural feature.

RNase D is thought to catalyze the reaction that generates the 3' end of the tRNA. All the tRNAs are capable of forming a base-paired stem-loop structure near their 3' ends, which probably serves as a recognition signal for the enzyme. The enzyme may also recognize the sequence —CCA, which becomes the 3' end of all mature tRNAs.

Processing of Bacterial-Cell rRNA Molecules

There are seven copies of the rRNA genes on the *E. coli* genome. Each one is organized as an operon containing the 5S, 16S, and 23S sequences. These

---

Fig. 5-11 Processing of bacterial-cell rRNA primary transcripts. (*a*) rRNA gene consisting of 16S, 23S, and 5S rRNAs and interspersed tRNA genes. (*b*) Stem-loop structures formed by the primary transcript, which loop out the 16S and 23S rRNA molecules. The likely sites of nucleolytic cleavage by RNase-III are shown. (*c*) The processing of the 5S rRNA by RNase M5. [Reproduced, with permission, from the *Annual Review of Biochemistry*, Vol. 48 (Ref. 48). Copyright © 1979 by Annual Reviews Inc.]

(a) *E. coli* rRNA gene

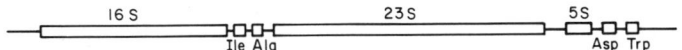

(b) Secondary structure of transcript of rRNA gene

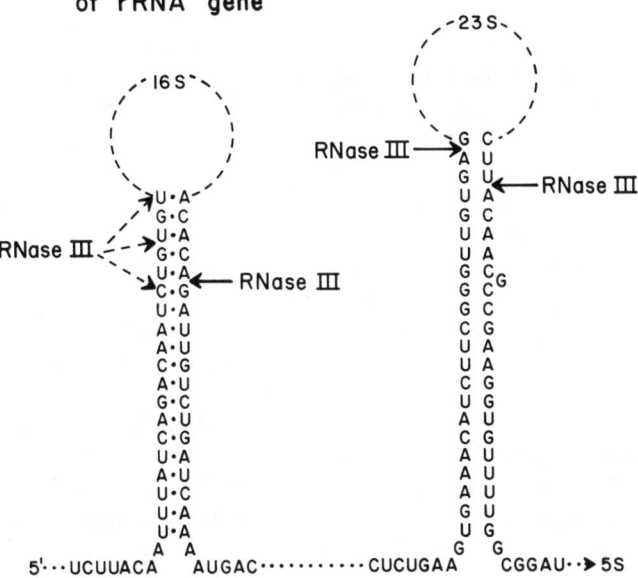

(c) Processing of 5S rRNA in *B. subtilis*

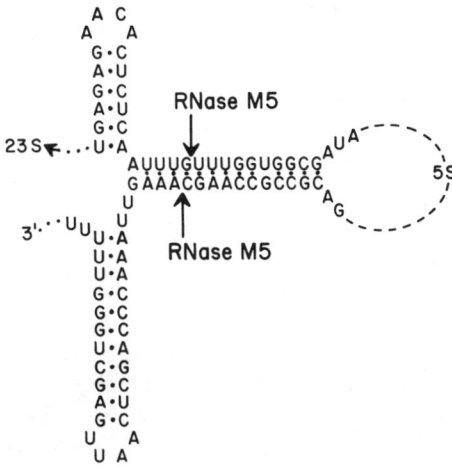

sequences are not contiguous but are separated by spacer DNA and DNA that codes for tRNA molecules (Fig. 5-11a). The seven ribosomal operons are dispersed on the genome, and the tRNA found between the ribosomal RNA sequences varies from operon to operon.

The primary transcript of the operon is usually about 5000 nucleotides in length and has been isolated and sequenced. RNase-III together with a number of other enzymes are responsible for the processing of these primary transcripts. Processing commences while transcription is still in progress, and the entire 5000-nucleotide primary transcript can only be isolated from mutants deficient in RNase activity. There is extensive base-pairing of the primary transcript to generate secondary structures that bring sites to be cleaved into close proximity. For example, both the 16S and the 23S rRNA segments have complementary flanking sequences that generate loops containing the entire 16S or 23S rRNA as shown in Fig. 5-11b. RNase-III cleaves the double-stranded RNA to generate intermediate transcripts that are trimmed down to mature rRNAs.

The processing of the small 5S rRNA molecule by RNase M5 has been studied extensively in *B. subtilis*. This enzyme also reacts with the base-paired stem of a stem-loop RNA secondary structure that loops out the 5S rRNA from the larger rRNA transcript (Fig. 5-11c). RNase M5 is a dimeric molecule with dissimilar subunits α and β. The β subunit has a high affinity for RNA, whereas the α subunit does not bind to RNA. The likely reaction mechanism involves the binding of β to the helical RNA segment, followed by the binding of α to β to generate an active enzyme that cleaves the helix to process the precursor into 5S rRNA. Thus, this is another example of an enzyme that recognizes RNA secondary structure rather than RNA sequences. It also illustrates that bacterial-cell RNases can process large molecules without the need for them to be organized into complex RNP particles. In total, about 10 RNase enzymes are found in bacterial cells, and they are capable of carrying out all the cell's processing needs.

THE TRANSPORT OF mRNA FROM NUCLEUS TO CYTOPLASM

Transport

Intranuclear RNA processing reduces the size of the primary RNA transcript to about 1000–2000 nucleotides in length, at which point it is a mature mRNA molecule consisting of a 5' nontranslatable leader sequence, a central sequence that codes for a protein, and a 3' nontranslated sequence to which is often attached a poly-A tail. Either during, or immediately after, processing the mRNA is translocated to a nuclear pore complex ready for export to the cytoplasm. Translocation is thought to involve the matrix that terminates at the pore complexes [see Preobrazhensky and Spirin (17) and Heinrich et al. (18)]. The size of the matured mRNP approaches or exceeds the mean diameter of the pores, and there is a marked structural reorganization of the mRNA as it passes into the cytoplasm. This is an energy-requiring process and appears to involve one or more covalent modifications (phosphorylation) or pore-complex proteins. Virtually all of the proteins associated with the mRNA in the nucleus are retained within the organelle with the exception of one protein ($M_r = 110,000$), which is believed to exit the nucleus along with the mRNA (50).

mRNP Particles in the Cytoplasm

Messenger RNA molecules in the cytoplasm are usually organized into several discrete pools. The molecules are translated by ribosomes that may be bound to the endoplasmic reticulum or may be "free" in the cytoplasm. However, not all mRNAs are translated immediately, and these molecules remain in the cytoplasm either as distinct mRNP particles or in association with 80S monoribosomes. Thus at least five different pools of mRNA can be distinguished:

(a) mRNA in free mRNP particles
(b) mRNA bound to monoribosomes
(c) mRNA bound to "free" polysomes
(d) mRNA bound to membrane-associated polysomes
(e) mRNA in the process of degradation

It is difficult to quantitate the amounts in each pool because of difficulties in isolating each as a discrete entity. In a recent report, McMullen et al. (51) placed 65% of the polyadenylated mRNAs on the polysomes ("free" and membrane-bound), 9% attached to 80S monoribosomes, and 27% as free mRNP in mouse ascites tumor cells.

There have been other reports that as much as 40-50% of the mRNA in the cytoplasm exists as free mRNP and is not being actively transcribed. Indeed, the kinetics of labeling (52) and RNA hybridization data (53) is not consistent with these particles being merely transport vehicles for mRNA from nucleus to cytoplasm. Moreover, it appears that there are some RNA species in these particles that are *never* translated, indicating that the selection of mRNAs from the mRNP pool for translation may be an important step in the regulation of gene expression.

Free mRNP particles are heterogeneous in size, sedimenting at 20S-70S in sucrose gradients. Generally, smaller particles contain smaller mRNAs, and larger particles contain larger mRNAs. The protein-to-RNA ratio is about 3:1, but estimates of the number of associated proteins have varied from 3 to 23 depending on the tissue and method of isolation. More recent reports, particularly those using UV light to cross-link the proteins to the RNA (54, 55), suggest that there are about six to nine proteins with molecular weights in the ranges 34,000-36,000; 45,000-48,000; 60,000-63,000; 68,000; 73,000-78,000; 98,000; and 105,000-115,000. The 105,000-115,000 protein appears to be the one that accompanies the RNA upon its release from the nucleus. The 78,000 protein is associated with the poly-A tail of the mRNA molecule in cross-linking experiments. There have also been reports (56, 57) that small RNA molecules, other than tRNA molecules, are associated with these particles and may, along with the proteins, be involved in the regulation of the expression of the mRNAs.

The mRNAs in the free mRNP particles constitute a distinct subset of total cellular mRNA sequences. For example, in ascites tumor cells (53) there are about 8500 different mRNAs being actively transcribed on the polysomes, but only 500 different sequences (even though they make up 27% of the mass) present in the mRNP fraction. Other studies have also shown that some mRNA sequences are present mostly on the ribosomes, whereas other sequences are restricted to the RNPs. Although the restricted mRNAs usually cannot be translated *in vitro* when presented as mRNP, they have high translation activity when the proteins (and small RNAs) are removed. Thus there is considerable indirect evidence that the selection of mRNAs for translation is regulated. (This will be discussed in more detail in the section on the regulation of protein synthesis in the next chapter.)

The complexity of regulation at this level is becoming apparent from studies directed toward an understanding of the subcellular organization of the translation machinery (58, 59). It now appears that the cytoplasm has a highly organized cyoskeletal architecture and that polyribosomes originally considered to be "free" in the cytoplasm are attached to this network, and their location within the cell may determine which mRNAs they translate. Moreover, membrane-bound polysomes apparently translocate from one subcellular location to another. For example, mRNAs synthesized in response to the drug phenobarbital exit the nucleus of liver cells and become associated with polysomes on the outer nuclear membrane (see Fig. 3-3), and these *polysomes* translocate to the part of the endoplasmic reticulum where translation will take place. These new concepts in cell organization and intracellular translocation add extra dimensions to the regulation of gene expression through mRNA availability.

Degradation of mRNA

The concentration of mRNA determines the rate of synthesis of the protein it codes for. As already described, a considerable amount of work has been put into an understanding of how the rates of synthesis of mRNAs are modulated. However, mRNA concentration can also be altered by changing the rate of degradation of the molecule, a process about which we know virtually nothing. There is a great heterogeneity of mRNA turnover in eukaryotic cells; some mRNAs have half-lives of 15–20 min whereas other appear to be very stable. It is not known whether this reflects an intrinsic property of each mRNA or whether the association (or lack) of certain proteins determines its half-life. Since degradation follows first-order kinetics, it should be a random process and nonspecific unless the mRNAs are protected from the degradative enzymes. In addition, many cells contain ribonuclease inhibitor proteins, and these could be of regulatory significance. This is one area of mRNA metabolism that needs much more attention.

REFERENCES

1. S. M. Berget, C. Moore, and P. A. Sharpe (1977). Spliced segments at the 5' terminus of adenovirus-2 late mRNA. *Proc. Natl. Acad. Sci. USA* **74**, 3171–3175.
2. L. I. Chow, R. Gélinas, T. R. Broker and R. J. Roberts (1977). An amazing sequence arrangement at the 5' ends of adenovirus-2 messenger RNA. *Cell* **12**, 1–8.
3. B. W. O'Malley, J. P. Stein, S. L. C. Woo, A. Dugaiczyk, J. F. Catterall, and E. C. Lai (1979). A comparison of the sequence organisation of the chicken ovalbumin and ovomucoid genes. In R. Axel, T. Maniatis, and C. F. Fox (Eds.). *Eukaryotic Gene Regulation*, Vol. 14, ICN-UCLA Symposium on Molecular and Cellular Biology. Academic Press, New York, pp. 281–299.
4. P. Chambon, C. Benoîst, R. Breathnach, M. Cochet, F. Gannon, P. Gerlinger, A. Krust, M. LeMeur, J. P. LePennec, J. L. Mandel, K. O'Hare, and F. Perris (1979). Structural organisation and expression of ovalbumin and related chicken genes. In T. R. Russell, K. Brew, H. Faber, and J. Schultz (Eds.). *From Gene to Protein: Information Transfer in Normal and Abnormal Cells*, Vol. 16, Miami Winter Symposium. Academic Press, New York, pp. 55–78.
5. T. Maniatis, E. T. Butler III, E. F. Fritsch, R. C. Hardison, E. Lacy, R. M. Lawn, R. C. Parker and C.-K. S. Shen (1979). The linkage arrangement of mammalian β-like globin genes. In R. Axel, T. Maniatis, and C. F. Fox (Eds.). *Eukaryotic Gene Regulation*, Vol. 14, ICN-UCLA Symposium on Molecular and Cellular Biology. Academic Press, New York, pp. 317–333.
6. D. C. Lee, G. S. McKnight, and R. D. Palmiter (1980). The chicken transferrin gene. *J. Biol. Chem.* **255**, 1442–1450.
7. E. J. Gubbins, R. A. Maurer, M. Lagrimini, C. R. Erwin, and J. E. Donalson (1980). Structure of the rat prolactin gene. *J. Biol. Chem.* **255**, 8655–8662.
8. C. P. Ordahl, S. M. Tilghman, C. Ovitt, J. Fomwald, and M. T. Larger (1980). Structure and developmental expression of the chick α-actin gene. *Nuc. Acid Res.* **8**, 4989–5005.
9. R. J. MacDonald, M. M. Crerar, W. F. Swain, R. L. Pictet, G. Thomas, and W. J. Rutter (1980). Structure of a family of rat amylase genes. *Nature* **287**, 117–122.
10. W. Lindenmaier, M. C. Nguyen-Huu, R. Lurz, M. Stratmann, N. Blin, T. Wurtz, H. J. Hauser, A. E. Sippel, and G. Schütz, (1979). Arrangement of coding and intervening sequences of chicken lysozyme gene. *Proc. Natl. Acad. Sci. USA* **76**, 6196–6200.
11. J. H. Nunberg, R. J. Kaufman, A. C. Y. Chang, S. N. Cohen and R. T. Schimke (1980). Structure and genomic organisation of the mouse dihydrofolate reductase gene. *Cell* **19**, 355–364.
12. B. Lewin (1980). *Gene Expression*, Vol. 2. John Wiley, New York, pp. 801–802.
13. D. R. Roop, M.-J. Tsai, and B. W. O'Malley (1980). Definition of the 5' and 3' ends of transcripts of the ovalbumin gene. *Cell* **19**, 63–68.
14. R. Williamson (1980). The processing of HnRNA and its relation to mRNA. In L. Goldstein and D. M. Prescott (Eds.). *Cell Biology: A Comprehensive Treatise*, Vol. 3. Academic Press, New York, pp. 547–562.
15. C. A. Holland, S. Mayrand, and T. Pederson (1980). Sequence complexity of nuclear and messenger RNA in HeLa cells. *J. Mol. Biol.* **138**, 755–778.
16. P. R. Wilkes and G. D. Birnie (1981). The appearance of new polyadenylated nuclear RNA sequences during rat liver regeneration. *Nuc. Acid Res.* **9**, 2021–2035.
17. A. A. Preobrazhensky and A. S. Spirin (1978). Informosomes and their protein components: the present state of knowledge. *Rec. Prog. Nuc. Acid Res.* **21**, 1–38.
18. P. C. Heinrich, V. Gross, W. Northemann, and M. Scheurlen (1978). Structure and function of nuclear ribonucleoprotein complexes. *Rev. Physiol. Biochem. Pharmacol.* **81**, 102–104.
19. W. J. Venrooij and D. B. Janssen (1978). HnRNP particles. *Mol. Biol. Rep.* **4**, 3–8.
20. T. E. Martin, J. M. Pullman, and M. D. McMullen (1980). Structure and function of nuclear and cytoplasmic ribonucleoprotein complexes. In D. M. Prescott and L. Goldstein (Eds.). *Cell Biology: A Comprehensive Treatise*, Vol. 4. Academic Press, New York, pp. 137–174.

21. H. Busch (Ed.) (1981). *The Cell Nucleus; Nuclear Particles, Part A*, Vol. 8. Academic Press, New York.
22. F. Puvion-Dutilleul, E. Puvion, and W. Bernhard (1978). Visualization of nonribosomal transcriptional complexes after cortisol stimulation of isolated rat liver cells. *J. Ultrastructure Res.* **63**, 118-131.
23. A. L. Beyer and O. L. Miller, Jr. (1980). Ribonucleoprotein structures in nascent hnRNA is nonrandom and sequence-dependent. *Cell* **20**, 75-84.
24. K. Maundrell and K. Scherrer (1979). Characterisation of pre-messenger RNA containing nuclear ribonucleoprotein particles from avian erythroblasts. *Eur. J. Biochem.* **99**, 225-238.
25. C. Brunel and M.-N. Lelay (1979). Two-dimensional analysis of proteins associated with heterogeneous nuclear RNA in various animal cell lines. *Eur. J. Biochem.* **99**, 273-283.
26. R. Herman, G. Zieve, J. Williams, R. Lenk, and S. Penman (1976). Cellular skeletons and RNA messages. *Prog. Nuc. Acid. Res. Mol. Biol.* **19**, 379-401.
27. J. M. Blanchard, C. Brunnel, and P. Jeanteur (1977). Characteristics of an endogenous protein kinase activity in ribonucleoprotein structures containing heterogeneous nuclear RNA in HeLa cell nuclei. *Eur. J. Biochem.* **79**, 117-131.
28. A. J. Shatkin (1976). Capping of Eukaryotic mRNAs. *Cell* **9**, 645-653.
29. R. P. Perry and D. E. Kelley (1976). Kinetics of formation of 5' terminal caps in mRNA. *Cell* **8**, 433-442.
30. F. M. Rottman (1978). Methylation and polyadenylation of heterogeneous nuclear and messenger RNA. In B. F. C. Clark (Ed.). *International Review of Biochemistry: Biochemistry of the Nucleic Acids*, Vol. 17. University Park Press, Baltimore, pp. 45-73.
31. M. Revel and Y. Groner (1978). Post-transcriptional and translational controls of gene expression in eukaryotes. *Ann. Rev. Biochem.* **47**, 1079-1126.
32. J. E. Darnell, Jr. (1979). Steps in processing of mRNA, implications for gene regulation. In T. R. Russell, K. Brew, H. Farber, and J. Schultz (Eds.). *From Gene to Protein: Information Transfer in Normal and Abnormal Cell*, Vol. 16, Miami Winter Symposium. Academic Press, New York, pp. 207-228.
33. G. Brawerman (1981). The role of the poly(A) sequence in mammalian messenger RNA. *CRC Crit. Rev. Biochem.* **10**, 1-38.
34. M. Fitzgerald and T. Shank (1981). The sequence 5'-AAUAAA-3' forms part of the recognition site for polyadenylation of late SV40 mRNAs. *Cell* **24**, 251-260.
35. B. C. Trapnell, P. Tolstoshev, and R. G. Crystal (1980). Secondary structures for splice junctions in eukaryotic and viral messenger RNA precursors. *Nuc. Acid Res.* **8**, 3659-3672.
36. V. Murray and R. Holliday (1979). Mechanism for RNA splicing of gene transcripts. *FEBS Lett.* **106**, 5-7.
37. J. Rogers and R. Wall (1980). A mechanism for RNA splicing. *Proc. Natl. Acad. Sci. USA* **77**, 1877-1879.
38. M. R. Lerner, J. A. Boyle, S. M. Mount, S. L. Wolin, and J. A. Steitz (1980). Are SnRNPs involved in splicing? *Nature* **283**, 220-224.
39. R. Reddy and H. Busch (1981). UsnRNA's of nuclear SnRNPs. In H. Busch (Ed.). *The Cell Nucleus*, Vol. 8. Academic Press, New York, pp. 261-306.
40. M.-J. Tsai, A. C. Ting, J. L. Nordstrom, W. Zimmer, and B. W. O'Malley (1980). Processing of high molecular weight ovalbumin and ovomucoid precursor RNAs to messenger RNA. *Cell* **22**, 219-230.
41. R. L. Erikson (1980). The expression of animal virus genes. In L. Goldstein and D. M. Prescott (Eds.). *Cell Biology: A Comprehensive Treatise*, Vol. 3. Academic Press, New York, pp. 265-301.
42. O. Hagenbüchle, M. Tori, U. Schibler, R. Bovey, P. K. Wellauer, and R. A. Young (1981). Mouse liver and salivary gland α-amylase mRNAs differ only in 5' non-translated sequences. *Nature* **289**, 643-646.

REFERENCES

43 R. A. Young, O. Hagenbüchle, and U. Schibler (1981). A single mouse α-amylase gene specifies two different tissue-specific mRNAs. *Cell* **23**, 451-458.
44 R. P. Perry (1976). Processing of RNA. *Ann. Rev. Biochem.* **45**, 605-629.
45 H. Busch (1978). The current excitement about gene controls of nuclear rDNA. *Life Sciences* **23**, 2543-2554.
46 S. Altman (1978). Transfer RNA biosynthesis. In B. F. C. Clark (Ed.). *Biochemistry of the Nucleic Acids II*, Vol. 17, International Review of Biochemistry. University Park Press, Baltimore, pp. 19-44.
47 H. S. Kang, R. C. Ogden, G. Knapp, C. L. Peebles, and J. Abelson (1979). Structure of yeast tRNA precursors containing intervening sequences. In R. Axel, T. Maniatis, and C. F. Fox (Eds.). *Eucaryotic Gene Regulation*, Vol. 14, ICN-UCLA Symposium on Molecular and Cellular Biology. Academic Press, New York, pp. 69-84.
48 J. Abelson (1979). RNA processing and the intervening sequence problem. *Ann. Rev. Biochem.* **48**, 1035-1069.
49 D. Apirion and P. Gegenheimer (1981). Processing of bacterial RNA. *FEBS Lett.* **125**, 1-9.
50 A. Schweiger and G. Kostka (1980). Identification of a 110,000 molecular weight protein associated with heterogeneous nuclear RNA and messenger RNA in rat liver cells. *Exp. Cell Res.* **125**, 211-219.
51 M. D. McMullen, P. H. Shaw, and T. E. Martin (1979). Characterization of Poly(A$^+$) RNA in free messenger ribonucleoprotein and polysomes of mouse taper ascites cells. *J. Mol. Biol.* **132**, 679-694.
52 A. Mauron and G. Spohr (1978). Kinetics of synthesis of cytoplasmic messenger-like RNA not associated with ribosomes in HeLa cells. *Eur. J. Biochem.* **82**, 619-625.
53 A. J. Kinniburgh, M. D. McMullen, and T. E. Martin (1979). Distribution of cytoplasmic poly(A)$^+$ RNA sequences in free messenger ribonucleoprotein and polysomes of mouse ascites cells. *J. Mol. Biol.* **132**, 695-708.
54 A. J. M. Wagenmakers, R. J. Reinders, and W. J. Van Venrooij (1980). Cross-linking of mRNA to proteins by irradiation of intact cells with ultraviolet light. *Eur. J. Biochem.* **112**, 323-330.
55 J. R. Greenberg (1980). Proteins crosslinked to messenger RNA by irradiating polyribosomes with ultraviolet light. *Nuc. Acid Res.* **8**, 5685-5701.
56 W. Northemann, E. Schmelzer, and P. C. Heinrich (1980). Characterisation of 20-S and 40-S non-polysomal cytoplasmic messenger ribonucleoprotein particles from rat liver. *Eur. J. Biochem.* **112**, 451-459.
57 A. Vincent, O. Civelli, K. Maundrell, and K. Scherrer (1980). Identification and characterization of the translationally repressed cytoplasmic globin messenger-ribonucleoprotein particles. *Eur. J. Biochem.* **112**, 617-633.
58 M. Cervera, G. Dreyfuss, and S. Penman (1981). Messenger RNA's translated when associated with the cytoskeletal framework in normal and VSV-infected HeLa cells. *Cell* **23**, 113-120.
59 F. J. Gonzalez and C. B. Kasper (1981). Sequential translocation of two phenobarbital-induced polysomal messenger ribonucleic acids from the nuclear envelope to the endoplasmic reticulum. *Biochemistry* **20**, 2292-2298.

CHAPTER SIX

Control of the Translation of Messenger RNA in Eukaryotes and Prokaryotes

The mechanism of protein synthesis is essentially the same in all living cells. Amino acids are linked together to form a polypeptide chain in a sequence determined by the nucleotide sequence of an individual mRNA molecule. The transfer of information is mediated by small tRNA molecules, each of which can (i) carry an amino acid and (ii) recognize a unique trinucleotide sequence of messenger RNA. In all cells, synthesis takes place on a specific subcellular structure—the ribosome.

Although the principal reaction of protein synthesis, the covalent condensation of two amino acids, is quite simple, an extraordinarily complex three-dimensional network of interacting protein and RNA molecules is necessary to ensure rapid and accurate translation. In eukaryotic cells, for example, approximately 80 proteins and at least 25 different RNA molecules are involved in the synthesis of each polypeptide chain. The identity of virtually all of these molecules has been established in the past 10 to 20 years, but we still know relatively little about how they interact and even less about the regulation of their activity.

A detailed description of the mechanism of protein synthesis will not be given—the books of Lewin (1) and Weissbach and Pestka (2) and the review of Hershey (3) should be consulted for background information. In the first part of this chapter recent advances in our knowledge of the components of the system, together with developments in our understanding of the mechanisms of protein synthesis in both bacterial and eukaryotic cells, is described. The second part of the chapter is devoted to a discussion of the regulatory role that translation plays in gene expression.

In bacterial cells there is no doubt that transcription is the critical regulatory step except in the expression of some phage proteins. Transcription and translation are tightly coupled in prokaryotes to the point where, in attenuation, the entire translational apparatus is used to mediate transcriptional control. In animal cells, the regulatory importance of the translation step in the production of proteins once appeared firmly established. However, recent developments in methodology for measuring the concentration of individual mRNAs has shifted the emphasis away from translation and toward pretranslational regulation of mRNA availability. The chapter concludes with a discussion of these pretranslational mechanisms.

THE PROTEIN-SYNTHESIZING MACHINERY

Ribosomes: The Sites of mRNA Translation

The physical characteristics of bacterial and animal-cell ribosomes are summarized in Table 6-1. In both cell types the monomer is composed of two dissimilar subunits, each of which has a major rRNA species and as many as 50 associated proteins. All of these proteins have been characterized by 2D gel electrophoresis and are listed in several reviews (1, 4, 5). In general they have molecular weights in the 20,000–30,000 range and are characteristically different from mRNP and HnRNP particle proteins. The proteins associate with the rRNA molecules during nuclear processing, and mature subunits are exported to the cytoplasm. In rapidly growing cells the ribosomal proteins constitute 30% of total cell protein.

Little is known of the function of the ribosomal proteins in terms of which are structural proteins and which may participate in translation reactions. The particles are too large and too complex to study by X-ray diffraction, but some progress has been made using indirect techniques (chemical cross-linking, neu-

Table 6-1 Physical Properties of Ribosomes from Bacterial and Animal Cells

	E. coli			Mammalian cell		
	small subunit	large subunit	monomer	small subunit	large subunit	monomer
Sedimentation coeff. (S)	30	50	70	40	60	80
Dimensions (Å)	80 × 160	140 × 170	200 × 170	100 × 190	180 × 200	220 × 240
Number of proteins	21	34	54	30	45–50	~80 (200*)
RNA species	16S	23S + 5S	16S + 5S + 23S	18S	5S 5.8S 28S	18S 5S 5.8S 28S
M_r of particle	0.9×10^6	1.8×10^6	2.7×10^6	1.5×10^6	3×10^6	4.5×10^6

* Mitochondrial and chloroplast ribosomes contain their own proteins, so as many as 200 different proteins may be ribosome-associated in higher eukaryotes.

tron diffraction, and immunoelectron microscopy), and the approximate location of many of the proteins has now been established (2, 4). It is thought that the two ribosomal subunits combine in such a way that a groove approximately 100 Å in length runs between them. The active areas of the ribosome, the A (aminoacyl tRNA) and P (peptidyl tRNA) binding sites, project into this groove, and mRNA is pulled through it one codon at a time.

Rapid RNA-sequencing techniques have been applied to the long rRNAs with the result that the entire sequence of the 16S bacterial rRNA is now known. Moreover, a computer-generated secondary structure has been proposed by Noller and Woese (6) that involves extensive hydrogen-bonding to form numerous stem-loop structures. This secondary structure may now be used as a model framework upon which to hang the associated proteins.

The sequence at the 3' end of the 16S–18S (small subunit) rRNA is strictly conserved in all cells, as shown in Fig. 6-1a (7). This sequence is capable of generating an intramolecular base-paired stem-loop structure (8, 9) and appears to be important in ribosome function since it can also base-pair with the 5S rRNA component of the large ribosomal subunit in both bacterial and animal

(a) 18S rRNA sequence homology

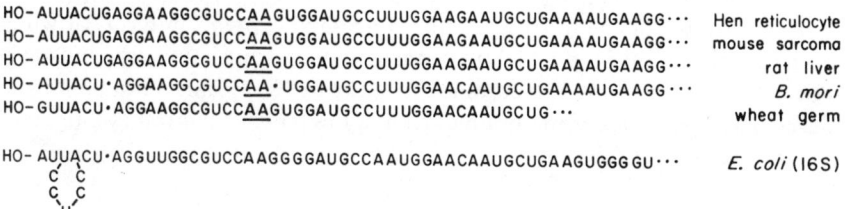

(b) Base pairing to mRNA and 5S rRNA

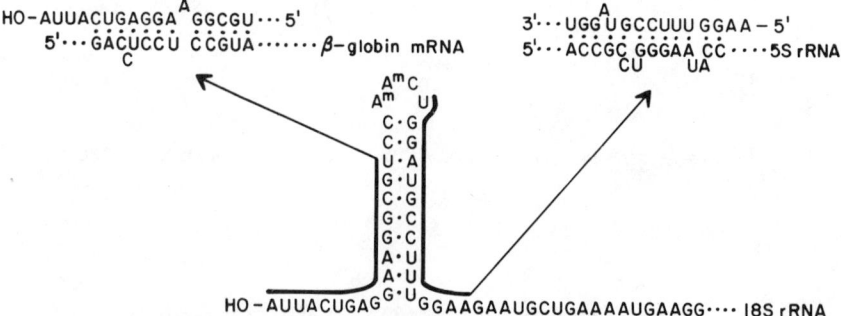

Fig. 6-1 The highly conserved nucleotide sequence at the 3' end of 18S rRNAs in eukaryotic cells. In (a) sequences from different sources are compared with each other and with the 16S rRNA of *E. coli*. The methylated adenyl residues are underlined. In (b) the base-paired stem-loop structure of this sequence is given, showing the regions of complementarity with β-globin mRNA and with the 5S rRNA of the large subunit. [Adapted from Fig. 2 of Ref. 7, with permission.]

cells (Fig. 6-1*b*). It is thought that this reaction is important in the binding together of the subunits to generate a translationally functional monomer. Moreover, in both bacterial and animal cells the 3' end of the 16S–18S rRNA molecule can also base-pair with mRNA upstream from the translation-start codon (see Figs. 6-1*b* and 6-3).

There are two important differences between bacterial and animal-cell ribosomes. Under appropriate conditions bacterial-cell ribosomes synthesize the unusual nucleotides ppGpp and pppGpp (see Fig. 6-8), which are the mediators of the stringent response. These guanine nucleotide derivatives are responsible for pleiotypically slowing-down biosynthesis when there is a lack of amino acylated tRNAs. The ribosomes from animal cells cannot do this. Some of the proteins of eukaryotic cell ribosomes can be covalently modified by phosphorylation [see the review of Kramer and Hardesty (10), and also Thomas et al. (11)], but the significance of this is not yet apparent, since no change in ribosome function can be detected.

Transfer RNA

Transfer RNA molecules are small (~80 nucleotides), multifunctional molecules that carry the appropriate amino acid to the A site of the ribosome during chain initiation and elongation. They have attracted a lot of attention recently as models for protein–nucleic acid interaction studies since, during the course of protein synthesis, they must be able to interact with aminoacyl tRNA synthetase, peptidyl transferase, initiation and elongation factors, and GTPase, as well as ribosomes, mRNA, and, under certain circumstances, ppGpp (2, 12). All the

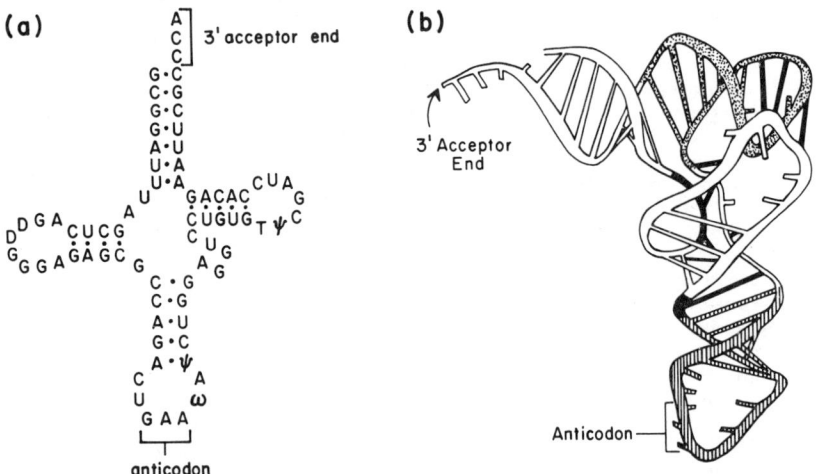

Fig. 6-2 Yeast tryptophanyl tRNA, showing the conventional base-paired structure (*a*) and the proposed tertiary structure (*b*). The modified codons are D = dihydrouridine, ω = guanosine derivatives, ψ = pseudouridine. [The tertiary structure is reproduced with the permission of Professor A. Rich from Ref. 13.]

molecules of tRNA have essentially the same structure as shown in Fig. 6-2. The molecule is represented on the left as a conventional, base-paired, secondary structure, and on the right the proposed three-dimensional structure is given (12, 13). Although all tRNAs have common structural features that permit them to be recognized by peptidyl transferase and to bind to the tRNA-binding sites of the ribosome, they also have specific structural elements so that only the correct tRNA is recognized by each aminoacyl tRNA synthetase.

The degeneracy of the genetic code permits some amino acids to use more than one codon (Table 6-2). Indeed, there are only two amino acids with a single codon, whereas some amino acids have as many as six codons. Thus in many cells there are 50–60 tRNA molecules, but since there is usually only *one* aminoacyl tRNA synthetase for *each amino acid*, most of these tRNAs are not used. However, there is evidence in eukaryotes that the cells of different tissues use different sets of tRNAs. Moreover, during development the cells of one tissue may switch codon usage. This is of regulatory significance since it may alter the rates of translation of mRNAs that use rare codons.

Messenger RNA

Each mRNA molecule carries the information that determines the amino acid sequence of one or more proteins. In eukaryotic cells all the mRNAs are monocistronic (i.e., they code for only one polypeptide chain), whereas many bacterial mRNAs are polycistronic. The message part of the molecule is clearly

Table 6-2 Degeneracy of the Genetic Code in Bacterial and Eukaryotic Cells

Amino acid	Codons					
methionine	AUG					
tryptophan	UGG					
asparagine	AAU	AAC				
aspartic acid	GAU	GAC				
cysteine	UGU	UGC				
glutamic acid	GAA	GAG				
glutamine	CAA	CAG				
histidine	CAU	CAC				
lysine	AAA	AAG				
phenylalanine	UUU	UUC				
tyrosine	UAU	UAC				
isoleucine	AUU	AUC	AUA			
alanine	GCU	GCC	GCA	GCG		
glycine	GGU	GGC	GGA	GGG		
proline	CCU	CCC	CCA	CCG		
threonine	ACU	ACC	ACA	ACG		
valine	GUU	GUC	GUA	GUG		
arginine	CGU	CGC	CGA	CGG	AGA	AGG
leucine	CUU	CUC	CUA	CUG	UUA	UUG
serine	UCU	UCC	UCA	UCG	AGU	AGC
translation stop	UAA	UAG	UGA			

defined; translation starts at an AUG codon (or sometimes GUG in bacterial cells) and stops at a translation-stop codon (UAA, UAG, or UGA). The mRNA molecule must also carry the information enabling it to bind at the appropriate site on the small ribosomal subunit during the initiation process. In addition, eukaryotic-cell mRNAs may also carry information that determines their intracellular sites of translation, their efficiencies of translation, and their rates of degradation.

Bacterial-cell mRNAs have a short leader sequence that precedes the AUG codon. This sequence is relatively well-conserved in those bacterial-cell mRNAs that have been studied (their very rapid turnover makes isolation of intact molecules very difficult), and it can base-pair to the highly conserved 3' end of the 16S rRNA as shown in Figs. 6-3a and 6-3b. It is believed that this interaction serves to stabilize the mRNA–small-ribosomal-subunit interaction and to properly align the AUG codon to ensure a correct start to translation (9, 14).

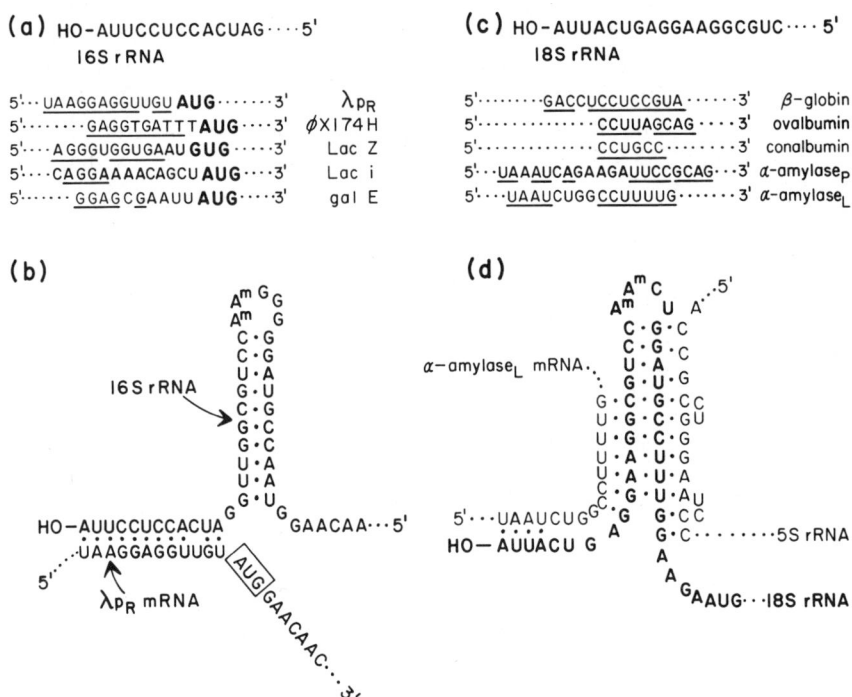

Fig. 6-3 Sequence complementarity between small subunit rRNA and mRNAs in (a) bacterial cells and (c) eukaryotic cells. The sequence of the 3' terminal end of the rRNA is shown at the top, and the sequences of the 5' leaders of selected mRNAs are shown underneath. The nucleotides that base-pair are underlined. In (b) the stem-loop structure of the bacterial 16S rRNA is shown base-paired to mRNA. The translation start codon is shaded. In (d) the equivalent structure for 18S rRNA is given, showing the mutually exclusive base-pairing capability with liver α-amylase mRNA and 5S rRNA.

CONTROL OF MESSENGER RNA TRANSLATION

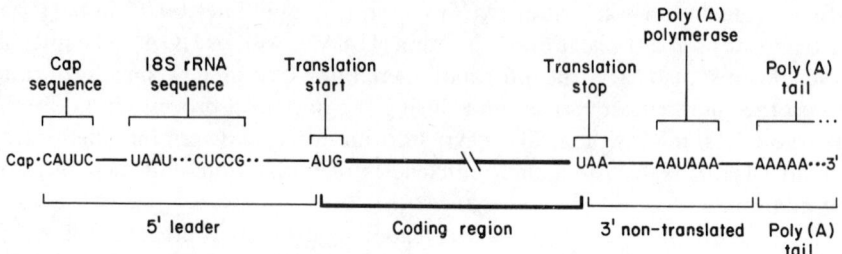

Fig. 6-4 Diagrammatic summary of the important sequences of a typical animal-cell mRNA.

The length of leader sequences in eukaryotic cell mRNAs varies considerably, from as few as 30 nucleotides to more than 200 and, as indicated in Fig. 6-3c, parts of the leaders can base-pair with the 3' terminal region of the 18S small subunit rRNA. However, rather than being adjacent to the AUG codon this complementary sequence is some 30 to 60 nucleotides upstream (and from <10 to ~150 nucleotides from the 5' cap). This variability appears to rule out the participation of mRNA-18S rRNA base-pairing in the selection of a correct translation-start site, but it may participate in the binding of mRNA to the small ribosomal subunit and the subsequent interaction of this complex with the large subunit. As indicated in Fig. 6-3d, mRNAs can base-pair with the 18S rRNA 3' stem loop in a way that may destabilize it and permit its interaction

Table 6-3 The Soluble Factors Involved in the Mechanism of Translation in Prokaryotes and Eukaryotes[a]

	PROKARYOTES		EUKARYOTES	
		M_r		M_r
Initiation factors	IF—1	9,000	eIF—1	19,000
	IF—2	70,000	eIF—2	122,000
	IF—3	21,000	eIF—3	500,000
			eIF—4A	50,000
			eIF—4B	80,000
			eIF—4C	19,000
			eIF—4D	17,000
			eIF—4E	24,000
			eIF—5	150,000
Elongation factors	EF—G	80,000	EF—1	50,000
	EF—T_S	40,000	EF—1β	50,000
	EF—T_U	40,000	EF—2	100,000
Termination factors	RF—1	44,000	RF—1	50,000
	RF—2	47,000		
	RF—3	46,000		

[a] Note that the initiation factors for eukaryotic cells are prefixed by "e." M_r = molecular weight ratio.

with the 5S rRNA of the large subunit to complete the formation of the translation-initiation complex.

Eukaryotic cell mRNAs also have a 3' nontranslated sequence following the translation-stop codon. This is also of variable length and in many cases is followed by a poly-A tail 50–200 nucleotides in length. A generalized eukaryotic-cell mRNA molecule is illustrated in Fig. 6-4. This model summarizes the sequence aspects of the molecule, but it must be emphasized that it also has a three-dimensional structure (see Fig. 6-9) and is associated with several proteins, even during the process of translation. It is likely that the higher orders of RNA structure and the proteins with which it associates play an important role in the regulation of translation.

Soluble Factors

A number of proteins, found either in the cytoplasm or in a 0.5 M salt wash of ribosomes, play important roles in protein synthesis in prokaryotes and eukaryotes. These factors, which act catalytically at either the initiation, elongation, or termination steps of translation (14–16), are listed in Table 6-3. The function of each factor, to the extent that it is known, will be described in the next section. Although these proteins have received the most attention it is becoming apparent that other proteins, particularly some of the more tightly bound ribosomal proteins (see Ref. 17 for details), also have important functions.

THE MECHANISM OF PROTEIN SYNTHESIS

Under optimal conditions protein synthesis proceeds rapidly and efficiently with a protein of $M_r = 50,000$ being synthesized in 1–2 min. Since the synthesis of each molecule of a protein of this size consumes 2000 GTP(+ATP) molecules, it is a major point of commitment for the cell. Thus in rapidly growing bacterial cells more than 80% of the energy generated is consumed in protein synthesis. The cell must choose carefully the kinds of proteins and the amount of each it synthesizes.

The mechanism of protein synthesis will be considered in three stages: initiation, elongation–translocation, and termination. The recent reviews by Hershey (3) and Richter and Isono (17) should be consulted for more details on each stage. The initiation step is the most complicated step and appears to be rate-limiting.

Initiation of Translation

The mechanism of the initiation of translation is more completely understood in bacteria than in any eukaryotic cell, chiefly because fewer factors are involved (1, 14, 17, 18). A major problem in reaching an understanding of the roles played by the various initiation factors is their reaction cooperativity.

That is, they facilitate each other's reactions and "all or none" biochemical responses are seldom seen, making the interpretation of results very difficult.

The cyclic nature of the reactions of the three bacterial initiation factors is shown in Fig. 6-5. The factors act catalytically to bring the three main elements in the reaction mechanism—70S ribosomes, messenger RNA, and initiator aminoacyl tRNA—together to produce a translation system capable of elongation.

The first reaction, ①, is the dissociation of inactive 70S monosomes into subunits catalyzed by IF-3. The initiation of protein synthesis can only take place on the 30S subunit, and IF-3 remains bound to prevent reassociation. IF-1 and IF-2 rapidly bind to this 30S–IF-3 complex (reaction ②). The binding of the three factors appears to be cooperative in that the binding of any one factor is stabilized when the other two are present. GTP also binds at this stage. It is not known whether the initiation factors bind to RNA or to subunit proteins.

The order of binding of mRNA and the initiator aminoacyl tRNA is still unknown. Figure 6-5 depicts initiator aminoacyl tRNA binding (reaction ③) before mRNA (reaction ④), but these may well be reversed. The net result, however, is the formation of the 30S–mRNA-initiation complex. IF-3 dissociates from the complex when the initiator aminoacyl tRNA binds, whereas IF-1 and IF-2 remain bound. The roles of the initiation factors are not fully resolved, but it seems that IF-2 and IF-3 play a key role in the recognition and binding of mRNA such that the AUG codon is properly aligned. IF-3, along with IF-1, also appears to be involved in stabilizing the initiator aminoacyl tRNA.

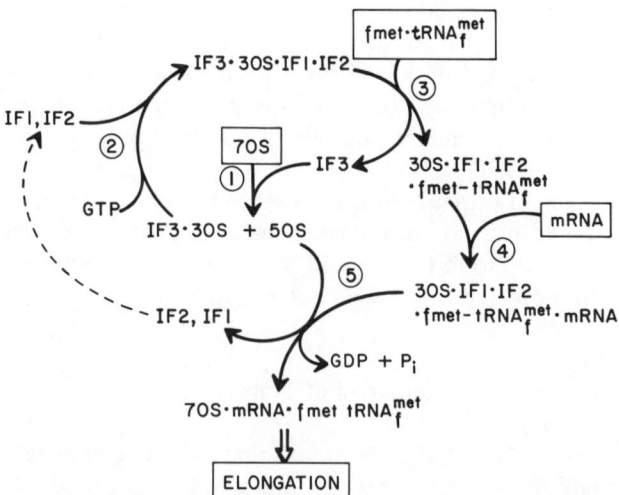

Fig. 6-5 The initiation-factor cycles in bacterial-cell protein synthesis. The sequence of the reactions is described in the text.

In addition to the three initiation factors another protein, ribosomal protein S1, is also involved in the interaction between mRNA and the 30S subunit. This protein probably interacts with the hydrogen-bonded mRNA–16S rRNA structure described in Fig. 6-3b. Unlike the initiation factors, which function catalytically and nonspecifically, S1 has some discriminatory properties. It facilitates the binding of some mRNAs and inhibits the binding of others. Several other such proteins, called interference factors, are also known to exist (14), but their physiological significance is far from understood. However, their presence raises an important question—are all the ribosomes the same? It is generally assumed that ribosomes are nondiscriminatory and are available to translate any mRNA. However, if such interference factors are bound they become a heterogeneous population, and this must be considered in any discussion on the regulation of protein synthesis.

The 30S initiation complex interacts with a free 50S subunit to form the 70S monomer–mRNA-initiator aminoacyl tRNA complex (reaction ⑤). When the 50S subunit binds, GTP is hydrolyzed and IF-2 is released (possibly as an IF-2·GDP complex). IF-1 is considered to be responsible for catalyzing this association of the large subunit to complete the initiation complex, and it is then released. At this stage the complex is ready for the start of chain elongation.

Although the mechanism of initiation in eukaryotic cells is still poorly understood (see Refs. 18 and 19 for recent reviews), some differences from the mechanism in bacteria are apparent (Fig. 6-6). Thus 80S ribosomal monomers are dissociated by the combined actions of eIF-3 and eIF-4c that remain bound to the 40S small subunit. Methionyl tRNA$_i^{met}$ and GTP bind to this small-subunit complex to generate a stable complex. The formation of this latter complex requires eIF-2, and it is the rate-limiting step in the initiation process. It also appears to be the most highly regulated step.

The cooperative interaction of eIF-3 with eIFs 4A, 4B, and 4E is essential for mRNA binding to the small subunit complex (the function of eIF-4D is unknown). One of these proteins, eIF-4E, has a high affinity for the 5' cap of eukaryotic cell mRNAs and may be responsible for selecting the 5' end of the molecule (in addition to possible interactions between mRNA leader and the 3' end of the 18S rRNA molecule). Because of the limited amount of information on the structure of ribosomal subunits and mRNP particles the nature of their interaction is not understood.

Until recently it was thought that translation always started at the AUG codon nearest to the 5' end of the mRNA molecule and that the small-subunit complex bound onto the leader segment of the mRNA near the cap and translocated to this codon (20). However, translation of the α-amylase mRNAs (21) does not start at the 5' proximal AUG codon; it starts at the second one some 50 nucleotides downstream. Thus the manner in which the ribosome locates the *correct* translation start site remains unknown.

Factors eIF-4C (and possibly eIF-4D) and eIF-5 are involved in the recognition and binding of the 60S large ribosomal subunit to generate the 80S monomer–mRNA-initiator aminoacyl tRNA complex. GTP is hydrolyzed dur-

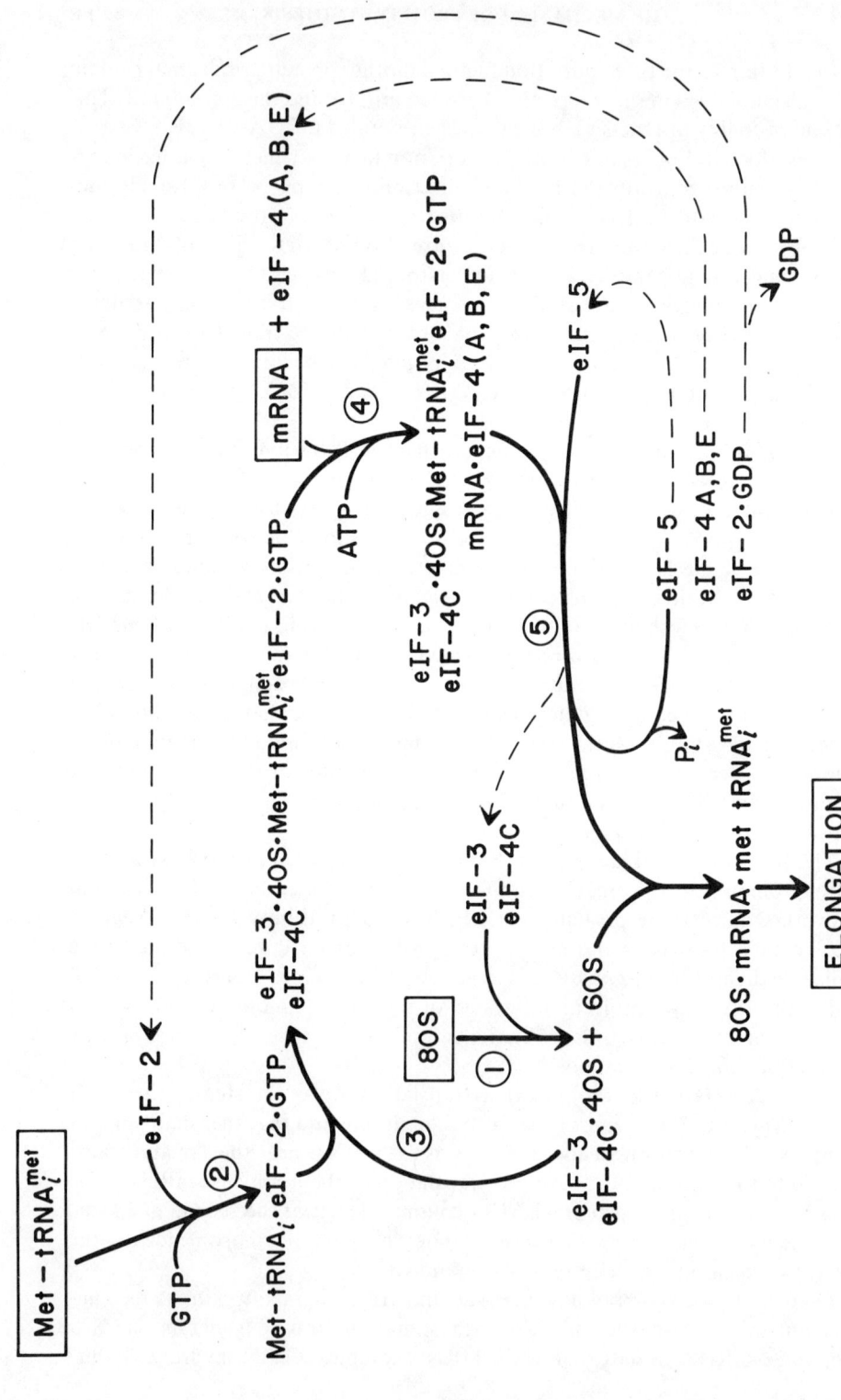

Fig. 6-6 The initiation factor cycles in eukaryotic cell protein synthesis. The cycles are described in the text.

ing 60S particle-binding, and all initiation factors are released (eIF-2 as an eIF-2·GDP complex). It is likely that the 5S rRNA component of the large subunit pairs with the 3' end of 18S rRNA of the small subunit (Figs. 6-1 and 6-3). This completes the formation of the translation-initiation complex.

Elongation and Translocation

In bacterial cells the dissociation of IF-2 upon formation of the 70S–mRNA-initiator aminoacyl tRNA complex (reaction ⑤ in Fig. 6-5) exposes the A binding site to the aminoacyl tRNA molecule complementary to the codon adjacent to AUG. Before binding, the incoming aminoacyl tRNA molecule interacts with elongation factor EF-T and with GTP. EF-T is actually a mixture of two proteins, EF-T$_s$ (heat stable) and EF-T$_u$ (heat unstable), and the aminoacyl tRNA reacts with the latter as shown in Fig. 6-7. GTP is hydrolyzed as the complex binds to the A site of the ribosome and EF-T$_u$ · GDP dissociates. EF-T$_s$ acts catalytically to generate fully functional EF-T$_u$, and another elongation cycle can take place with the next aminoacyl tRNA.

Eukaryotes use a similar cyclic mechanism for aminoacyl tRNA binding with factors EF-1 and EF-1β being equivalent to EF-T$_u$ and EF-T$_s$, respectively. EF-1 is an interesting protein since it seems to exist in some cells in aggregated forms with M_r up to 1×10^6. These multimers are inactive, and monomeric EF-1 protein ($M_r = 50,000$) has to be released to be active.

Peptidyl transferase, which is an enzyme bound to the large ribosomal subunit in all cells, catalyzes the formation of the peptide bond between the amino acid of the aminoacyl-tRNA at the A site and the peptide chain held at the P site. Peptide bond formation transfers the growing peptide chain from the tRNA molecule bound to the P site to the one bound at the A site. The deacylated tRNA dissociated from the P site, and the newly formed peptidyl-tRNA translocates from A to P. At the same time the mRNA molecule moves relative to the ribosome, bringing the next codon into the A site, and chain elongation continues. Under optimum conditions about 20 amino acids are added to the growing polypeptide chain every second.

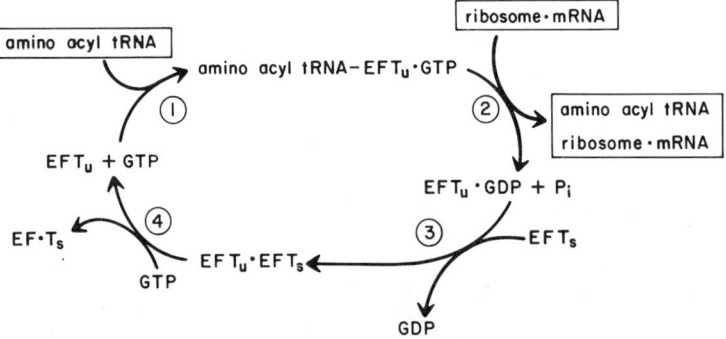

Fig. 6-7 The elongation-factor cycle in *E. coli.*

Translocation is mediated by EF-G in bacterial cells and EF-2 in eukaryotic cells. GTP is consumed during this process in all cells. The observation that EF-2 can be inhibited by ADP ribosylation (see Chapter 7 for details of this modification) is of regulatory significance. Moreover, diphtheria toxin, which has ADP-ribosylating activity, inhibits protein synthesis in cells in which it is present by this mechanism. The mechanisms of translation and elongation are discussed in more detail by Miller and Weissbach (15) and Friesen et al. (26).

Termination

When one of the nonsense codons UAA (ocher), UAG (amber), or UGA (opal) becomes aligned with the A site no aminoacyl tRNA can bind, and this signals the completion of elongation. The newly formed polypeptide chain is then released. Bacterial cells use three release factors: RF-1 recognizes codons UAA and UAG, RF-2 recognizes UAA or UGA, whereas RF-3 is less specific. In eukaryotic cells there is only one releasing factor RF-1. Termination is an energy-requiring process with one molecule of GTP being hydrolyzed per polypeptide chain released. The monoribosome is released from the mRNA and becomes available for a new round of protein synthesis.

REGULATION OF ENZYME SYNTHESIS AT THE TRANSLATIONAL LEVEL IN PROKARYOTES

Messenger RNA molecules transcribed from bacterial genomes are translated almost immediately and degraded within 1–2 min. Since the efficiency of translation is very high there is little or no regulation at the translational level in these cells. However, the mRNAs of some phage are quite stable, and there is evidence for the regulation of their translation.

Translational Control in Phage-Infected Cells

Bacteriophage usually shut off the translation of host cell mRNAs and use the host cell's translational machinery to synthesize their own proteins (9, 22). Phage such as F2, R17, and MS2 contain a single polycistronic molecule (3000–4000 nucleotides long) that codes for maturation protein, coat protein, and RNA replicase, but the proteins are not synthesized in equal amounts. Coat protein is synthesized in greater amounts than the other two.

The RNA molecule has three AUG codons, and translation can start at any one of them. However, the tertiary structure of the RNA molecule is such that only the AUG codon for the coat protein is easily accessible. As ribosomes bind and translate the coat-protein mRNA they induce conformational changes in the RNA molecule such that the initiation codons for the other two messenger sequences become available.

Furthermore, late in the replicative cycle the synthesis of replicase ceases abruptly. Apparently, phage coat-protein molecules bind to the RNA molecule at a location between the site of termination of coat-protein synthesis and the initiation of replicase synthesis (the RNA molecule is arranged: 5'-maturation protein–coat protein–replicase-3'). Binding of coat protein prevents translation of the replicase mRNA. A similar example has been found for gene 32 of the bacteriophage T4, which also seems to be regulated by feedback inhibition by its own protein product.

Tight Coupling between Transcription and Translation

Under optimal conditions the rate of polypeptide chain elongation in bacterial cells is 17–18 amino acids per second, and the rate of transcription is approximately 60 nucleotides per second [see Maaløe, (23)]. Thus the cells match the rate of production of messenger RNAs to the rate at which they can be translated, the rate of translation being dependent upon the availability of amino acids. This tight coupling between transcription and translation is the result of several levels of interaction between the two processes. Of these levels attenuation affords the most elegant example. In this mechanism (see Chapters 2 and 9 for details) the whole translational apparatus is used to regulate the expression of the genes of the operons for amino acid biosynthesis. Aminoacylated and unacylated tRNA molecules play a key role in this mechanism.

A more generalized, coupled response of the transcriptional and translational machinery occurs during stringency (17, 24). Under normal conditions the relative amounts of the various components of the translational apparatus—tRNA, rRNA, mRNA, and amino acid pools—are maintained at certain optimal levels. When one or more components become rate-limiting the cell responds by readjusting the levels of other components to afford maximum economy. For example, when certain amino acids become limiting the cell responds by decreasing the synthesis of the stable RNA species, tRNA and rRNA, until the shortfall can be made up. This mechanism is mediated by unacylated tRNA molecules, which, in the absence of their amino acids, bind to the A site of the ribosome and catalyze the biosynthesis of guanosine tetraphosphate, ppGpp (and also the pentaphosphate pppGpp), by the reaction shown in Fig. 6-8a. This reaction is catalyzed by a ribosomal protein ($M_r = 77,000$) known as the stringent factor (see Refs. 17 and 25 for more details).

The changes in ppGpp level can be quite dramatic, as shown in Fig. 6-8b for cells maintained in medium containing 0.2% acetate and shifted into a rich medium. The rich medium provides a plentiful supply of amino acids, the level of ppGpp drops to almost zero within 1 min, and the rates of RNA and protein synthesis increase rapidly (26). The levels of ppGpp eventually rise to establish a new steady state. The mechanism of action of these unusual nucleotides is not known. Prokaryotes contain only one polymerase enzyme that transcribes all RNA species, yet ppGpp prevents the transcription of rRNA and tRNA without interfering with mRNA transcription.

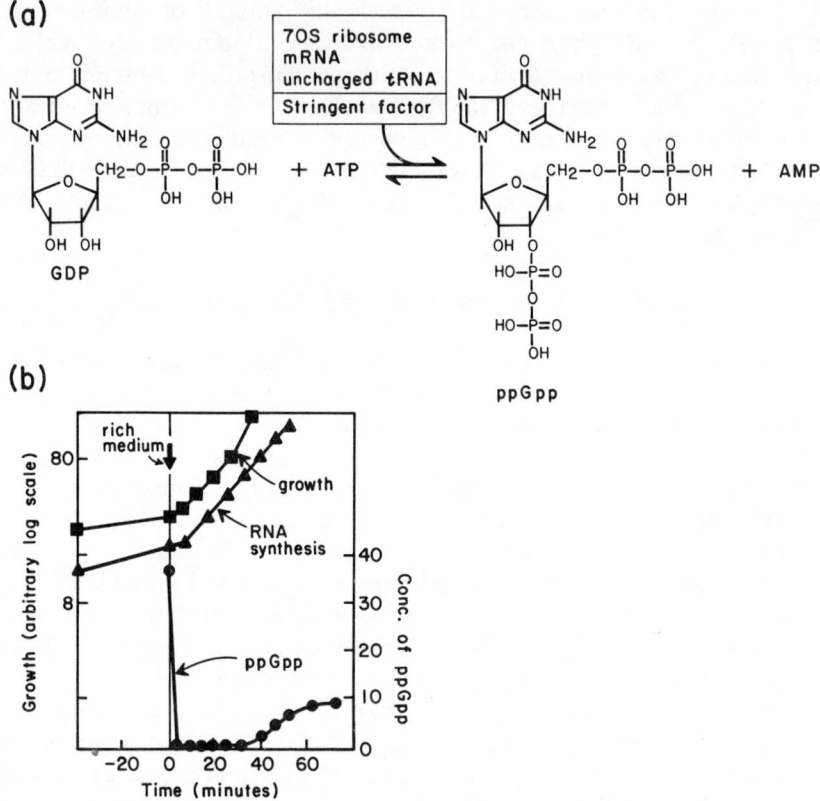

Fig. 6-8 The formation of guanosine tetraphosphate from GDP and ATP catalyzed by the 70S ribosome stringent factor (a). Changes in the concentration of ppGpp when *E. coli* cells are shifted from an amino acid-poor medium to a rich medium are shown in (b). [Reproduced, with permission, from Fig. 1 of Ref. 26. Copyright © 1975 The American Society of Biological Chemists, Inc.]

There appear to be several other, as yet poorly characterized, lines of communication between transcription and translation. For example, an increase in the concentration of fmet-tRNA$_f^{met}$ leads to an increase in RNA polymerase activity to produce more mRNA for initiation-complex formation. Moreover, a buildup of free IF-2 leads to an increase in rRNA synthesis to generate more ribosomes for translation. As described in Chapter 2 some of the initiation and elongation factors can activate RNA polymerase and increase transcription (see Ref. 27 for more details).

REGULATION OF ENZYME SYNTHESIS AT THE TRANSLATIONAL LEVEL IN EUKARYOTES

Until recently, gene expression was considered to be regulated at either the transcriptional level or the translational level. That is, either the rate of transcription of the gene was altered or the rate of translation of the mRNA was altered. The regulation of gene expression at the transcriptional level has been shown to occur in eukaryotes, as we discussed extensively in Section 1. Translational control also seems to exist. For example, in reticulocytes the synthesis of globin stops abruptly when the concentration of heme is lowered. Since these cells are anucleate, translational control appears the most likely mechanism. Mechanistically, this means either the rate of initiation is changed or the rate of chain elongation is changed (the concentration of the mRNA remaining constant). However, since all messenger RNAs in the cell are translated by the same machinery it is not immediately obvious how selective changes in gene expression could be effected at the translation level.

There are several model systems for translational control currently being studied, which are described in this section. In addition, some consideration is given to the regulation of mRNA availability since this post-transcriptional (pretranslational) level of control is emerging as an important new regulatory mechanism of gene expression.

Regulation of Translation at the Initiation Step

The formation of the ternary complex between eIF-2, met-tRNA$_i^{met}$, and GTP is considered to be the rate-limiting step in the initiation process, and there is some evidence that this is also a regulatory step. Studies on the reticulocyte lysate, on the effects of viruses on host-cell protein synthesis, and on the antiviral effects of the interferons all point to this step (10, 28–31). Moreover, all of these diverse effectors appear to act via a common mechanism—the phosphorylation of eIF-2.

Reticulocytes, anucleate precursors of red blood cells, are actively engaged in the synthesis of α- and β-globin, which combine with the cofactor heme to generate hemoglobin. The facts that the cells are easily isolated, there is no transcriptional control, and *in vitro* systems are fully active have made them a popular model system for translational control studies (28–30). Globin synthesis only takes place in the presence of heme. In the absence of heme, protein synthesis stops because of the accumulation of an inhibitory protein called the *heme-controlled repressor, HCR* (it is also sometimes referred to as the heme-controlled inhibitor, HCI). Under normal conditions HCR exists in an inactive form but is converted to an active form as the concentration of heme falls. The mechanism by which it occurs is not clear [a recent proposal (28, 30) that it is activated by a cAMP-dependent protein kinase has been discounted (29, 31)], but since HCR has been found to possess endogenous protein kinase activity it may activate itself by autophosphorylation (32).

Active HCR specifically phosphorylates eIF-2 on the $M_r = 38,000$ small subunit and it is believed that the inhibition of protein synthesis is due to a failure of phosphorylated eIF-2 to function correctly in initiation. Such an inhibitory mechanism is entirely nondiscriminatory, and the initiation of all proteins is inhibited to the same extent, but since globin synthesis represents more than 90% of total protein synthesis it may be an acceptable regulatory mechanism in reticulocytes.

The interferons are a group of glycoproteins produced by cells in response to viral infection (10). Their synthesis is triggered by the double-stranded RNA of the viral genome (and by synthetic double-stranded RNA in laboratory experiments), and as the concentration of interferon increases viral-induced protein synthesis is inhibited. Host-cell protein synthesis is also inhibited but to a lesser degree. Detailed studies on the mechanism involved (see Refs. 10, 18, 28, and 31) reveal that it is remarkably similar to the action of HCR. Interferon induces the synthesis of a protein kinase, which, in the presence of dsRNA, phosphorylates the small subunit of eIF-2. This protein kinase, termed *dsRNA-activated inhibitor (DAI)*, is not the same as HCR [M_r(HCR) = 80,000, M_r(DAI) = 68,000]. For example, it is ribosome-bound rather than being free in the cytoplasm. It is cyclic nucleotide-independent and appears to be activated by autophosphorylation (33).

The inhibition of the initiation of protein synthesis is just one facet of the antiviral effect of interferons. These proteins also inhibit chain elongation by an unknown mechanism that precedes the inhibition of initiation. In addition, interferon induces the synthesis of an endonuclease that preferentially degrades the viral messenger RNAs and an unusual trinucleotide, pppApApA, which is a powerful inhibitor of protein synthesis (10).

Protein kinases similar to HCR and DAI have been found to exist in a number of normal cells (18, 28–31), including rat liver, wheat germ, and brine shrimp embryos (*Artemia salina*), which suggests that this mechanism may be occurring widely. All of these kinases phosphorylate the small subunit of eIF-2. However, there has been no direct demonstration that the phosphorylation of eIF-2 prevents its participation in the process of initiation. Indeed, it has been shown that phosphorylated eIF-2 can *initiate* ternary complex formation (34), which creates considerable confusion over this role for eIF-2 in the regulation of initiation.

eIF-2 Cycling and the Regulation of Initiation

Initiation factor eIF-2 acts catalytically and is released upon completion of the initiation complex (Fig. 6-6). Since there appeared to be a correlation between the phosphorylation of eIF-2 by HCR or DAI and the inhibition of protein synthesis it was assumed that covalent modification of the initiation factor prevented its participation in ternary complex formation. However, it has now been shown that purified eIF-2 in the phosphorylated form can participate in ternary complex formation and be recycled (34), ruling out this modification as the *direct* cause of its inability to participate in initiation.

More recently (35) another form of covalent modification of eIF-2 has been described involving the redox state of sulfhydryl residues of the protein. The protein must be in a fully reduced form to actively catalyze the initiation of translation. When sulfhydryl groups become oxidized the protein loses 85% of its activity by being unable to release GDP during recycling or bind GTP during ternary complex formation. It is proposed that phosphorylation traps eIF-2 in the oxidized form, preventing its participation in the initiation process leading to an inhibition of translation.

Further work should clarify the roles of oxidation/reduction and phosphorylation in the regulation of initiation. This is important since it is one of the few examples where the redox state of a protein could have a regulatory function. Moreover, if found to be generally applicable, it indicates that the redox status of the cell is a great determinant in the overall rate of protein synthesis. However, it must be stressed that this is a *nonspecific mechanism* for changing the rate of total protein synthesis. Since the mechanism *does not even require the participation of mRNA*, it cannot function in selective changes in gene expression.

Regulation of Translation at the Elongation Step

Codon degeneracy could be exploited to regulate the rate of translation of specific mRNAs at the elongation step (36, 37). The rate of elongation of some polypeptides is discontinuous since, when a rare codon occurs, the system must wait for the appropriate aminoacyl tRNA, which may be at a very low concentration. These are referred to as *modulating aminoacyl-tRNAs*, and mRNAs such as globin mRNA that are translated very efficiently tend to avoid such codons. For example, all four codons for glutamic acid are CAG and not CAA. The aminoacyl tRNA that recognizes CAG binds to the ribosome much more efficiently than the one recognizing CAA. Since the relative proportions of the iso-accepting tRNAs for each amino acid vary from tissue to tissue, the same mRNAs may be translated much more efficiently in one tissue than in another. Moreover, the relative proportions of the tRNAs change within the cells of the same tissue during differentiation and may contribute to the changes in the spectrum of proteins synthesized.

mRNA Structure and the Regulation of Translation

Any model for the regulation of gene expression at the level of translation must be able to explain how the rate of translation of one (or a small number of) mRNA(s) can be altered without any change in the rates of translation of thousands of other mRNAs. It is almost inevitable that the mRNAs themselves determine this aspect of their translatability via the nucleotide sequence, or sequence-directed structural conformations, of the leader portion of the molecule. Thus each mRNA has a characteristic efficiency of translation reflected in the ability of the molecule to interact with the small ribosomal subunit during the initiation process. This aspect of translational control has been discussed extensively in a recent review by Steitz (9). The emphasis of her review is on the structural aspects of bacteriophage mRNA translation, but recent studies with eukaryotic mRNAs (38) suggest that similar mechanisms may operate in all cells.

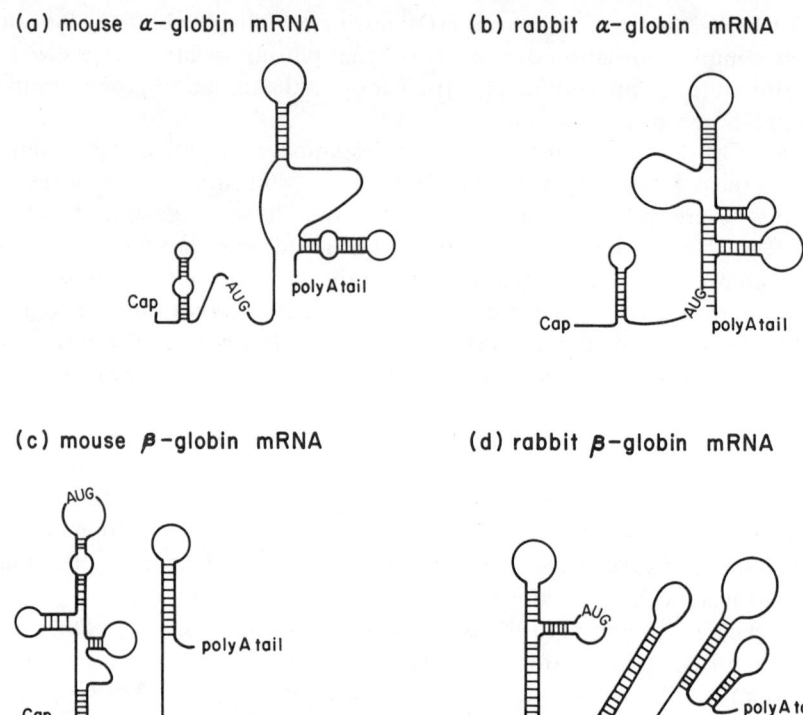

Fig. 6-9 Proposed secondary structures of the α- and β-globins of mouse and rabbit showing the position of the AUG translation-start codon. [Redrawn from Pavlakis et al. (38).]

Models for the secondary structure of the α- and β-globins of mouse and rabbit have been generated by computer analysis of the products of digestion of the molecules with single and double-strand specific RNases [these nucleases can detect the likely regions of stem-loop formation (38)]. Such models show that the AUG codons of the β-globins, which are preferentially translated, are in very exposed locations, whereas the translation-start codons for the α-globins are either base-paired or otherwise less accessible (Fig. 6-9). The structure of the mRNA is also likely to affect its interaction with the large and small ribosomal subunits during initiation. Moreover, translatability will also be affected by the proteins (and RNA molecules) bound to the mRNA, which may mask or expose ribosome binding sites and the AUG codon. It is clearly going to be a formidable task to understand the initiation of translation in these structural terms.

Summary of the Role of mRNA in Determining its Own Rate of Translation

Although the various soluble, catalytic factors play important roles in coordinating the activities of the components of the translational apparatus it is unlikely that modulation of their activity can lead to selective changes in the phenotype of the cell. It is much more likely that selectivity is determined at the mRNP level by either the mRNA itself or by the association (dissociation or modification) of binding proteins or small RNA molecules. Some of the latter molecules are known to be powerful inhibitors of translation, but their regulatory significance is not yet understood (39, 40). (The structure and some properties of RNP particles were described in Chapter 5.)

This conclusion is supported by measurements of the concentration of individual mRNAs during enzyme induction where, in *every* case examined so far, the rate of enzyme synthesis is paralleled, or preceded, by an increase in concentration of its mRNA. These changes which are discussed in considerable detail in Chapter 12, leave little doubt that transcriptional control is the major level for the regulation of gene expression in eukaryotes. However, there is evidence that the induction of some enzymes requires a transcriptional event and a post-transcriptional event(s) to ensure that the product of transcription is efficiently processed, transported, and made available for translation. For example, a number of enzymes induced by steroid hormones (see Table 12-2) have an absolute requirement for cAMP at some transcriptional or post-transcriptional site(s), which ensures that their mRNAs are made available for translation. cAMP-dependent and cAMP-independent protein kinase activities have been detected on mRNP particles (41–43). As more is learned about the fate of mRNA molecules from the time they leave the nucleus to the time they become polysome-associated it is likely that a major new post-transcriptional, pretranslational site for the regulation of gene expression will emerge.

REFERENCES

1. B. Lewin (1974). *Gene Expression*, Vol. I, *Bacterial Genomes*. John Wiley, London, pp. 38–228.
2. H. Weissbach and S. Pestka (Eds.) (1977). *Molecular Mechanisms of Protein Biosynthesis*. Academic Press, New York.
3. J. W. B. Hershey (1980). The translational machinery: Components and mechanisms. In D. M. Prescott and L. Goldstein (Eds.). *Cell Biology: A Comprehensive Treatise*, Vol. 4, Academic Press, New York, pp. 1–68.
4. R. A. Garrett (1979). The structure, assembly, and function of ribosomes. *Int. Rev. Biochem.* **25**, 121–177.
5. I. G. Wool (1979). The structure and function of eukaryotic ribosomes. *Ann. Rev. Biochem.* **48**, 719–754.
6. H. F. Noller and C. R. Woese (1981). Secondary structure of the 16S ribosomal RNA. *Science* **212**, 403–411.
7. A. A. Azed and N. J. Deacon (1980). The 3′-terminal primary structure of five eukaryotic 18S rRNAs determined by the direct chemical method of sequencing. The highly conserved sequences include an invariant region complementary to eukaryotic 5S rRNA. *Nuc. Acid Res.* **8**, 4365–4376.
8. E. Darzynkiewicz, K. Nakashima, and A. J. Shatkin (1980). Base-pairing in conserved 3′ end of 18S rRNA as determined by posoralen photoreaction and RNase sensitivity. *J. Biol. Chem.* **255**, 4973–4975.

9   J. S. Steitz (1979). Genetic signals and nucleotide sequences in messenger RNA. In R. F. Goldberger (Ed.). *Biological Regulation and Development*, Vol. 1. Plenum Press, New York, pp. 349–399.

10 G. Kramer and B. Hardesty (1980). Regulation of eukaryotic protein synthesis. In R. F. Goldberger and L. Goldstein (Eds.). *Cell Biology: A Comprehensive Treatise*, Vol. 4. Academic Press, New York, pp. 69–105.

11 G. Thomas, M. Siegmann and J. Gordon (1979). Multiple phosphorylation of ribosomal protein S6 during transition of quiescent 3T3 cells into early G_1, and cellular compartmentalization of the phosphate donor. *Proc. Natl. Acad. Sci. USA* **76**, 3952–3956.

12 B. F. C. Clark (1979). Structure and function of tRNA. In J. E. Celis and J. D. Smith (Eds.). *Nonsense Mutations and tRNA Suppressors*. Academic Press, London, pp. 1–46.

13 G. J. Quigley, A. H. J. Wang, N. C. Seeman, F. L. Suddath, A. Rich, J. L. Sussman, and S. H. Kim (1975). Hydrogen bonding in yeast phenylalanine transfer RNA. *Proc. Natl. Acad. Sci. USA* **72**, 4866–4870.

14 M. Revel (1977). Initiation of messenger RNA translation into protein and some aspects of its regulation. In H. Weissbach and S. Pestka (Eds.). *Molecular Mechanisms of Protein Biosynthesis*. Academic Press, New York, pp. 245–321.

15 D. L. Miller and H. Weissbach (1977). Factors involved in the transfer of aminoacyl tRNA to the ribosome. In H. Weissbach and S. Pestka (Eds.). *Molecular Mechanisms of Protein Biosynthesis*. Academic Press, New York, pp. 323–373.

16 N. Brot (1977). Translocation. In H. Weissbach and S. Pestka (Eds.). *Molecular Mechanisms of Protein Biosynthesis*. Academic Press, New York, pp. 375–411.

17 D. Richter and K. Isono (1977). The mechanism of protein synthesis—initiation, elongation and termination in translation of genetic messages. *Curr. Top. Microbiol. Immunol.* **76**, 83–125.

18 M. Revel and Y. Groner (1978). Post-transcriptional and translational controls of gene expression in eukaryotes. *Ann. Rev. Biochem.* **47**, 1079–1126.

19 R. Jagus, W. F. Anderson, and B. Safer (1981). The regulation of initiation of mammalian protein synthesis. *Prog. Nuc. Acid Res. Mol. Biol.* **25**, 127–185.

20 M. Kozak (1978). How do eukaryotic ribosomes select initiation regions in messenger RNA? *Cell* **15**, 1109–1123.

21 O. Haganbüchle, M. Tosi, U. Schiller, R. Bovery, P. K. Wellauer, and R. A. Young (1981). Mouse liver and salivary gland α-amylase mRNAs differ only in 5′ non-translated sequences. *Nature* **289**, 643–646.

22 H. F. Lodish (1976). Translational control of protein synthesis. *Ann. Rev. Biochem.* **45**, 39–72.

23 O. Maaløe (1979). Regulation of the protein-synthesizing machinery—ribosomes, tRNA factors, and so on. In R. F. Goldberger (Ed.). *Biological Regulation and Development*, Vol. 1, *Gene Expression*. Plenum Press, New York, pp. 487–542.

24 R. Cortese (1979). The role of tRNA in regulation. In R. F. Goldberger (Ed.). *Biological Regulation and Development*, Vol. 1. *Gene Expression*. Plenum Press, New York, pp. 401–432.

25 D. P. Meirlich (1978). Regulation of bacterial growth, RNA, and protein synthesis. *Ann. Rev. Microbiol.* **32**, 393–432.

26 J. D. Friesen, N. P. Fül, and K. von Meyenburg (1975). Synthesis and turnover of basal level guanosine tetraphosphate in *Escherichia coli*. *J. Biol. Chem.* **250**, 304–309.

27 R. Lathe (1978). RNA polymerase of *Escherichia coli*. *Curr. Topics Microbiol. Immunol.* **83**, 37–91.

28 S. Ochoa and C. de Haro (1979). Regulation of protein synthesis in eukaryotes. *Ann. Rev. Biochem.* **48**, 549–580.

29 T. Hunt (1979). The control of protein synthesis in rabbit reticulocytes. In T. R. Tussell, K. Brew, H. Faber, and J. Schultz (Eds.). *From Gene to Protein: Information Transfer in Normal*

REFERENCES

and Abnormal Cells, Vol. 16, Miami Winter Symposium. Academic Press, New York, pp. 321-346.

30. S. Ochoa (1979). Regulation of protein synthesis. *CRC Critical Reviews Biochemistry* **7**, 7-22.
31. S. A. Austin and M. J. Clemens (1980). Control of the initiation of protein synthesis in mammalian cells. *FEBS Lett.* **110**, 1-7.
32. R. Fagard and I. M. London (1981). Relationship between phosphorylation and activity of heme-regulated eukaryotic initiation factor 2α kinase. *Proc. Natl. Acad. Sci. USA* **78**, 866-870.
33. H. Grosfeld and S. Ochoa (1980). Purification and properties of the double-standard RNA-activated eukaryotic factor 2 kinase from rabbit reticulocytes. *Proc. Natl. Acad. Sci. USA* **77**, 6526-6530.
34. R. Benne, M. Salimans, H. Goumans, H. Amesz, and H. O. Voorma (1980). Regulation of protein synthesis in rabbit reticulocyte lysates. *Eur. J. Biochem.* **104**, 501-509.
35. R. Jagus and B. Safer (1981). Activity of eukaryotic initiation factor 2 is modified by processes distinct from phosphorylation. *J. Biol. Chem.* **256**, 1324-1329.
36. H. Chantrenne (1978). Regulation of protein synthesis at translation in eukaryotes. A brief review. In J. Dumont and J. Nunez (Eds.). *Hormones and Cell Regulation*, Vol. 2. Elsevier/North Holland, Amsterdam, pp. 1-13.
37. L. A. Osterman (1979). Participation of tRNA in regulation of protein biosynthesis at the translational level in eukaryotes. *Biochimie* **61**, 323-342.
38. G. N. Pavlakis, R. E. Lockard, N. Vamvakopoulos, L. Rieser, U. L. Raj Bhandary, and J. N. Vournakis (1980). Secondary structure of mouse and rabbit α- and β-globin mRNAs: Differential accessibility of α and β initiator AUG codons towards nucleases. *Cell* **19**, 91-102.
39. I. C. Bathhurst, R. K. Craig, and P. N. Campbell (1980). Inhibition of cell-free protein synthesis by low-molecular-weight nuclear polyadenylate—containing ribonucleic acid species isolated from the lactating guinea pig. *Biochem. J.* **186**, 561-570.
40. A. K. Mukherjee, C. Guha, and S. Sarkar (1981). The translational inhibitory 10S cytoplasmic ribonucleoprotein of chicken embryonic muscle is distinct from messenger ribonucleoproteins. *FEBS Lett.* **127**, 133-138.
41. J. Bag and B. H. Sells (1980). Presence of cyclic-AMP-independent protein kinase activity in RNA-binding proteins of embryonic chicken muscle. *Eur. J. Biochem.* **106**, 411-424.
42. J. Cardelli and H. C. Pitot (1980). Characterization of protein kinase activity associated with rat liver polysomal messenger ribonucleoprotein particles. *Biochemistry* **19**, 3164-3169.
43. R. E. Moore and R. K. Sharma (1980). Endogenous substrate for cyclic AMP-dependent protein kinase in adrenocortical polyadenylated messenger ribonucleoproteins. *Nature* **210**, 1137-1139.

CHAPTER SEVEN

Regulation of Enzyme Activity by Post-Translational Modification

Not all newly synthesized enzyme molecules are catalytically active when they dissociate from the ribosome. Moreover, some enzymes may be active but not physiologically functional until they are translocated to a specific location within the cell. For these molecules some kind of post-translational modification is essential to permit full expression of their activity. Thus, gene expression must be considered in terms of these post-translational alterations, and the regulation of enzyme synthesis must acknowledge regulatory mechanisms that may operate at this level.

There are two kinds of post-translational modification; proteolytic cleavage of the polypeptide chain and covalent modification of the amino acid residues of the polypeptide. Proteolytic cleavage is *irreversible* and is used in three different ways: liberation of an active enzyme from an inactive proenzyme or zymogen, fixation of an enzyme or protein into a cell membrane, and removal of the enzyme or protein from the cell by degrading it. Covalent modifications of amino acid residues, on the other hand, are *reversible* and are used to modulate the *activity* of the gene product. The latter modifications are the basis of a major level of metabolic regulation intermediate between the acute regulation of metabolism by allosteric interactions and the chronic regulation of metabolism by changes in enzyme synthesis. Consequently, the discussion of this aspect of post-translational modification is brief and limited to a few examples sufficient to illustrate the mechanisms involved and their fundamental importance in regulating enzyme activity. Similar mechanisms are also important factors in transcriptional activation (Chapters 3 and 4) and in information transfer (Chapter 11).

POST-TRANSLATIONAL MODIFICATIONS INVOLVING PROTEOLYSIS

Activation of Enzymes by Proteolytic Cleavage

This kind of post-translational modification occurs mostly in eukaryotic cells and usually involves enzymes that are activated following secretion from the cell. Only one or two intracellular enzymes have been found to be activated by this mechanism. The best known examples of enzymes synthesized and secreted as zymogens are the digestive proteases of the pancreas, and the most significant enzyme to be activated by an intracellular proteolytic event is the calcium-dependent, protease-activated protein kinase.

Four enzymes secreted by the pancreatic acinar cells—chymotrypsin, trypsin, and carboxypeptidases A and B—are synthesized as chymotrypsinogen, trypsinogen, and procarboxypeptidases A and B, respectively, and stored in granules (1, 2). The granules are secreted into the lumen of the duodenum, where the intestinal enzyme enterokinase cleaves a hexapeptide from the N-terminal end of trypsinogen. The activated trypsin molecules then catalyze the conversion of more molecules of trypsinogen to trypsin and also proteolytically activate the other zymogens. Thus the enzymes only become active outside the cell, thereby avoiding the potentially lethal problem of having an active, nonspecific protease loose in the cytoplasm.

A small number of very specific proteases are found in the cytoplasm of a number of eukaryotic cells. For example, the glycolytic enzyme fructose diphosphatase can be modified by proteolytic cleavage (3), and there is also a protein kinase activated by a calcium-dependent protease (4, 5). The latter reaction is particularly interesting since it may mediate information transfer from calcium ions to intracellular targets. Moreover, since calcium fluxes are usually transient (see Chapter 11), irreversible proteolytic activation of a protein kinase may be a mechanism for converting this transient into a longer-lasting effect. For example, phosphorylase kinase is "activated" by a calcium-dependent protease, rendering it sensitive to cyclic nucleotides.

The fact that relatively few enzymes are activated by proteolytic cleavage does not detract from the importance of proteases in the regulation of biochemical processes, particularly the multiprotein systems such as blood coagulation and complement. Further details of this important area of metabolic control can be found in the book edited by Reich, Rifkin, and Shaw (6) and the review by Holzer and Heinrich (7).

Translocation of Enzymes across Membranes

Messenger RNA molecules emanating from the nucleus code not just for cytoplasmic proteins but for proteins that are integral components of most subcellular organelles and membranes, and also for proteins that are exported from the cell. Most of these proteins and enzymes are only physiologically functional when integrated into their specific subcellular location. Thus for these proteins the final act of gene expression is the translocation of each newly synthesized polypeptide to its specific site of action.

These proteins must become integrated into, or translocated through, one or more membranes. For example, cytochrome *c* passes entirely through the outer-mitochondrial membrane and becomes incorporated into the inner-mitochondrial membrane, whereas albumin is translocated across the membrane of the endoplasmic reticulum into the lumen of this organelle, from which it is secreted into the bloodstream. Other proteins, such as ATPase and cytochrome p_{450}, become incorporated into the membrane of the endoplasmic reticulum but do not pass entirely through it. Bacterial cells also secrete many proteins through their cell membranes. In recent years a number of theories and hypotheses have been proposed (see Refs. 8–11 for reviews), most notably the "signal hypothesis" developed by Blobel and his colleagues (10), which proposes a common mechanism of protein transfer across membranes.

The signal hypothesis proposes that mRNAs coding for these proteins interact with cytoplasmic ribosomes, and translation commences. The first 20 to 30 amino acids that are polymerized are very hydrophobic; for example, cytochrome p_{450} has 10 leucine residues in the first 20 amino acids, together with methionine, proline, serine, and alanine (see Table II of Ref. 11 for other similar sequences). This amino acid sequence acts as a signal, directs the translation complex to a specific membrane location, and anchors the ribosome to the membrane. *Thus membrane-bound polysomes are actually held there by the proteins they are translating.* The signal sequence has a high affinity for the hydrophobic regions of the membrane and either becomes integrated into it or passes straight through it. The remainder of the polypeptide chain follows as it is synthesized, and the whole protein is either integrated into the membrane of, or secreted into the lumen of, the endoplasmic reticulum (or other subcellular organelle, or even outside the cell in the case of bacterial proteins such as penicillinase and α-amylase). The signal sequence is usually proteolytically cleaved during the translocation process by an enzyme, signal peptidase. This ensures that the protein is *irreversibly* transferred to its new location.

Virtually all proteins secreted by eukaryotic cells are synthesized as precursor molecules containing a signal sequence that directs the nascent polypeptide chain to the endoplasmic reticulum and facilitates its transfer into the lumen, from where it is transported to the golgi apparatus and eventually secreted. Moreover, enzymes such as the lysosomal hydrolases are directed to the golgi apparatus, where they are packaged into vesicles that mature into lysosomes. In this way the cell avoids the synthesis in the cytoplasm of active degradative enzymes. It is now known that all these proteins undergo an additional covalent modification, glycosylation, with carbohydrate residues being added to the polypeptide chain as it enters the lumen of the endoplasmic reticulum (12). These residues tag the protein for export or for incorporation into a specific cellular organelle. For example, the lysosomal enzymes are tagged with mannose-6-phosphate.

This model for protein secretion and translocation requires the presence of a structure in the membrane with which the signal sequence interacts. In recent work Walter and Blobel (13) have isolated a multiprotein complex (six proteins of M_r = 72,000, 68,000, 54,000, 19,000, 14,000, and 9000) from membranes that has some of the properties of the proposed translocator.

Not all proteins that cross membranes or are components of membranes contain signal sequences. Fumarase (synthesized in the cytoplasm and translocated to the mitochondrial matrix), adenylate cyclase (translocated to the plasma membrane), and pyruvate oxidase (a component of bacterial membranes) are typical examples. It has been proposed (9) that these enzymes exist in two conformations: in the ionic environment of the cytoplasm they are folded in such a way that polar amino acid residues are exposed, whereas when they come into contact with membranes they refold, exposing hydrophobic residues that facilitate incorporation into the membrane.

POST-TRANSLATIONAL COVALENT MODIFICATION OF AMINO ACID RESIDUES

At least 13 of the 20 amino acids that are commonly incorporated into proteins can undergo some kind of covalent modification that alters their reactivity (14). Some of these reactions occur in both prokaryotes and eukaryotes, whereas others appear to be more or less restricted to one cell type. However, it is not known whether this is due to a lack of study or to the certain absence of that particular reaction. For example, protein modification by phosphorylation was long considered to be restricted to eukaryotic cells, but it is now established (15, 16) that prokaryotes do indeed possess protein kinase activity. The various kinds of covalent modification are listed in Table 7-1, and each is described briefly below. Particular emphasis is placed on phosphorylation because of its fundamental importance in the regulation of gene expression and in the regulation of the activity of the gene products, particularly in eukaryotic cells.

Phosphorylation

In eukaryotic cells many proteins, including some 20 to 30 enzymes, have been found to undergo covalent modification by phosphorylation (17–20), and, as mentioned above, this form of covalent modification has now been confirmed in bacterial cells. Several amino acid residues can be phosphorylated by either a

Table 7-1 The Various Forms of Covalent Modifications Involved in the Regulation of Enzyme Activity

Covalent modification	Eukaryotes	Prokaryotes
phosphorylation	+	+
acetylation	+	?
adenylylation	?	+
uridylylation	?	+
ADP-ribosylation	+	+
methylation	+	+
oxidation of —SH groups	+	?

REGULATION BY POST-TRANSLATIONAL MODIFICATION

Table 7-2 The Amino Acid Residues That Can Be Modified by Either Phosphorylation or Phosphoramidation Reactions Catalyzed by Protein Kinases

Amino acid residue	linkage
serine	P—O
threonine	P—O
tyrosine	P—O
aspartic acid	P—O
glutamic acid	P—O
lysine	P—N
arginine	P—N
histidine	P—N

P—O or a P—N linkage as shown in Table 7-2. Strictly speaking, the formation of P—N bonds should be referred to as *phosphoramidation*, since the whole of the phosphate group is not transferred (Fig. 7-1). Serine and threonine are the two amino acids most widely phosphorylated, with the other residues only being phosphorylated under more specific circumstances. Phosphorylation of these residues is likely to induce conformational changes in the protein molecule because the phosphate (phosphoramidate) group introduces both charge changes and steric hindrance. Such a *conformation-induced change of state* is the basis of the action of phosphorylation and other forms of covalent modification.

The phosphorylation reaction is carried out by a group of ATP(GTP):phosphotransferases commonly referred to as protein kinases. Since new protein

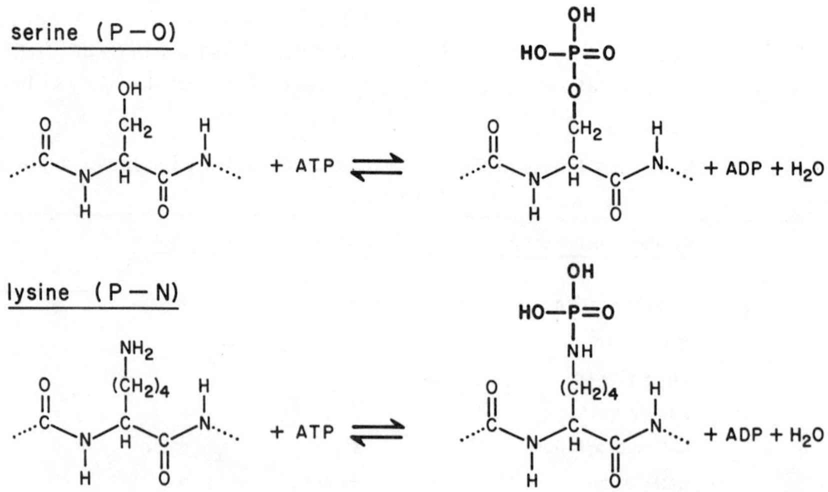

Fig. 7-1 Protein kinase-catalyzed covalent phosphorylation of serine, and phosphoramidation of lysine residues of proteins.

kinases are still being discovered (21) it is difficult to classify the enzymes into discrete groups. In general, there are two functional categories: those that are activated as a result of a change in an extracellular stimulus (e.g., a hormone) and those that respond to purely intracellular signals (Table 7-3). Only the well-characterized kinases are listed here, but there have been many more reports of protein kinase activities that remain to be studied in more detail.

Protein kinases participate in the transfer of information from regulatory signals to the key enzymes, or other proteins, of metabolic pathways. They are part of a system by which the cell obtains a coordinated response from a number of diverse cellular processes, usually in response to an extracellular stimulus. The protein kinases that initially respond are those activated by the second messengers (cAMP, cGMP, or calcium ions), which increase in response to the extracellular stimulus. These kinases either phosphorylate a target pro-

Table 7-3 Some of the Well-Characterized Protein Kinases and the Signals That Activate Them

Kinases responding to extracellular signals

Kinase	Second messenger\activator
cAMP – PK I	cAMP
cAMP – PK II	cAMP
cGMP – PK	cGMP
Ca^{2+}\CDR – PK	Ca^{2+} and \or CDR
Nuclear kinases	cAMP\CDR\Ca^{2+}

Kinases responding to intracellular signals

Kinase	Signal
HCR	hemin
phosphorylase kinase	cAMP-PK \ Ca^{2+}-CDR
synthetase kinase 1	cAMP–PK
synthetase kinase 2	CDR
synthetase kinase 3	—
AcetylCoA carboxylase kinase	cAMP-PK
fatty acid synthetase kinase	"
HMG CoA reductase kinase	"
pyruvate dehydrogenase kinase	"
myosin light chain kinase	Ca^{2+} – CDR
Nuclear kinases	—

tein directly or they activate a second kinase that then phosphorylates a specific protein.

The coordinated response of the enzymes of energy metabolism to a change in the insulin:glucagon ratio in the bloodstream is an excellent example of this level of control (Fig. 7-2). The enzymes responsible for energy storage (the conversion of glucose to glycogen and lipid) are all inhibited by phosphorylation, whereas the enzymes responsible for energy mobilization (glucose formation by glycogen breakdown and gluconeogenesis, and lipid conversion to free fatty acids) are all activated by phosphorylation. Moreover, all these enzymes are either phosphorylated directly by cAMP-dependent protein kinase (there are two separate cAMP-dependent kinases, but which one is involved is not known) or are phosphorylated by kinases that are *activated* by a cAMP-dependent kinase (Table 7-3). In general, the enzymes that exist as large, multi-subunit complexes (phosphorylase, glycogen, synthetase, fatty acid synthetase, etc.) are activated by the latter mechanism, whereas the smaller enzymes may be phosphorylated directly. When there is a fall in the concentration of blood glucose the insulin:glucagon ratio also falls and the concentration of cAMP in the liver rises, leading to an inhibition of the enzymes of energy storage and an activation of the enzymes of glucose production, which restores the level of glucose in the blood. Thus the activities of 12 enzymes are coordinated to satisfy a change in one extracellular stimulus.

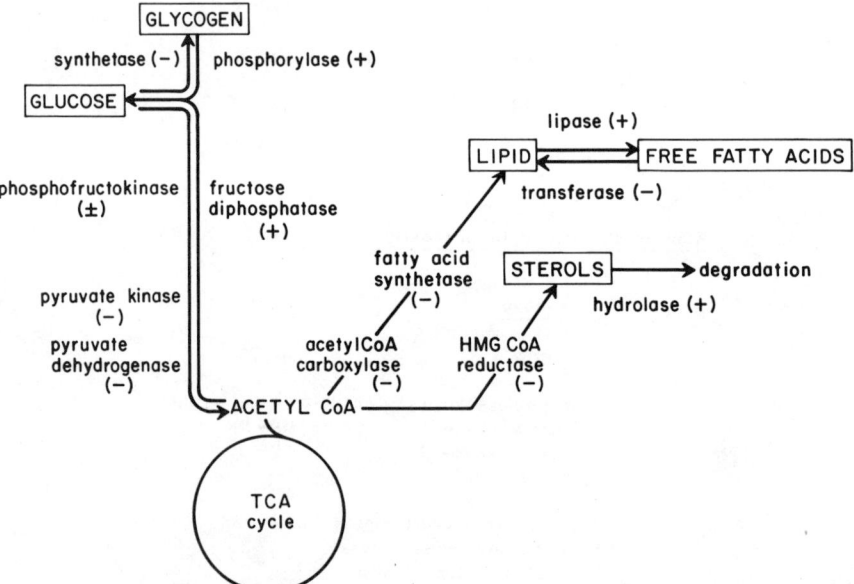

Fig. 7-2 The regulation of energy storage and energy mobilization by phosphorylation of key enzymes. (+) indicates enzyme activation by phosphorylation, (−) indicates enzyme inhibition by phosphorylation. All of these enzymes are phosphorylated in response to an increase in cAMP.

Phosphorylation overrides the intracellular, allosteric control signals, committing the cell to a more long-term change in metabolism. If the extracellular stimulus persists it leads to a chronic readjustment of the metabolism of the cell by inducing increases in enzyme synthesis. The role of second messengers and protein kinases in the regulation of enzyme synthesis is discussed in more detail in Chapters 11 and 12, taking into account the roles of covalent modifications in gene expression that were described in Section 1.

Phosphate (phosphoramidate) groups are removed enzymatically from proteins by the action of phosphoprotein phosphatases. Until recently it was assumed that these enzymes were relatively nonspecific and that the kinases were principally responsible for regulating cycles of phosphorylation and dephosphorylation. However, some recently discovered properties of phosphatases (17, 22, 23) indicate they play an active role in covalent modification. For example, certain phosphatases specifically dephosphorylate all the proteins phosphorylated by a particular protein kinase. Thus phosphatase C dephosphorylates the protein substrates of cAMP-dependent protein kinases. Moreover, some of these phosphatases are under complex regulatory control.

Adenylylation and Uridylylation

Tyrosine residues can be covalently modified by the addition of either the adenylyl groups from ATP or the uridylyl group from UTP (Fig. 7-3a) catalyzed by adenylyltransferase and uridylyltransferase, respectively. This modification has been found predominantly in bacterial cells, and the most characterized system is that of glutamine synthetase in *E. coli* (23, 24) (see Fig. 7-3b).

Glutamine synthetase consists of 12 subunits, each of which can be adenylylated. Upon adenylylation the subunit becomes inactive, thus the overall activity of the enzyme is determined by the fraction of subunits that remain active (deadenylylated). The *removal* of adenylyl groups is stimulated by a tetrameric protein P_{II}, each subunit of which can be uridylylated (Fig. 7-3b). Uridylylation of P_{II} activates the molecule, and it stimulates the deadenylylation of glutamine synthetase. The transferases that catalyze the adenylylation and uridylylation are also under quite complex allosteric control, making this one of the most highly regulated metabolic systems found in any cell. The physiological consequences of this complex system are discussed in more detail in Chapter 8.

ADP-Ribosylation

Arginine, lysine, glutamic acid, and phosphorylated serine residues of both prokaryotes and eukaryotes can be modified by the covalent addition of the ADP-ribose moiety of nicotinamide adenine dinucleotide (Fig. 7-4) (25-27). Moreover, in eukaryotic cells (but apparently not in bacterial cells) the ADP ribose can be polymerized (Fig. 7-4b) to form a chain of residues attached to the protein. As many as 100 residues have been found attached to some proteins by a reaction catalyzed by poly(ADP ribose)synthetase.

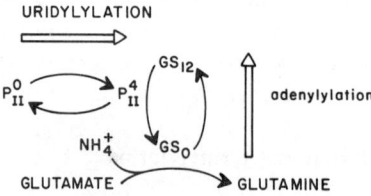

Fig. 7-3 Covalent modification of tyrosine residues by adenylylation and uridylylation (*a*) and the roles of uridylylation and adenylylation in the activation of glutamine synthase (*b*). P_{II} catalyzes the deadenylylation of glutamine synthetase. GS_0 is fully deadenylylated glutamine synthetase, and GS_{12} is fully adenylylated glutamine synthetase.

At the present time the physiological significance of this form of covalent modification has not been firmly established. Enzymes that have been shown to undergo ADP-ribosylation include RNA polymerase and adenylate cyclase in both prokaryotes and eukaryotes and the nuclear Ca^{2+}—Mg^{2+}-dependent endonuclease in eukaryotes. A number of nonenzyme proteins also undergo this kind of covalent modification, particularly histones H1 and H2B (28). Moreover, histone H1 can be poly(ADP-ribosylated), and it is possible that the polymer cross-links H1 molecules, locking the DNA into a higher order of chromatin structure (see Chapter 4). The observation that phosphoserine, but not serine, can be ADP-ribosylated is particularly interesting since histone H1 (and other proteins) can be phosphorylated on a number of serine residues. Thus there may be interactions between the different forms of covalent modification that enhance the regulatory capacity of the cell.

Methylation, Acetylation, and Changes in Sulfydryl Redox State

Carboxymethyltransferases transfer methyl groups from S-adenosyl methionine to aspartic acid or glutamic acid residues (O-methylation) and to histidine, lysine, or arginine residues (N-methylation), as shown in Fig. 7-5*a*. These enzymes have been found in both prokaryotes and eukaryotes. A number of proteins, particularly membrane-bound proteins, have been shown to undergo cycles of methylation–demethylation (29–31), but as yet there are no known examples of *enzymes* that are modified in this way.

Acetylation of amino groups (lysine, N-terminal residues) or hydroxyl groups (serine) is another form of covalent modification that has received relatively little attention. The reaction catalyzed is shown in Fig. 7-5b. Although no enzymes have been found to undergo cycles of acetylation–deacetylation, this form of covalent modification appears to be an important part of the mechanism of transcriptional activation. The extensive acetylation of histones that accompanies gene transcription is described in Chapter 4.

The sulfydryl group of cysteinyl residues can exist in two states (Fig. 7-5c) that can have a profound effect on protein structure—when oxidized the groups participate in intramolecular disulfide bridge formation, when reduced these bonds are broken. It is becoming apparent that this form of covalent modification is used by cells to modulate the activity of a number of enzymes and proteins. For example, the state of oxidation of eIF-2 determines its participation in the initiation of protein synthesis (Chapter 6).

The best examples of the regulation of enzyme activity through changes in the redox state of sulfydryl groups come from plant cells. Chloroplast enzymes

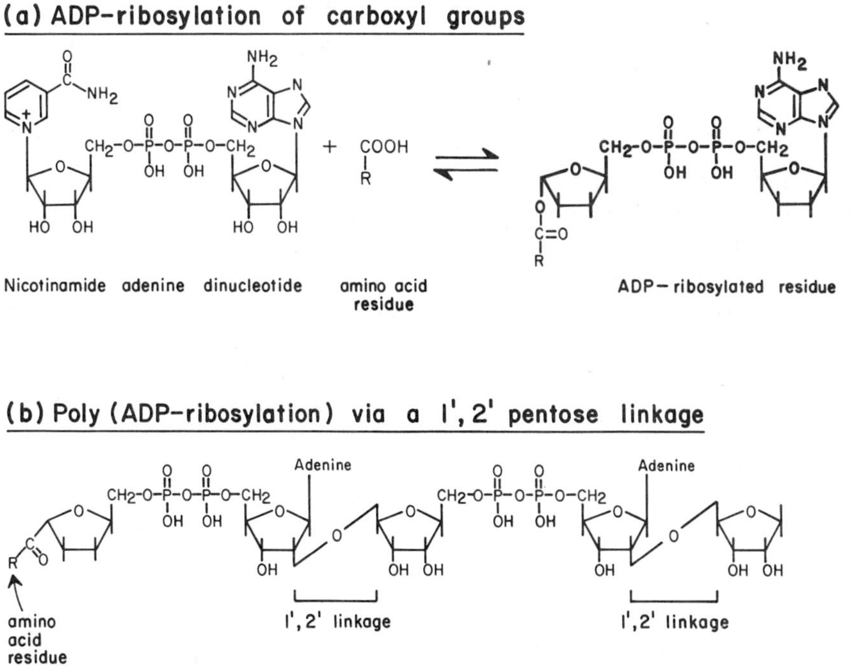

(a) ADP-ribosylation of carboxyl groups

(b) Poly (ADP-ribosylation) via a 1′, 2′ pentose linkage

Fig. 7-4 ADP-ribosylation of amino acid residues (*a*). Nicotinamide adenine dinucleotide is the donor, and carboxyl residues are usually modified. Poly(ADP-ribosylation) occurs through a 1′,2′ pentose linkage (*b*).

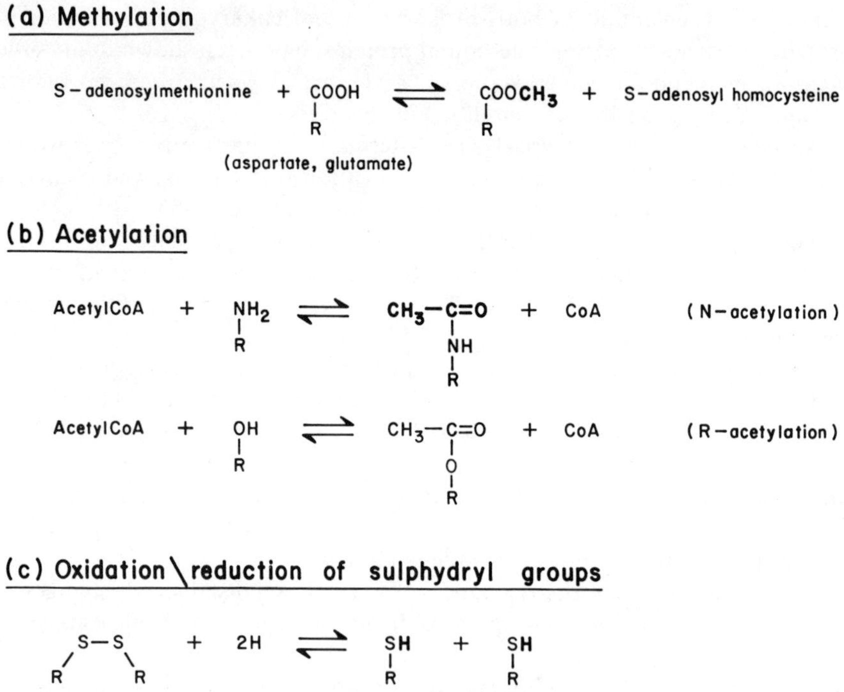

(a) Methylation

S–adenosylmethionine + COOH–R (aspartate, glutamate) ⇌ COOCH$_3$–R + S–adenosyl homocysteine

(b) Acetylation

AcetylCoA + NH$_2$–R ⇌ CH$_3$–C(=O)–NH–R + CoA (N–acetylation)

AcetylCoA + OH–R ⇌ CH$_3$–C(=O)–O–R + CoA (R–acetylation)

(c) Oxidation\reduction of sulphydryl groups

R–S–S–R + 2H ⇌ SH–R + SH–R

Fig. 7-5 Summary of covalent modifications involving methylation, acetylation, and sulfydryl group oxidation/reduction (see text).

such as fructose 1,6-biphosphatase and sedoheptulose 1,7-biphosphatase are inactive in the dark, but their activities are rapidly increased in the light (32). This activation has been found to be accompanied by a reduction of disulfide groups. Thus in the oxidized state the enzymes are inactive, and in the reduced state they are active. A cascade of reactions catalyzes the light-dependent reduction as shown in Fig. 7-6. In the presence of light ferredoxin is reduced by chlorophyll, and this, in turn, reduces thioredoxin, a reaction catalyzed by thioredoxin reductase. Reduced thioredoxin participates in a number of biological oxidation–reduction reactions, including activation of chloroplast enzymes (ascorbate- or glutathione-dependent oxidation inactivates the enzymes in the dark). Thus ferredoxin and thioredoxin act as second messengers for the extracellular stimulus—light. Thioredoxin is present in many other eukaryotic cells and participates in a number of redox reactions in what is likely to emerge as an important form of covalent modification.

Covalent Modification and Enzyme Cascades

The modulation of enzyme activity by covalent modification of amino acid residues has two important physiological functions. The first, already described,

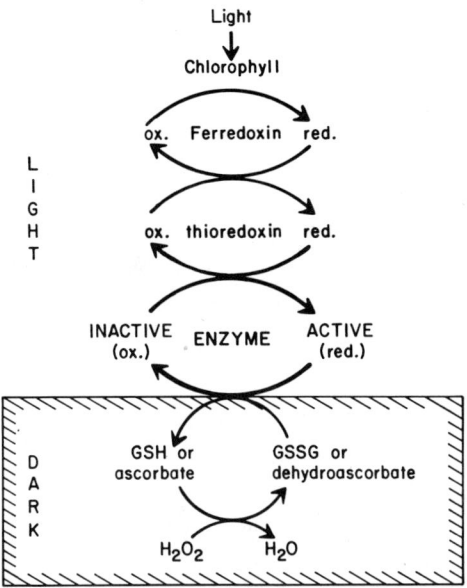

Fig. 7-6 Covalent modifications of sulfydryl groups of chloroplast enzymes during light-induced activation and dark-induced inactivation. Light-activated chlorophyll is the source of reducing equivalents, and ferredoxin and thioredoxin are intermediate carriers. The enzymes modified in this way are described in the text.

is the role it plays in the integration of cellular responses to satisfy a change in an exogenous signal (hormonal or neural). The second is to increase the sensitivity of the regulatory system to permit major changes in cellular metabolism to be effected by small changes in effector concentration. This latter concept has been developed by Stadtman and his colleagues (23, 24) and is best illustrated by the series of reaction cascades involved in the activation of hepatic (or muscle cell) glycogen phosphorylase in response to a fall in the concentration of blood glucose (Fig. 7-7a).

The physiological importance of this phenomenon is readily apparent. A very small number of glucagon molecules (released from the pancreas in response to the fall in blood glucose) bind to cell surface receptors and activate an equal number of adenylate cyclase molecules. Each molecule of adenylate cyclase may then synthesize hundreds or thousands of molecules of cAMP, which can activate many molecules of cAMP-dependent protein kinase. Each active catalytic subunit of this kinase phosphorylates and activates many molecules of phosphorylase *b* kinase, each of which, in turn, can activate many molecules of phosphorylase, leading to a massive breakdown of glycogen. Thus a cascade *amplifies the effect of the original signal*, and each additional step incorporated into a cascade leads to an *exponential* increase in amplification (Fig. 7-7b). One cascade cycle amplifies the signal approximately 300-fold whereas four cycles amplify the signal by a factor of 10^{10} for only a *twofold* increase in the concentration of initial effector.

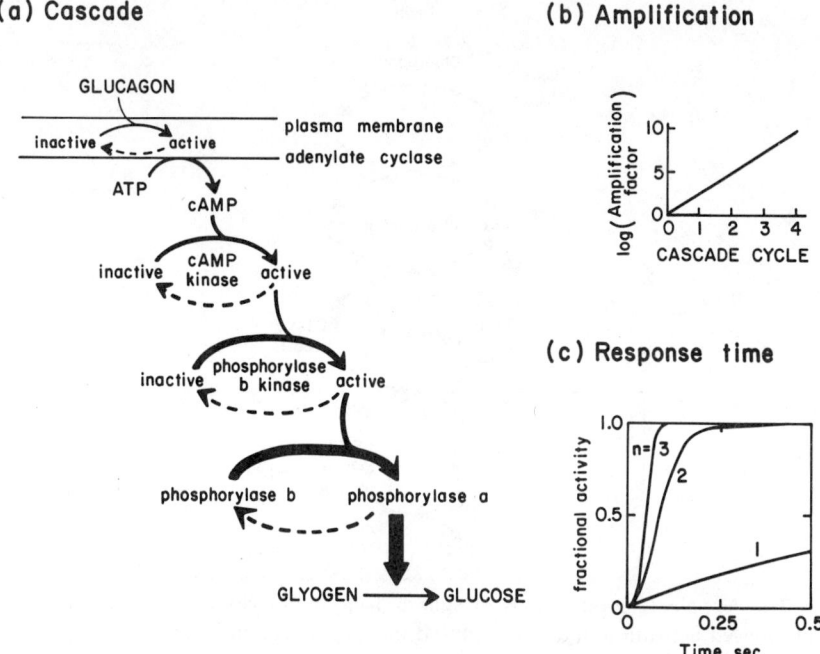

Fig. 7-7 The physiological importance of enzyme cascades in metabolic regulation. The glycogen phosphorylase cascade is illustrated in (*a*). The relationship between the number of cycles in a cascade and the amplification factor is shown in (*b*)—note the log scale. The relationship between the time taken to achieve maximal activity and the number of cycles in a cascade (*n*) is shown in (*c*). [Figures (*b*) and (*c*) are reproduced from Figs. 8 and 9 of Ref. 24, with permission.]

Cascades also offer a number of other major advantages. They permit large changes in rate to be effected by very low concentrations of effector. Moreover, the response time is greatly decreased, as illustrated in Fig. 7-7c, with each additional reaction in a cascade. Furthermore, in a cascade involving several enzymes the total number of effectors that can interact with the system is greatly increased. For example, as many as 40 effectors have been shown to influence the bicyclic glutamine synthetase cascade of *E. coli* (Fig. 7-3b).

ENZYME DEGRADATION

The concentration of an enzyme is determined by its relative rates of synthesis and degradation. Although we know a great deal about the molecular mechanisms involved in the regulation of enzyme synthesis, we know virtually nothing about the mechanisms of enzyme degradation, despite its potential importance. However, as described in Chapter 1, it is the rate of degradation of an enzyme that determines the kinetics of change during enzyme induction. This section describes the mechanism of enzyme protein degradation, to the extent that it is known, and deals with the question of the specificity of the process. Several recent reviews are available for background information (33–36).

The Mechanism of Enzyme Degradation

Enzymes are degraded by proteolytic cleavage of the polypeptide chain. In eukaryotes much of cellular protein catabolism takes place in lysosomes where several enzymes, cathepsins A, B1, C, and D, are capable of degrading proteins entirely to amino acids. Bacterial cells do not possess a definite organelle, but proteins subject to degradation appear to be concentrated into a particular part of the cell before being degraded. Protease activity has been detected in these regions of the bacterial cell.

Despite considerable effort, the pathway by which proteins are degraded is still not known. The major stumbling block appears to be the inability to establish *in vitro* model systems suitable for detailed analytical studies. This suggests there is some structural component of the cell, difficult to recreate *in vitro*, that is involved in degradation. A curious aspect of the degradative process is the requirement for energy, despite the fact that proteolytic cleavage of peptide bonds is a highly exergonic process. The energy may be required to pump the protein molecules into lysosomes or into the degradative area of bacterial cells. A recent study (37), however, implicates ATP (possibly via an adenylylation reaction) in the covalent binding of an ATP-dependent proteolysis factor to susceptible proteins. Once tagged these proteins are attacked by cellular proteases. In another study, a microsome-associated protein that inactivates proteins by oxidation of their sulfydryl groups prior to degradation has been isolated (38). Although much further work is needed, both of these studies implicate reactions that select the proteins that are to be degraded before the polypeptide chain is cleaved.

Specificity of the Process

For the degradative process to play a role in the regulation of gene expression, mechanisms must exist for the rate of degradation of specific proteins to be changed without altering the rates of degradation of all other proteins. At the present time it has not been convincingly demonstrated that such mechanisms exist. It appears that each protein has an intrinsic susceptibility to degradation that is determined, in part, by its size, isoelectric point, degree of hydrophobicity, and solubility (in general, large, insoluble, hydrophobic molecules are the most susceptible to degradation). All of these characteristics are determined by the amino acid sequence of the protein (and ultimately by the nucleotide sequence of the gene).

However, the cell does appear to have some facility for modifying this intrinsic degradability for a number of proteins, or protein groups. For example, tryptophan oxygenase is completely stabilized by its substrate (Fig. 1-2). Moreover, phosphorylation of pyruvate kinase inactivates it and makes it 10 times more susceptible to proteolytic degradation. Additionally, enzymes using pyridoxal phosphate as cofactor are susceptible to a specific group of serine proteases in the absence of cofactor (33, 34), but are stabilized in its presence. These observations indicate that an active enzyme is much less susceptible to degradation than an inactive one, but the generality of this has not been established.

In conclusion, it must be said that protein degradation does not appear to play a major role in specifically modulating the concentration of *individual* gene products. Indeed, virtually all the adaptable enzymes discussed in Chapter 12 are modulated by changing their rate of synthesis (Tables 12-2, 12-3) rather than by altering their rates of degradation.

REFERENCES

1 H. Neurath (1975). Limited proteolysis and zymogen activation. In E. Reich, D. B. Rifkin, and E. Shaw (Eds.). *Proteases and Biological Control.* Cold Spring Harbor Laboratory, pp. 51–64.
2 P. Cohen (1976). *Control of Enzyme Activity.* Halsted Press (John Wiley), New York, pp. 20–31.
3 S. Pontremoli, E. Melloni, and A. DeFlora (1974). Conversion of neutral to alkaline fructose 1,6-bisphosphatase by proteolytic mechanisms: Structure–function relationships. In E. H. Fisher, E. G. Krebs, H. Neurath, and E. R. Stadtman (Eds.). *Metabolic Interconversion of Enzymes.* Springer-Verlag, New York, pp. 285–300.
4 Y. Takai, A. Kishimoto, M. Inoue, and Y. Nishizuka (1977). Studies on a cyclic nucleotide-independent protein kinase and its proenzyme in mammalian tissues: I. Purification and characterization of an active enzyme from bovine cerebellum. *J. Biol. Chem.* **252**, 7603–7609.
5 M. Inoue, A. Kishimoto, Y. Takai, and Y. Nishizuka (1977). Studies on a cyclic nucleotide-independent protein kinase and its proenzyme in mammalian tissues: II. Proenzyme and its activation by calcium-dependent proteases from rat brain. *J. Biol. Chem.* **252**, 7610–7616.
6 E. Reich, D. B. Rifkin, and E. Shaw (1975). *Proteases and Biological Control.* Cold Spring Harbor Laboratory.
7 H. Holzer and P. C. Heinrich (1980). Control of proteolysis. *Ann. Rev. Biochem.* **49**, 63–91.
8 S. Bar-Nun, G. Kreibich, M. Adesnik, L. Alterman, M. Negiski, and D. Sabatini (1980). Synthesis and insertion of cytochrome P-450 into endoplasmic reticulum membranes. *Proc. Natl. Acad. Sci. USA* **77**, 965–969.
9 W. Wickner (1979). The assembly of proteins into biological membranes: The membrane trigger hypothesis. *Ann. Rev. Biochem.* **48**, 23–45.
10 G. Blobel (1980). Intracellular protein topogenesis. *Proc. Natl. Acad. Sci. USA* **77**, 1496–1500.
11 B. D. Davis and P.-C. Tai (1980). The mechanism of protein secretion across membranes. *Nature* **283**, 433–438.
12 J. E. Rothman (1981). The Golgi Apparatus: Two organelles in tandem. *Science* **213**, 1212–1219.
13 P. Walter and G. Blobel (1980). Purification of a membrane-associated protein complex required for protein translocation across the endoplasmic reticulum. *Proc. Natl. Acad. Sci. USA* **77**, 7112–7116.
14 F. Wold (1981). *In vivo* chemical modification of proteins (post-translational modification). *Ann. Rev. Biochem.* **50**, 783–814.
15 M. Manai and A. J. Cozzone (1979). Analysis of the protein-kinase activity of *Escherichia coli* cells. *Biochem. Biophys. Res. Commun.* **91**, 819–826.

REFERENCES

16 J. Y. J. Wang and D. E. Koshland, Jr. (1981). The identification of distinct protein kinases and phosphatases in the prokaryote *Salmonella typhimurium*. *J. Biol. Chem.* **256**, 4640–4648.

17 E. G. Krebs and J. A. Beavo (1979). Phosphorylation–dephosphorylation of enzymes. *Ann. Rev. Biochem.* **48**, 923–959.

18 M. Weller (1979). *Protein Phosphorylation: The Nature, Function, and Metabolism of Proteins Which Contain Covalently Bound Phosphorus.* Pion Ltd., London.

19 P. Cohen, E. Embi, G. Foulkes, G. Hardie, G. Nimmo, D. Hylatt, and S. Shenolikar (1979). The role of protein phosphorylation in the coordinated control of intermediary metabolism. In T. R. Russell, K. Brew, H. Faber, and J. Schultz (Eds.). *From Gene to Protein: Information Transfer in Normal and Abnormal Cells*, Vol. 16, Miami Winter Symposium. Academic Press, New York, pp. 463–481.

20 T.-S. Huang, J. R. Feramisco, D. B. Glass, and E. G. Krebs (1979). Specificity considerations relevant to protein kinase activation and function. In T. R. Russell, K. Brew, H. Faber and J. Schultz (Eds.). *From Gene to Protein: Information Transfer in Normal and Abnormal Cells*, Vol. 16, Miami Winter Symposium. Academic Press, New York, pp. 449–461.

21 M. Sikorska, J. P. MacManus, P. R. Walker, and J. F. Whitfield (1980). The protein kinases of rat liver nuclei. *Biochem. Biophys. Res. Commun.* **93**, 1196–1203.

22 E. Y. C. Lee, J. H. Aylward, R. L. Mellgren, and S. D. Killilea (1979). Protein phosphatase C: Properties, specificity and structural relationship to a larger holoenzyme. In T. R. Russell, K. Brew, H. Faber, and J. Schultz (Eds.). *From Gene to Protein: Information Transfer in Normal and Abnormal Cells*, Vol. 16, Miami Winter Symposium. Academic Press, New York, pp. 483–499.

23 P. B. Chock, S. G. Rhee, and E. R. Stadtman (1980). Interconvertible enzyme cascades in cellular regulation. *Ann. Rev. Biochem.* **49**, 813–843.

24 P. B. Chock and E. R. Stadtman (1979). Covalently interconvertible enzyme cascade and metabolic regulation. In D. E. Atkinson and C. F. Fox (Eds.). *Modulation of Protein Function*, Vol. 13, ICN-UCLA Symposia on Molecular and Cellular Biology. Academic Press, New York, pp. 185–202.

25 H. Hilz and P. Stone (1976). Poly(ADP-ribose) and ADP-ribosylation of proteins. *Rev. Physiol. Biochem. Pharmacol.* **76**, 1–58.

26 K. Ueda, O. Hayaisha, M. Kawaichi, N. Ogata, K. Ikai, J. Oka, and H. Okayama (1979). Poly (ADP-ribose) and ADP-ribosylation of proteins. In D. E. Atkinson and C. F. Fox (Eds.). *Modulation of Protein Function*, Vol. 13, ICN-UCLA Symposia on Molecular Biology. Academic Press, New York, pp. 47–64.

27 M. E. Smulson and T. Sugimura (Eds.) (1980). *Novel ADP-Ribosylations of Regulatory Enzymes and Proteins.* Elsevier/North-Holland, New York.

28 H. Hilz, P. Adamietz, R. Bredehorst, and K. Wielckers (1979). ADP-ribosylation of nuclear proteins. *Adv. Enz. Reg.* **17**, 195–211.

29 S. M. Panasenko and D. E. Koshland, Jr. (1979). Methylation and demethylation in the bacterial chemotactic system. In D. E. Atkinson and C. F. Fox (Eds.). *Modulation of Protein Function*, Vol. 13, ICN-UCLA Symposia on Molecular Biology. Academic Press, New York, pp. 273–284.

30 C. Gagnon and S. Heisler (1979). Protein carboxyl-methylation: role in exocytosis and chemotaxis. *Life Sciences* **25**, 993–1000.

31 M. S. Springer, M. F. Goy, and J. Adler (1979). Protein methylation in behavioural control mechanisms and in signal transduction. *Nature* **280**, 279–284.

32 B. B. Buchanan (1979). Thioredoxin and enzyme regulation in photosynthesis. In D. E. Atkinson and C. F. Fox (Eds.). *Modulation of Protein Function,* Vol. 13, ICN-UCLA Symposia on Molecular Biology. Academic Press, New York, pp. 93–111.

33 R. T. Schimke and N. Katanuma (Eds.) (1975). *Intracellular Protein Turnover.* Academic Press, New York.

34 N. Katanuma (1975). Regulation of intracellular enzyme levels by limited proteolysis. *Rev. Physiol. Biochem. Pharmacol.* **72**, 83–104.
35 A. L. Goldberg and A. C. St. John (1976). Intracellular protein degradation in mammalian cells. Part 2. *Ann. Rev. Biochem.* **45**, 747–803.
36 F. J. Ballard (1977). Intracellular Protein Degradation. In P. N. Campbell and W. N. Aldridge (Eds.). *Essays in Biochemistry*, Vol. 13. The Biochemical Society, London, pp. 1–37.
37 A. Hershko, A. Ciechanover, H. Heller, A. L. Haas, and I. A. Rose (1980). Proposed role of ATP in protein breakdown: Conjugation of proteins with multiple chains of the polypeptide of ATP-dependent proteolysis. *Proc. Natl. Acad. Sci. USA* **77**, 1783–1786.
38 G. L. Francis and F. J. Ballard (1980). Enzyme inactivation via disulphide-thiol exchange as catalysed by rat liver membrane protein. *Biochem. J.* **186**, 581–590.

SECTION THREE

Regulation of Enzyme Synthesis in Bacterial Cells

Prokaryotic cells are able to live in an environment composed of very simple compounds. From a source of nitrogen and a source of carbon they can produce energy and synthesize all of the cellular constituents needed to support growth. In this section we describe some of these metabolic pathways and the regulation of the synthesis of the enzymes involved. Catabolic pathways are described in Chapter 8, and biosynthetic pathways are described in Chapter 9.

Maximum growth is achieved by minimizing the number of proteins synthesized, since as much as 80% of all carbon and energy goes into protein biosynthesis. The cells have evolved feedback mechanisms that switch off the transcription of the genes for the enzymes and proteins that are not needed in a given environment. But, at the same time, the cells retain the capacity to adapt quickly to *changes* in the composition of their environment. Unlike animal cells, the bacterial cells switch *all* of the enzymes in a pathway on or off, and the genes coding for the enzymes of a single pathway are usually organized into an operon that facilitates the coordination of their expression.

Two different mechanisms regulate the transcription of the genome. The first mechanism, of which there are a number of variations, involves the placement of a protein molecule, the repressor, onto the DNA of the genome so that it prevents, by steric hindrance, the binding of RNA polymerase. Both catabolic and biosynthetic operons have been shown to be regulated in this way. The presence or absence of the repressor on the genome is determined by the concentration of substrate or end product (or a closely related metabolite). The second mechanism, found to regulate only pathways of amino acid biosynthesis, involves a complex series of reactions, mediated by the translational apparatus, which causes the dissociation of a transcriptionally active RNA polymerase from the operon. This is the mechanism of attenuation.

REGULATION OF ENZYME SYNTHESIS

These mechanisms are well suited to the pathways they control. The regulation of *catabolic* pathways is relatively simple; each substrate induces the synthesis of the enzymes required for its own catabolism. This mechanism of induction produces large amounts of enzymes quickly and transforms the substrate into metabolites that can readily enter general metabolic pathways. Some *biosynthetic* pathways are also controlled in a simple way, with their enzymes being synthesized until the concentration of end product reaches an optimal level. When this level is reached enzyme synthesis is repressed. The more complex mechanism of attenuation serves to match the *rate* of biosynthesis of an amino acid with the *rate* at which it is consumed in protein synthesis.

Although individual metabolites and substrates can effect changes in gene expression, the cell has evolved a series of higher-level controls that *coordinate* the expression of various operons. The catabolite activator protein and glutamine synthetase are two molecules that mediate control at this higher level. They are described in Chapter 8.

Finally, this section concludes with a description of the regulation of the synthesis of the enzymes of histidine metabolism. The biosynthesis and catabolism of the amino acid are considered in relation to the metabolic demands of the cell, illustrating the various mechanisms of gene expression in bacteria.

CHAPTER EIGHT

Catabolic Pathways

Bacterial cells, particularly enteric bacteria such as *E. coli* and *S. typhimurium*, are faced with a constantly changing environment in which the types and amounts of carbon and nitrogen sources fluctuate wildly. To survive and proliferate the cells must be able to derive as much energy as possible from the available nutrients. Moreover, when availability changes the cells must adapt to the new carbon and nitrogen sources.

The first clues to the way in which bacterial cells adapt came from work carried out during the 1940s [reviewed by Monod (1)]. This work established that a cell faced with a *mixture* of carbon sources catabolized them sequentially rather than simultaneously. A typical example is shown in Fig. 8-1 for cells growing on a mixture of glucose, sorbitol, and glycerol. During the first 2 or 3 h the cells grow rapidly, and all the glucose in the medium is consumed. Cell growth ceases temporarily and then resumes for about 2 h during which time all the sugar, sorbitol, is consumed. Finally, after a second brief period without growth the glycerol is consumed, but this carbon source supports only a small additional amount of growth, and the cells cease to proliferate when all these carbon sources are depleted. This phenomenon, known as *triauxic growth*, became the basis for all studies on the molecular biology of the regulation of enzyme synthesis in bacterial cells.

The lag periods in growth in Fig. 8-1 were interpreted as the time required to synthesize the enzymes necessary for the catabolism of the next substrate. This was confirmed in subsequent work (2, 3) on the diauxic growth of *E. coli* cells exposed to a mixture of glucose and lactose (Fig. 8-2a). The glucose is catabolized first, and then growth ceases for a short time. When growth resumes it is paralleled by an increase in the activity of β-galactosidase, the enzyme responsible for the catabolism of lactose. Figure 8-2b shows that the induction of this enzyme commences within a few minutes after the depletion of glucose. Moreover, the enzyme is not synthesized before this time even though lactose is present in the medium from the beginning of the experiment. Thus, the enzymes of some catabolic pathways are only synthesized when required, whereas others are expressed constitutively.

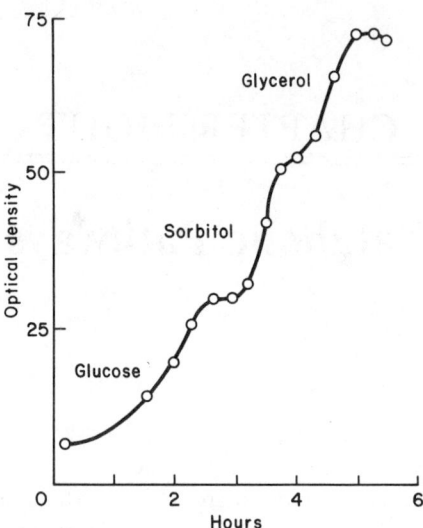

Fig. 8-1 Triauxic growth of *E. coli* on a mixture of glucose, sorbitol, and glycerol. Growth of the culture over the 6-hr period was measured by turbidity. The amount of growth supported by each carbon source is indicated. [Reproduced from Fig. 11 of Monod (1) with permission from *Growth*.]

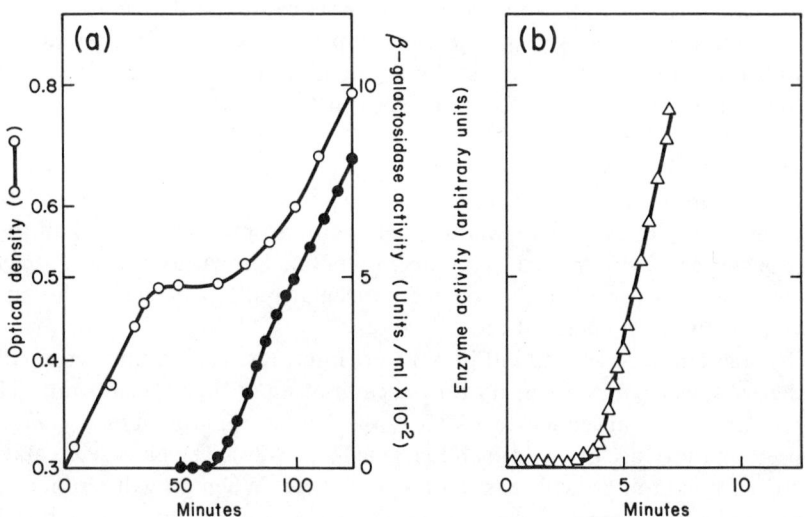

Fig. 8-2 Diauxic growth of *E. coli* on a mixture of glucose and lactose and the induction of β-galactosidase. In (*a*) growth was measured by turbidity and enzyme activity by *in vitro* assay. Note the difference in time scales between (*a*) and (*b*). The time course of β-glactosidase induction is shown in (*b*) for cells switching from glucose to galactose as in (*a*). [Figure (*a*) is reproduced from Fig. 1 of Epstein et al. (2) with permission of Academic Press. Figure (*b*) is reproduced from Fig. 1 of Pardee and Prestidge (3) with permission.]

INDUCIBLE AND NONINDUCIBLE PATHWAYS

Glucose is catabolized by Embden–Meyerhof glycolysis with the carbon being converted into (i) *biosynthetic intermediates* by the activities of the pentose phosphate pathway and the tricarboxylic acid cycle and (ii) *energy* by glycolysis and the tricarboxylic acid cycle (Fig. 8-3 and Ref. 4). The enzymes of these pathways are expressed constitutively (i.e., they are present in the cell at all times, although they are not all expressed to the same extent) since even when glucose is not present in the medium the cell requires most of these enzymes to permit the carbon from other sources to enter general metabolic pathways.

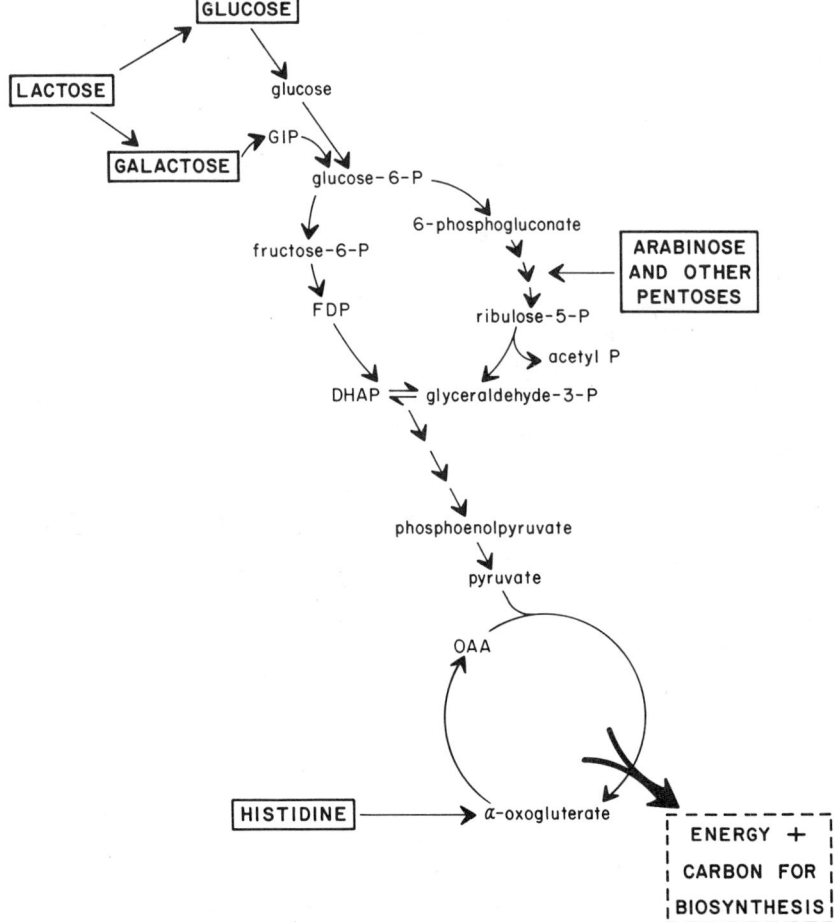

Fig. 8-3 The pathways of carbon catabolism in bacterial cells showing the points of entry of commonly used carbon sources. The operons regulating the inducible catabolic pathways are described in the text. The pathways expressed constitutively are Embden–Meyerhof glycolysis, the pentose phosphate pathway, and the tricarboxylic acid cycle.

Many carbon sources such as lactose, galactose, and a number of pentoses and trioses, as well as two-carbon sources such as pyruvate, can all be catabolized by bacterial cells. Moreover, certain amino acids, for example, histidine, provide carbon that can enter general metabolic pathways and support growth (Fig. 8-3). The enzymes catalyzing the reactions of these pathways are not expressed unless the particular carbon source is present in the environment and all carbon sources of a higher energy value (e.g., glucose) have been depleted. This affords the cell maximum economy in the operation of catabolic pathways.

The first part of this chapter describes the regulation of the synthesis of the enzymes of each pathway, commencing with a brief review of the *lac* operon and progressing to the more complex regulatory systems. In the second half of the chapter the way in which these pathways are coordinated to permit the bacterial cell to make optimum use of its environment is discussed, together with some comments on the integration of carbon and nitrogen catabolic pathways. Several recent reviews deal with various aspects of this subject (5–11).

OPERONS OF CATABOLIC PATHWAYS

The *lac* Operon

The *lac* operon is approximately 5200 nucleotides in length, mapping at about 8′ on the *E. coli* chromosome (Fig. 8-4). It codes for the proteins responsible for the transport of the disaccharide lactose into the cell and its catabolism to glucose and galactose. Glucose enters the Embden–Meyerhof pathway directly, whereas galactose must be further metabolized to glucose 1-phosphate before entering this pathway.

β-Galactoside permease is a transport protein (M_r = 30,000) that facilitates the movement of lactose across the cell membrane. Galactoside transacetylase (M_r = 32,000) also appears to be involved in the uptake process, but its exact function is not known. Inside the cell lactose is cleaved to glucose and galactose by β-galactosidase (M_r = 544,000), with allolactose being generated as a reaction intermediate (Fig. 8-4). The three enzymes are coded by three contiguous structural genes with the Z gene being transcribed first. Details of the genetic analysis of the *lac* (and other operons) can be found in Refs. 9 and 11.

Upstream from the 5′ end of the Z gene is the promoter–operator region. This region, of about 120 nucleotides, has been sequenced (see Fig. 2-7), and the interactions between RNA polymerase and *lac* repressor have been studied in detail (see Chapter 2). The concentration of the three enzymes is determined by the frequency with which RNA polymerase is allowed to bind to the promoter and form an initiation complex. The regulation of this reaction is elegantly simple.

In the absence of lactose the operon is not expressed, because the presence of repressor, tightly bound to its operator site, prevents RNA polymerase from binding to its promoter. The repressor is a tetrameric molecule (M_r = 600,000)

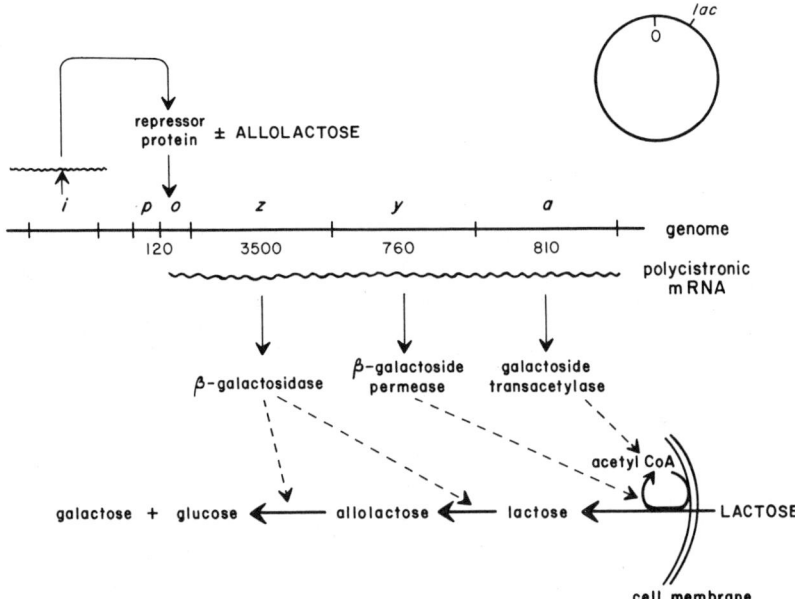

Fig. 8-4 The *lac* operon and the pathway of lactose catabolism. The position of the operon on the *E. coli* genome is shown in the inset. The numbers under each gene indicate their approximate lengths in base pairs. Transcription of the operon is from left to right.

synthesized constitutively at a low level (approximately 10 molecules/cell) from the *i* gene located immediately upstream from the *lac* promoter (Fig. 8-4). Although the operon is almost completely repressed under these circumstances there is sufficient "leakage" (i.e., basal level of transcription) to maintain a low concentration of β-galactosidase in the cell.

The introduction of lactose to the medium (or the depletion of glucose as in Fig. 8-2a) leads, after a few minutes delay, to a rapid synthesis of the enzymes of the operon (Fig. 8-2b). Lactose, itself, is not the inducer; the intermediate, allolactose, generated by the basal activity of β-galactosidase is the true inducer, and the delay in enzyme induction represents the time required for the concentration of allolactose to reach a critical level. At this level it binds to, and induces an allosteric transition in, the repressor molecule such that its affinity for DNA is markedly decreased and the allolactose–repressor complex dissociates from the operon. The promoter site is then free for RNA polymerase binding. As long as the concentration of allolactose remains above the critical level the operon may be transcribed. When its concentration falls, repressor reassociates with its operator binding site, and transcription of the operon declines. Thus the operon is under *negative* control—*it is only induced in the presence of a specific low-molecular-weight inducer.*

The key to the regulation of the expression of the operon is the dual role played by the repressor molecule. It interacts with a small intermediary metabolite, and it also binds, specifically, to the *lac* operon. This establishes a direct link for information transfer between the environment of the cell and the genome, with the actual transfer of information being mediated by a conformational change in the repressor protein.

Like most catabolic operons the expression of *lac*, even in the presence of allolactose (i.e., after repressor has dissociated from its operator), depends upon the binding of the CAP–cAMP complex to its binding site near the *lac* promoter (see Fig. 2-7). RNA polymerase has a very low affinity for the *lac* promoter in the absence of this complex. Since CAP–cAMP mediates a coordinated response of the cell to its external environment by interacting with a number of operons this facet of the expression of the *lac* operon will be deferred until a later section (see Ref. 12 for a review of the control of *lac* expression by CAP–cAMP).

The *gal* Operon

Galactose taken up from the medium, or produced intracellularly from lactose catabolism, is metabolized by different pathways in different bacterial strains. The pathway in *E. coli* is shown in Fig. 8-5 together with the organization of the *gal* operon (7, 13–15). In this series of reactions galactose is converted to glucose 1-phosphate by the actions of the three enzymes of the *gal* operon. A fourth enzyme, UDPG pyrophosphorylase, catalyzes the initial charging of UDPglucose and is coded by the *constitutively* expressed *gal U* gene located some distance from the operon.

The *gal* operon occupies about 3500 nucleotides, mapping at 17' on the *E. coli* chromosome. Expression of the operon is negatively controlled by a repressor (product of the *gal R* gene mapping at 61'), and galactose is the inducer. However, regulation of the expression of the operon differs in several respects from that of the *lac* operon. The operator, which binds *gal* repressor, is located *upstream* from the promoter region, rather than being between the promoter and the start of transcription as in *lac* (cf. Figs. 8-5 and 2-7). The binding site for CAP–cAMP lies between the operator and the promoter region (Fig. 8-5, inset). The presence of repressor at its operator (signifying no galactose in the medium) prevents CAP–cAMP from binding, and since RNA polymerase binding depends on CAP–cAMP the operon is not expressed. In the presence of galactose, *gal* repressor dissociates, and the operon is expressed when CAP–cAMP is also available.

The product of the *E* gene, UDP-galactose epimerase, also participates in the biosynthesis of the capsular polysaccharide, colonic acid, and it has been found that changes in the demand for colonic acid influence the expression of the *gal* operon, even in the absence of CAP–cAMP. Detailed studies of the control region (13–15) have revealed a second promoter, designated P_2 (P_1 is the promoter for the CAP–cAMP-dependent expression of the operon), to which RNA polymerase binds *in the absence of* CAP–cAMP. The Pribnow box for this promoter is located about 5 base pairs upstream (Fig. 8-5). Thus CAP–cAMP-

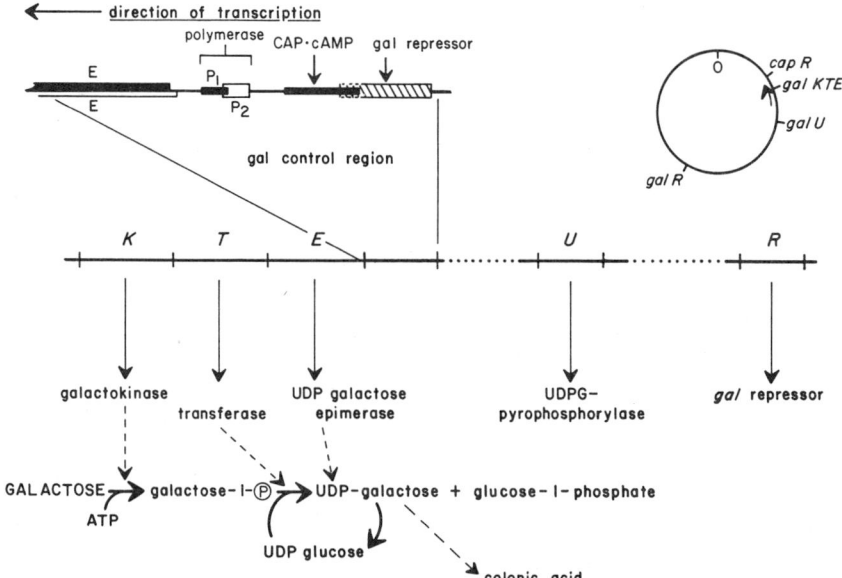

Fig. 8-5 The *gal* operon and the galactose catabolic pathway. The operon is transcribed from right to left; the structure of the control region is shown in more detail in the inset. P_1 and P_2 are CAP–cAMP-dependent and -independent promoters, respectively. The binding sites for CAP–cAMP and *gal* repressor are also shown. *Cap R* is the gene coding for the repressor of the colonic acid biosynthetic pathway. Regulation of expression of the operon is described in the text.

independent transcription of the operon can occur to ensure an adequate supply of colonic acid even under conditions of catabolite repression.

The product of the *cap R* gene has been found, by mutation studies, to also influence the expression of this operon (13) and is believed to behave as a repressor for capsular polysaccharide biosynthesis, but the operator binding site for this repressor has not been established. Thus, it is apparent that the expression of the *gal* operon depends upon *interactions between several proteins* and represents a considerable increase in complexity over the regulation of the *lac* operon.

The Catabolism of Arabinose

α-Arabinose is converted to D-xylulose by a pathway involving three enzymes. The pentose is first isomerized to L-ribulose, before undergoing phosphorylation by ribulokinase and epimerization to yield D-xylulose 5-phosphate, which can be readily catabolized by the pentose phosphate pathway and Embden–Meyerhof glycolysis (Figs. 8-2 and 8-6). The three enzymes are coded by the *ara BAD* operon. Two other genes *ara E* and *ara F* code for α-arabinose permease and the periplasmic binding protein, two proteins involved in arabinose transport

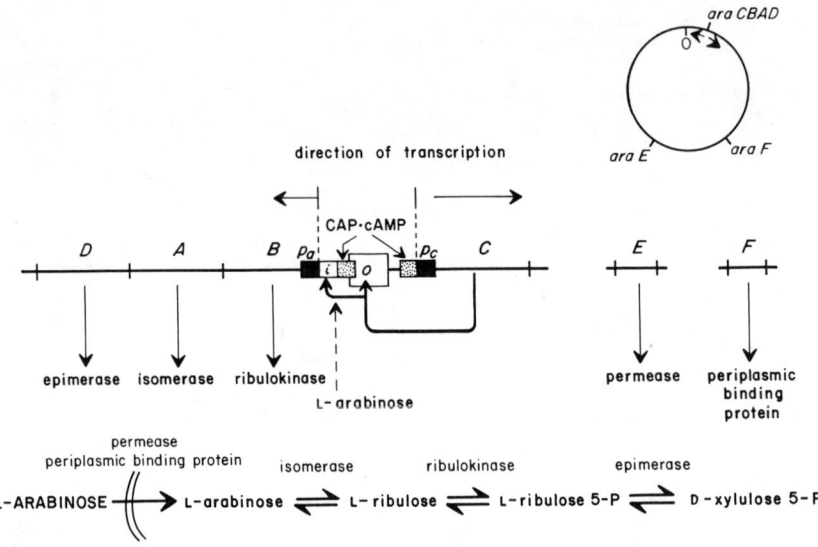

Fig. 8-6 The pathway of arabinose catabolism and the *ara* operon. Note the divergent transcription of *ara BAD* and *ara C*. The organization of the binding sites for regulatory molecules and RNA polymerases on the 140–160-base-pair segment of DNA between the genes is indicated. *o* = *ara BAD* operator, *i* = *ara BAD* inducer site to which the product of the *ara C* gene binds in the presence of L-arabinose. The binding sites are approximately to scale, but the structural genes are not.

into the cell. Although not contiguous with the rest of the operon, these two genes are also under negative control. The expression of the *ara BAD* operon has been studied in great detail (7, 16–19).

The three genes are transcribed into a polycistronic messenger RNA when the inducer α-arabinose is present. The operon is under the negative control of a repressor that is the product of the *ara C* (control) gene located near the regulatory region of the *ara BAD* operon (Fig. 8-6). Surprisingly, mutations in this repressor gene, instead of leading to constitutive expression of the operon by producing a defective repressor, actually *prevented* expression of the operon. Further work revealed that the *ara C* gene product functions *as both a repressor and a positive activator* of the operon. Moreover, the molecule has two distinct binding sites at the control region, moving from the operator to an initiator site in the presence of the inducer α-arabinose (Fig. 8-6).

The *ara C* gene is transcribed from the opposite strand to that of *ara BAD* (Fig. 8-6); moreover, the *ara C* gene is not expressed constitutively. It is repressed by its own gene product and is stimulated by CAP–cAMP. Recent studies on the sequence and organization of the control regions between the two genes (19) have revealed some details of the mechanism for the regulation of this operon. Thus, the 140–160 base pairs of DNA between the 3′ end of the *B* gene and the 5′ end of the *C* gene contain promoters for each gene as well as operator and initiator binding sites for the *ara C* gene product and two binding sites for CAP–cAMP (Fig. 8-6).

In the absence of inducer, residual transcription of the *C* gene provides enough repressor to bind to the *ara o* site and repress both *ara BAD* and virtually all *ara C* transcription. In the presence of α-arabinose, repressor dissociates from *ara o* and binds to *ara i*, where, in the presence of CAP–cAMP, it induces transcription of the operon. Dissociation of repressor from *ara o* allows CAP–cAMP to bind near *ara C* and the gene is transcribed, providing more product to further induce the operon. When the concentration of α-arabinose falls, this high concentration of *ara C* gene product becomes a repressor and rapidly shuts off transcription of the operon.

A number of other sugars, notably maltose and rhamnose, also appear to be catabolized by enzymes coded by operons controlled by *positive activation* as well as repression. This kind of mechanism permits both a rapid induction and a rapid repression of the operon and, in the case of the *α-arabinose* operon, depends upon the complex interaction of a number of proteins on a short segment of DNA between two genes. Since the binding sites for these proteins probably overlap, protein–protein interactions are likely to play a major role in the expression of this operon in addition to the protein–DNA interactions.

Histidine Utilization

Operon control in bacterial cells tends to be regarded as a transcriptional switch that either turns the operon *on* or turns it *completely off*. However, this seldom happens *in vivo*, and most operons are expressed to some extent even in the most heavily repressed situations. Indeed, some operons *must* be expressed at a basal level, and the *h*istidine *ut*ilization operon (*hut*) is a good example.

Histidine is catabolized by the bacterium *S. typhimurium* to glutamate, ammonia, and formamide. These compounds permit both the carbon and the nitrogen atoms of histidine to enter general metabolic pathways (Figs. 8-3 and 8-7). Four reactions are involved in histidine catabolism, and the genes coding for the enzymes are organized into two closely associated, but nevertheless discrete, operons (Fig. 8-7 and Refs. 6 and 20). The first two enzymes of the pathway, histidase and uroconase, are controlled by the *hut* operon to the right (PRQ, U, H), whereas the other two enzymes are controlled by the *hut* operon on the left (M, I, G, C). Both operons are transcribed in a clockwise direction from promoter–operator regions designated M and PRQ, respectively (Fig. 8-7). The expression of these operons involves a novel regulatory mechanism, known as *autogenous expression*, which is a simple, but effective modification of the negative control mechanism of the *lac* operon.

A single species of repressor protein binds to each operon, *ensuring that both operons are coordinately repressed*. Uroconate, the product of the first reaction of the catabolic pathway is the inducer and in its presence repressor dissociates from the operator of each operon and they are both expressed. A search for the gene responsible for repressor synthesis led to the discovery that it was part of

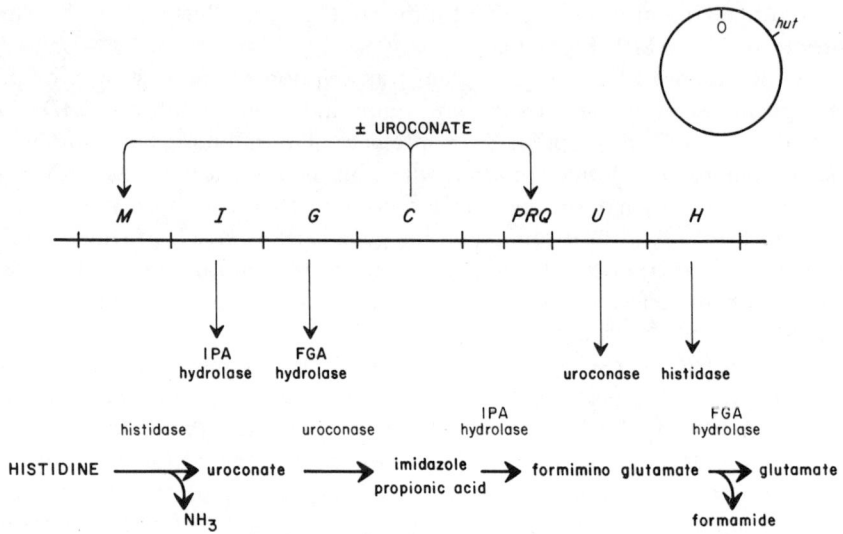

Fig. 8-7 The operon for histidine utilization (*hut*) and the pathway of histidine catabolism. M is the promoter for *hut IGC*, and PRQ is the promotoer for *hut UH*. The product of the *hut C* gene is a repressor protein, and uroconate is the inducer.

one of the operons. Thus the *hut C* gene of the operon on the left in Fig. 8-7 codes for repressor.

A residual activity of both operons is necessary; the operon on the left must synthesize enough repressor to repress both operons, and the operon on the right must direct the synthesis of sufficient histidase so that histidine can be converted to uroconate to induce the operon.

When the operons are expressed the enzymes for histidine catabolism are synthesized, and so is more repressor. However, as long as the concentration of uroconate remains high these repressor molecules do not bind to their operators, and the operons continue to be expressed. As the histidine is consumed the concentration of uroconate falls, free repressor molecules become available, and the operons are repressed.

Several other operons are also controlled autogenously [i.e., they induce the synthesis of their own repressor (6, 7, 18)]. This kind of mechanism has been described as a mechanism for the "cautious expression" (6) of operons in which overproduction of enzymes does not occur, particularly in situations where there are rapid fluctuations in the concentration of histidine and other catabolites in the cell's environment.

INTEGRATION OF THE PATHWAYS OF CATABOLISM

Glucose as a Preferred Source of Carbon and Energy and the Phenomenon of Catabolite Repression

The examples of di- and triauxic growth described earlier (Figs. 8-1 and 8-2) illustrate that when a mixture of carbon sources is available *glucose* is the preferred substrate. In fact, glucose is the carbon and energy source *par excellence*, and it is always metabolized first. Thus when glucose is present none of the operons discussed in this chapter are expressed, even though their inducers may be present in the environment.

This phenomenon was first observed in 1901 and became known as the "glucose effect" (see Refs. 1 and 21 for discussions of the early literature). When later work showed that other metabolites, such as glucose 6-phosphate, galactose, and gluconate, were capable of the same effect the more general term "catabolite repression" was introduced (21).

cAMP as a Mediator of Catabolite Repression

Little progress was made toward an understanding of the phenomenon of catabolite repression until after the discovery of cAMP in 1957 by Sutherland and Rall (22) and Cook, Lipkin, and Markham (23). This cyclic nucleotide, first discovered in eukaryotic cells and subsequently found in prokaryotes (24), was found to be inversely related to the concentration of the glucose upon which the cells were growing (25). As the concentration of glucose in the medium falls, the concentration of cAMP inside the cell increases.

A direct relationship between glucose and cAMP and a role for cAMP in catabolite repression were established by experiments in which cAMP was added to the medium of cells growing on glucose (26, 27). Before cAMP was added, all catabolic operons were repressed by the glucose present in the medium, and the *in vivo* concentration of cAMP was very low. When cAMP was added the cells took it up and the intracellular concentration rose to a point where, despite the presence of glucose, the catabolic operons were expressed and enzymes such as β-galactosidase and tryptophanase were synthesized.

Two independent lines of enquiry established that cAMP acted at the transcriptional level, at least in the case of the *lac* operon (28, 29). First, mutations that abolished the stimulatory effect of cAMP and also rendered the cells insensitive to catabolite repression mapped in the promoter–operon region of the operon. Second, it was shown by direct measurement (Fig. 8-8) that cAMP increased the concentration of β-galactosidase mRNA. Changes in the rate of mRNA degradation were ruled out, indicating that cAMP must be interacting at the transcriptional level.

The development of sensitive techniques for the detection of cAMP permitted direct measurements of the intracellular concentration of cAMP in

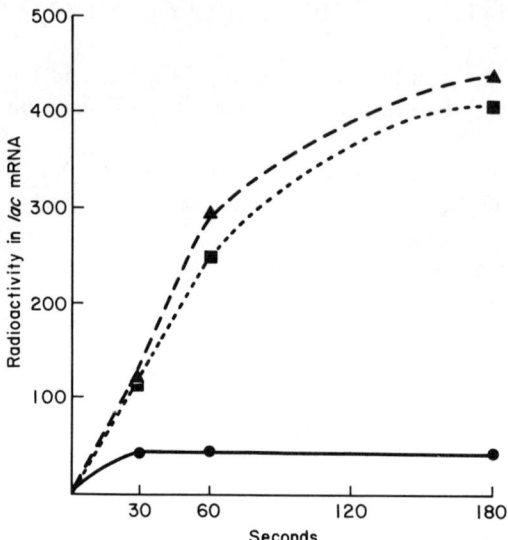

Fig. 8-8 The synthesis of β-galactosidase mRNA (measured by the incorporation of radioactive precursor) in *E. coli* cells in repressed cultures (●——●), in induced cultures (▲---▲), and in repressed cultures treated with cAMP (■- - -■). [Reproduced from Ref. 29 by permission of the American Society of Biological Chemists, Inc.]

glucose-starved and glucose-repressed cells. A typical example is shown in Fig. 8-9 (30). In this experiment *E. coli* cells were placed in a medium containing 3.5-mM glucose and allowed to grow. During the first 3 h of culture the cAMP concentration was barely detectable, but as soon as the glucose in the medium was depleted there was a 30-fold increase in the concentration of cyclic AMP over a period of about 15 min. This increased concentration of cyclic nucleotide was maintained as long as the cells were starved. The readdition of glucose to the medium caused an equally abrupt decrease in the cAMP level, and it remained low until the added glucose was consumed (Fig. 8-9).

The Catabolic Gene Activator Protein

Two classes of mutant that affect the mechanism of catabolite repression have been isolated. The first of these, described above, is located within the promoter–operator region of inducible operons. The second class maps at a distant site on the genome. This locus has been designated the *cap* (or *crp*) gene and codes for a protein, the catabolic gene activator protein (CAP), which is also called the cyclic AMP receptor protein (CRP). Following the discovery of this gene and its protein product a great deal of progress has been made toward an understanding of the molecular biology of catabolite repression. Moreover, this work has shown that negatively controlled, inducible, catabolic operons are also dependent upon a positive activation reaction catalyzed by CAP.

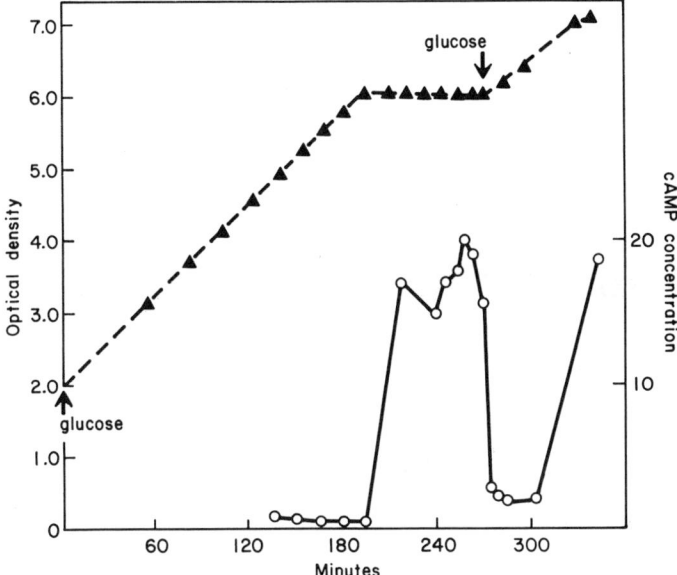

Fig. 8-9 The concentration of cAMP in *E. coli* cells cultured in the presence of 3.5-m*M* glucose. Growth (measured by turbidity, ▲——▲) continued until all glucose was consumed and then stopped until glucose was replenished. The concentration of cAMP (o——o) was high in the absence of glucose and low in its presence. [Reproduced from Fig. 3 of Ref. 30, with permission of the American Society for Microbiology.]

The role of CAP and cAMP in the regulation of the expression of operons such as *lac* and *hut* is summarized in Fig. 8-10. In the absence of glucose the concentration of cAMP increases, and the CAP–cAMP complex binds to its binding site on the promoter. If the inducer of the operon is also present the repressor–inducer complex dissociates from the operator, the binding of RNA polymerase is facilitated, and the operon is transcribed. When glucose is present, the concentration of cAMP decreases, and CAP dissociates. Even if the inducer of the pathway is present and it removes repressor from its operator binding site, the operon is not expressed, because of the lack of CAP–cAMP.

The arrangement of the regulatory elements in the control regions of the *gal* and *ara* operons is slightly different (Fig. 8-11). The binding site for CAP–cAMP is adjacent to, and upstream from, the RNA polymerase binding site as it is in *lac* and *hut*. However, the operator (site of repressor binding) overlaps the site of CAP–cAMP binding. Thus CAP–cAMP can only bind to the operon when inducer is present. Moreover, when CAP–cAMP is bound and the operon is being transcribed, a fall in inducer concentration necessitates a protein–protein interaction between repressor and CAP–cAMP to displace CAP–cAMP from the operon and permit repressor binding. Without this interaction the operon would continue to be transcribed in the absence of its catabolite.

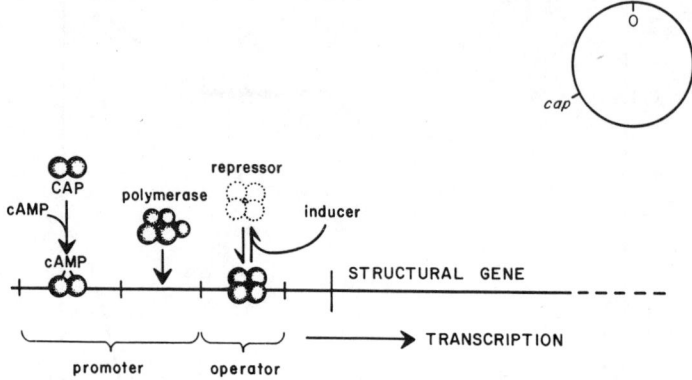

Fig. 8-10 The interactions between catabolic activator protein (CAP), RNA polymerase, repressor, and the two regulatory metabolites, cAMP and inducer, at the promoter-operator region of operons such as *lac* and *hut*. The map position of the *cap* gene that codes for CAP is shown in the inset.

The key element in catabolite repression is the bifunctional protein CAP, which transmits information on the energy status of the cell to a number of different operons. Like repressor, it can bind a small metabolite and also recognize, and bind to, a specific DNA sequence.

CAP–cAMP and Auxic Growth

If CAP–cAMP mediates catabolite repression the mechanism must include a means for discriminating between different catabolic operons, so that when glucose is depleted the next most efficient energy source is selected. For example, in cells growing on a mixture of glucose, lactose, and arabinose, will lactose or arabinose be metabolized after the glucose is depleted? This question has

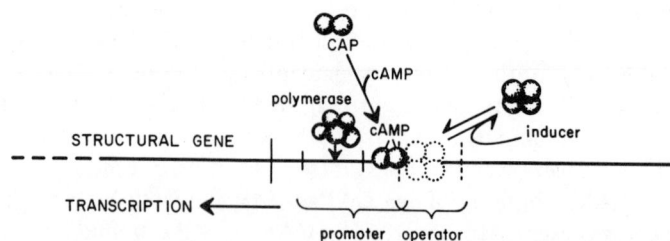

Fig. 8-11 The interactions between regulatory molecules in operons where the operator overlaps the site of binding of CAP–cAMP. Transcription of this operon is from right to left.

been partially answered by the observations (31, 32) that each operon is activated by different concentrations of cAMP. Thus half-maximal activation of the *lac* operon occurs when the concentration of cAMP reaches $3.8 \times 10^{-4} M$, whereas half-maximal activation of *ara* is not achieved until the concentration of cAMP reaches $8.6 \times 10^{-4} M$. It is proposed that the sensitivity of the operons to the concentration of cAMP establishes a hierarchy of catabolic substrates, and in the example cited above lactose would be catabolized after glucose and before arabinose.

However, it is far from obvious how this works, since it implies that different concentrations of cAMP induce a series of conformational changes in the CAP dimer, permitting it to bind to different operons. It has not yet been established that cAMP can do this, or that CAP is capable of recognizing a large number of different nucleotide sequences in this manner. This is just one example of a number of problems that have arisen since CAP–cAMP was invoked as a pleiotypic regulator of catabolite repression. Other problems are discussed below, along with some general comments on the metabolism and functions of cAMP.

cAMP and Catabolite Repression

cAMP is synthesized by adenylate cyclase and degraded by phosphodiesterase. Adenylate cyclase is coded by the *cya* gene; mutations in this gene, which result in an inability to synthesize cAMP, usually prevent the expression of catabolic operons. However, some cells have been found with *cya* mutations that *cannot* synthesize cAMP but *can* express catabolic operons (33), suggesting that molecules other than cAMP are involved in catabolite repression. Similarly, some cells with mutations in the *cap* (*crp*) gene that alter CAP production can still express catabolic operons (34). Moreover, certain other mutants are able to *repress* catabolic operons in the presence of their inducers and a high concentration of cAMP (35). Taken together, these observations cast considerable doubt on the *simple* CAP–cAMP model of catabolite repression.

There are indications that cAMP regulates other cellular processes involved in catabolite repression. For example, the cells with *cap* mutations discussed above were also found to be deficient in the synthesis of the termination factor rho. Thus the catabolic operons could only be expressed in those cells if, in addition to being deficient in CAP, they were also deficient in rho. This suggests that in some situations CAP–cAMP may activate catabolic genes by *preventing* rho-dependent termination (34).

There is little doubt that cAMP does play some role in catabolite repression and that it may act at more than one level. It is also clear that catabolite repression is a much more complex process than previously thought. Moreover, glucose may exert some of its effect by cAMP-independent mechanisms. Thus, glucose is thought to prevent expression of the *gal* operon by *inducer exclusion* (i.e., galactose is not transported into the cell when glucose is present) and not catabolite repression (36).

PATHWAYS OF NITROGEN SOURCE CATABOLISM

Escape from Catabolite Repression

There are metabolic situations in which repression of certain catabolic operons could pose a problem to the cell. Consider cells growing on a mixture of glucose and histidine as sources of carbon and nitrogen. Glucose is metabolized as the preferred source of energy since histidine is a much poorer substrate and the *hut* operon should be subject to catabolite repression. However, if this happened the cells would become starved of nitrogen, since histidine is the nitrogen source. Under these circumstances a mechanism exists to permit the operon to "escape" from catabolite repression and be expressed.

Nitrogen starvation, which results from repression of the *hut* operon, lowers the concentration of ammonia in the cell, and glutamine synthetase, an enzyme of central importance in nitrogen metabolism (see below), monitors this change. As the concentration of ammonia falls the fraction of the enzyme in the deadenylylated form increases (see Chapter 7 for a discussion of covalent modifications to this enzyme). Deadenylylated glutamine synthetase (GS_0) is now known to be a positive activator of gene transcription, and it is capable of the activation of the *hut* operon even in the presence of glucose. The enzyme interacts at a site similar to the CAP–cAMP-binding site of the operon and produces the necessary positive activation of the operon, so that both glucose and histidine can be catabolized simultaneously.

Operons governing the catabolism of other amino acids, such as tryptophan, asparagine, and proline, are also positively activated by deadenylylated glutamine synthetase. Moreover, the enzyme stimulates transcription of its own gene. However, other operons such as *lac* that do not catabolize nitrogen sources are not activated by GS_0.

The Central Role of Glutamine Synthetase in Cellular Nitrogen Metabolism

Although ammonia is the preferred source of nitrogen, most cells can also use a large number of nitrogen-containing organic compounds. All nitrogen sources are catabolized to either glutamate, glutamine, or ammonia, and these molecules serve as nitrogen sources in all biosynthetic reactions (see Ref. 37 for a recent review). Glutamine synthetase catalyzes the only reaction in the cell that involves all three key nitrogen-containing compounds:

$$\text{Glutamate} + NH_4^+ + ATP \rightleftharpoons \text{Glutamine} + ADP$$

The enzyme is under complex regulatory control, and many metabolites have been shown to modulate the degree of adenylylation of the enzyme. For example, glutamine—the product of the reaction—promotes adenylylation and is, therefore, a feedback inhibitor of its own synthesis. α-Oxogluterate, on the other hand, promotes deadenylylation and is a strong activator of the enzyme.

As many as 40 metabolites are thought to interact with the glutamine synthetase cascade system. Moreover, many of these metabolites also affect the transcription of the gene for the enzyme, and the enzyme's concentration can vary as much as 100-fold depending on the nitrogen source (37).

Glutamine synthetase is the product of the *gln A* gene mapping at about 85' on the *E. coli* genome. Deadenylylated glutamine synthetase is a positive activator of its own transcription. Thus when the enzyme is fully active (completely deadenylylated) it will stimulate the production of more enzyme molecules. As described above, the enzyme also acts as a positive activator for the catabolism of several amino acids as well as several other nitrogen-containing compounds such as putrescine and urea.

The enzyme is known to be a DNA-binding protein, but the site at which it interacts is not known. However, it is one more protein factor, along with CAP, repressor, and polymerase, that regulates the expression of certain critical operons that supply both carbon and nitrogen. These operons exhibit a considerable increase in complexity of control over the simple *lac* operon, and much work remains to be done to be able to fully understand these regulatory mechanisms.

The key role of glutamine synthetase in the regulation of catabolic operon transcription under conditions of nitrogen starvation and nitrogen excess is illustrated in Fig. 8-12. Under conditions of nitrogen starvation the concentration of α-oxogluterate increases, leading to the deadenylylation of glutamine synthetase (via inhibition of adenylylation, Fig. 8-12a). The deadenylylated

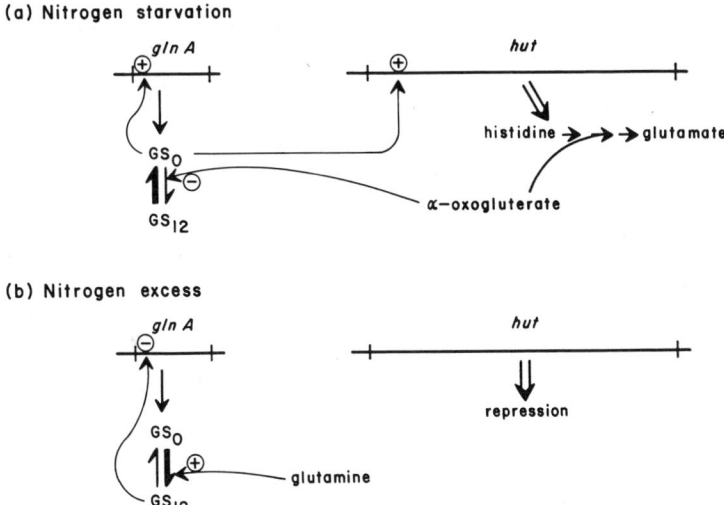

Fig. 8-12 The role of glutamine synthetase in the regulation of nitrogen balance. GS_0 = fully deadenylylated, fully active enzyme; GS_{12} = fully adenylylated, totally inactive enzyme. ⊕ = stimulation of covalent modification or increase in transcription; ⊖ = inhibition of covalent modification or decrease in transcription. These reactions are described in the text.

protein activates both *gln A* and the operons of nitrogen catabolism such as *hut*. The net result of these reactions is the production of glutamate and glutamine to overcome the nitrogen starvation.

Under conditions of nitrogen excess the glutamine: α-oxogluterate ratio increases, which promotes adenylylation of the enzyme. This form of the enzyme represses its own synthesis, the fall in the concentration of deadenylylated glutamine synthetase results in decreased transcription of catabolic operons, and further nitrogen production is curtailed (Fig. 8-12*b*).

These are just two examples of the critical role that the multifunctional glutamine synthetase molecule plays in the regulation of the synthesis of the enzymes of catabolic pathways in bacterial cells. In addition, there appears to be a connection between the catabolism of carbon and energy sources and the concentration of glutamine synthetase (37). This interaction remains poorly understood but indicates that the cell has a highly integrated system of regulatory controls that coordinate the expression of the operons that catabolize both nitrogen and carbon sources. Elucidating the mechanisms of these master control systems is a challenging problem for the future.

REFERENCES

1. J. Monod (1947). The phenomenon of enzymatic adaptation. *Growth* **11**, 223–289.
2. A. B. Pardee and L. S. Prestidge (1961). The initial kinetics of enzyme induction. *Biochim. Biophys. Acta* **49**, 77–88.
3. W. Epstein, S. Naono, and F. Gros (1966). Synthesis of enzymes of the lactose operon during diauxic growth of *E. coli*. *Biochem. Biophys. Res. Commun.* **24**, 588–592.
4. R. A. Cooper (1978). Intermediary metabolism of monossacharides by bacteria. *Int. Rev. Biochem.* **16**, 37–73.
5. F. Jacob and J. Monod (1961). Genetic regulatory mechanisms in the synthesis of proteins. *J. Mol. Biol.* **3**, 318–356.
6. B. Magasanik (1976). Classical and post-classical modes of regulation of the synthesis of degradative bacterial enzymes. *Prog. Nuc. Acid. Res. Mol. Biol.* **17**, 99–115.
7. R. F. Goldberger, R. G. Deeley, and K. P. Mullinix (1976). Regulation of gene expression in prokaryotic organisms. *Adv. Genetics* **18**, 1–67.
8. R. F. Goldberger (1979). Strategies of genetic regulation in prokaryotes. In R. F. Goldberger (Ed.). *Biological Regulation and Development*. Plenum Press, New York, pp. 1–18.
9. J. H. Miller and W. S. Reznikoff (Eds.) (1978). *The Operon*. Cold Spring Harbor Laboratory, (New York).
10. M. Rosenberg and D. Court (1979). Regulatory sequences involved in the promotion and termination of RNA transcripts. *Ann. Rev. Genet.* **13**, 319–353.
11. B. Lewin (1974). *Gene Expression,* Vol. I. John Wiley, London, pp. 272–408.
12. G. Carpenter and B. H. Sells (1975). Regulation of the lactose operon in *Escherichia coli* by cAMP. *Int. Rev. Cytol.* **41**, 29–58.
13. S.-S. Hua and A. Markovitz (1972). Multiple regulator gene control of the galactose operon in *Escherichia coli* K-12. *J. Bacteriol.* **110**, 1089–1099.
14. S. Adhya and W. Miller (1979). Modulation of the two promoters of the galactose operon of *Escherichia coli*. *Nature* **279**, 492–494.

REFERENCES

15. R. Dilauro, T. Taniguchi, R. Musso, and B. de Crombrugghe (1979). Unusual location and function of the operator in the *Escherichia coli* galactose operon. *Nature* **279**, 494–500.
16. G. Wilcox, J. Boulter, and N. Lee (1974). Direction of transcription of the regulatory gene *ara C* in *Escherichia coli* B/r. *Proc. Natl. Acad. Sci. USA* **71**, 3635–3639.
17. M. J. Casadaban (1976). Regulation of the regulatory gene for the arabinose pathway *ara C*. *J. Mol. Biol.* **104**, 557–566.
18. M. A. Savageau (1979). Autogenous and classical regulation of gene expression: A general theory and experimental evidence. In R. F. Goldberger (Ed.), *Biological Regulation and Development*. Plenum Press, New York, pp. 57–108.
19. S. Ogden, D. Haggerty, C. M. Storer, D. Kolodrubetz, and R. Schleif (1980). The *Escherichia coli* L-arabinose operon: Binding sites of the regulatory proteins and a mechanism of positive and negative regulation. *Proc. Natl. Acad. Sci. USA* **77**, 3346–3350.
20. D. C. Hagen and B. Magasanik (1973). Isolation of the self-regulated repressor protein of the *Hut* operons of *Salmonella typhimurium*. *Proc. Natl. Acad. Sci. USA* **70**, 808–812.
21. B. Magasanik (1961). Catabolite repression. *Cold Spring Harbor Symp. Quant. Biol.* **26**, 249–256.
22. E. W. Sutherland and T. W. Rall (1957). The properties of an adenine ribonucleotide produced with cellular particles, ATP, Mg^{2+} and epinephrine or glucagon. *J. Amer. Chem. Soc.* **79**, 3608.
23. W. H. Cook, D. Lipkin, and R. Markham (1957). The formation of a cyclic dihydrodiadenylic acid by alkaline degradation of adenosine 5' triphosphoric acid. *J. Am. Chem. Soc.* **79**, 3607–3608.
24. R. S. Makman and E. W. Sutherland (1963). Presence of cyclic 3',5'-adenosine phosphate (CA) in *E. coli. Fed. Proc.* **22**, 470.
25. R. S. Makman and E. W. Sutherland (1965). Adenosine 3',5'-phosphate in *Escherichia coli*. *J. Biol. Chem.* **240**, 1309–1314.
26. R. Perlman and I. Pastan (1968). Cyclic 3'5'-AMP: Stimulation of β-galactosidase and tryptophanase induction in *E. coli*. *Biochem. Biophys. Res. Commun.* **30**, 656–664.
27. A. Ullmann and J. Monod (1968). Cyclic AMP is an antagonist of catabolite repression in *Escherichia coli*. *FEBS lett.* **2**, 57–60.
28. I. Pastan and R. L. Perlman (1968). The role of the *lac* promoter locus in the regulation of β-galactosidase synthesis by cyclic 3',5'-adenosine monophosphate. *Biochemistry* **61**, 1336–1342.
29. H. E. Varmus, R. L. Perlman, and I. Pastan (1970). Regulation of *lac* messenger ribonucleic acid synthesis by cyclic adenosine 3'5'-monophosphate and glucose. *J. Biol. Chem.* **245**, 2259–2267.
30. M. J. Buettner, E. Sptiz, and H. V. Rickenberg (1973). Cyclic adenosine 3',5'-monophosphate in *Escherichia coli*. *J. Bacteriol.* **14**, 1068–1073.
31. J. T. Lis and R. Schleif (1973). Different cyclic AMP requirements for the induction of the arabinose and lactose operons of *Escherichia coli*. *J. Mol. Biol.* **79**, 149–162.
32. W. Epstein, L. B. Rothman-denes, and J. Hesse (1975). Adenosine 3':5'-cyclic monophosphate as a mediator of catabolite repression in *Escherichia coli*. *Proc. Natl. Acad. Sci. USA* **72**, 2300–2304.
33. A. Dessein, M. Schwartz, and A. Ullman (1978). Catabolite repression in *Escherichia coli* mutants lacking cyclic AMP. *Mol. Gen. Genet.* **162**, 83–87.
34. C. Guidi-Rontani, A. Danchin, and A. Ullmann (1980). Catabolite repression in *Escherichia coli* mutants lacking cyclic AMP receptor protein. *Proc. Natl. Acad. Sci. USA* **77**, 5799–5801.
35. B. L. Warner, R. Kodaira, and F. C. Neidhardt (1978). Regulation of *lac* operon expression: reappraisal of the theory of catabolite repression. *J. Bacteriol.* **136**, 947–954.
36. E. Joseph, A. Danchin, and A. Ullmann (1981). Regulation of galactose operon expression: Glucose effects and role of cyclic adenosine 3',5'-monphosphate. *J. Bacteriol.* **146**, 149–154.
37. B. Tyler (1978). Regulation of the assimilation of nitrogen compounds. *Ann. Rev. Biochem.* **47**, 1127–1162.

CHAPTER NINE

Biosynthetic Pathways

Prokaryotes are able to grow on a variety of simple compounds such as histidine or glucose plus ammonia. From these sources of carbon and nitrogen the cells synthesize all their amino acids, nucleotides, polynucleotides, and proteins. For optimal growth, biosynthetic pathways must be tightly coupled to the demand for their end product because any overproduction reduces the overall efficiency of the cell's metabolism. Moreover, when the end product is provided in the medium the cells must repress the enzymes of that pathway. Since as much as 80% of all the carbon and energy consumed by an actively growing cell is used in protein synthesis (1), regulation of the pathways of amino acid biosynthesis is particularly important. This was particularly well-demonstrated by Zamenhof and Eichhorn (2) for two strains of *B. subtilis* growing together in the presence of tryptophan. One strain was able to repress the synthesis of the enzymes for tryptophan biosynthesis, whereas the mutant strain synthesized these enzymes constitutively. In just 26 generations the cells capable of repressing the enzymes outnumbered the other cells by a ratio of $10^5:1$.

Most of the work on the regulation of the expression of the operons coding for the enzymes of biosynthetic pathways has been carried out on operons of amino acid biosynthesis. Some of these operons are controlled with repressors by mechanisms analogous to those regulating catabolic operons. However, some are not controlled by repressors at all, and from studies on these operons a major new regulatory mechanism emerged—attenuation. Attenuation appears to be geared specifically to the regulation of the operons of amino biosynthesis, although in principle it could control other operons. Unfortunately, operons controlling the enzymes of biosynthetic pathways, other than those for amino acids, have not been studied in sufficient detail to establish the mechanisms involved.

PATHWAYS OF AMINO ACID BIOSYNTHESIS

The biosynthesis of the 20 amino acids occurs in five major groups; the glutamate, serine, aspartate, pyruvate, and aromatic families, as shown in Fig. 9-1. The individual reactions of these pathways are summarized by Umbarger (3).

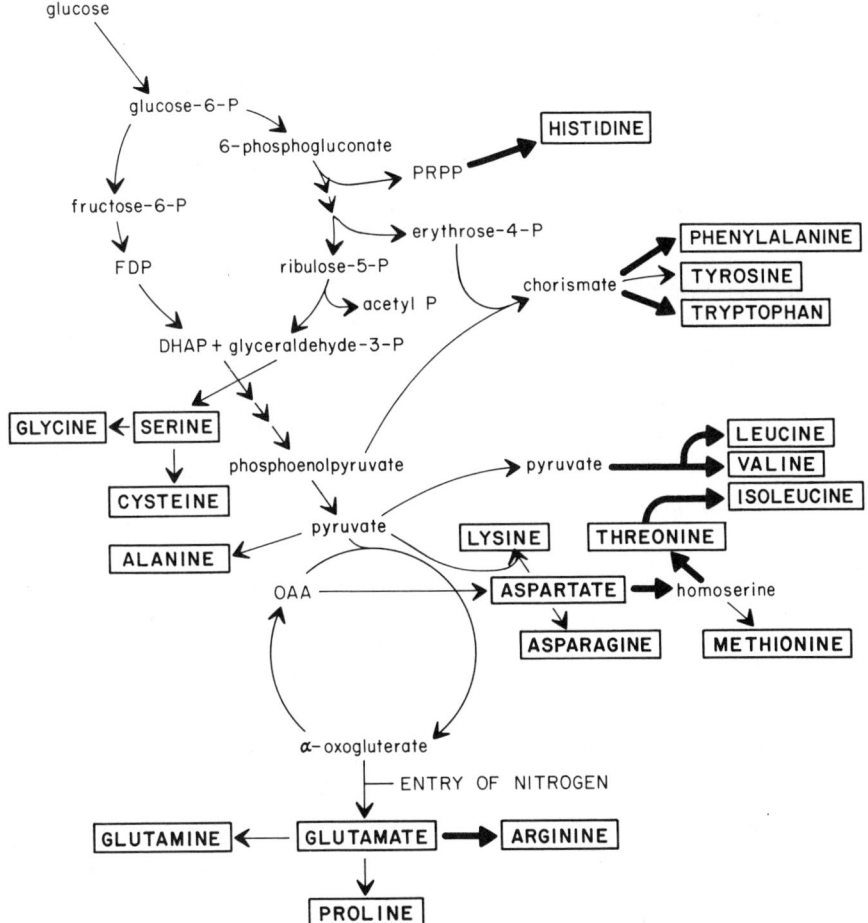

Fig. 9-1 Biosynthesis of the 20 amino acids. The amino acids are synthesized in five major groups from metabolites derived from general metabolic pathways (see also Fig. 8-3).

All of these amino acids can be synthesized from intermediates of the Embden-Meyerhof, pentose phosphate, and tricarboxylic acid cycle pathways. These are the pathways into which catabolic substrates are diverted (cf. Fig. 8-3).

The regulation of the synthesis of the enzymes for eight amino acids has been studied in detail (indicated by the heavier arrows in Fig. 9-1), each of which will now be discussed. In general, progress on the studies of the control of the operons for biosynthetic pathways lagged somewhat behind those of catabolic pahtways because repressor molecules, critical components of the Jacob and Monod model for operon control, could not be isolated (nor evidence of their existence detected genetically). However, the emergence of the concept of attenuation, coupled with advances in nucleic acid sequencing methodology, has led to a more complete understanding of the regulation of the operons of biosynthetic pathways.

OPERONS FOR PATHWAYS OF AMINO ACID BIOSYNTHESIS

Tryptophan Biosynthesis

Tryptophan is synthesized in *E. coli* and other bacterial cells from the glycolytic intermediate phosphoenolpyruvate and erythrose 4-phosphate (generated by the pentose phosphate pathway). Tyrosine and phenylalanine, the other aromatic amino acids, also share some of the same reactions (Fig. 9-2).

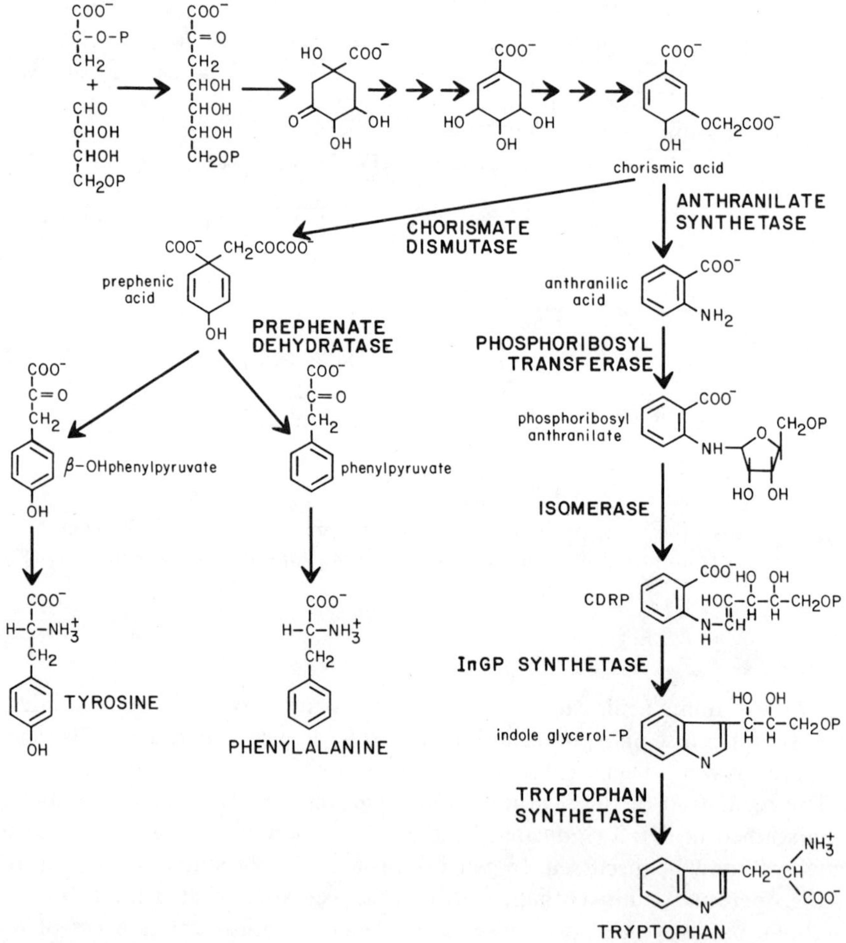

Fig. 9-2 Biosynthesis of the aromatic amino acid family from phosphoenolpyruvate and erythrose 4-phosphate.

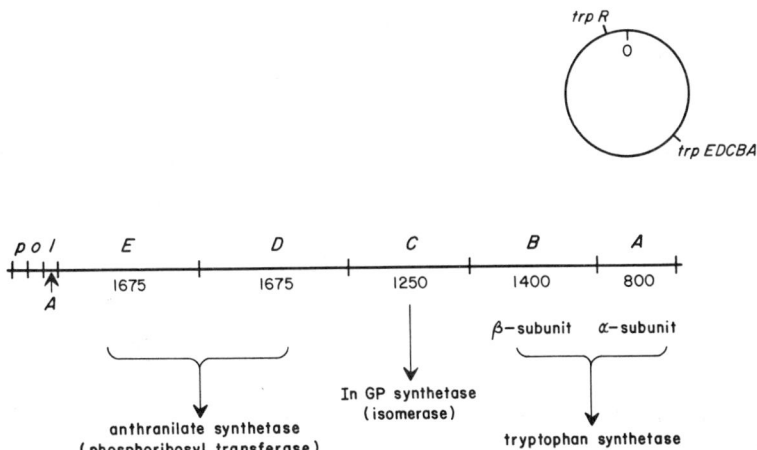

Fig. 9-3 The tryptophan operon of *E. coli*. *p*, *o*, and *l* are the promoter, operator, and leader DNA segments, respectively, and *A* is the site of attenuation. The length of each gene, in base pairs, is indicated.

The five reactions for the conversion of chorismic acid to tryptophan are catalyzed by three enzyme complexes. The genes coding for these three complexes are organized into an operon as shown in Fig. 9-3. [The *trp* operon differs somewhat in different bacterial strains (4) and the discussion here is restricted to that for *E. coli*.] The operon maps at 27' and occupies 7000 base pairs of the *E. coli* genome. It consists of five structural genes together with promoter–operator and leader regulatory regions. Genes *E* and *D* code for the tetrameric ($M_r = 240,000$) enzyme complex of phosphoribosyl transferase–anthranilate synthetase, which catalyzes the first two reactions of tryptophan biosynthesis. Gene *C* codes for the isomerase–InGP synthetase complex ($M_r = 45,000$), and genes *A* and *B* code for the α and β subunits of tryptophan synthetase ($M_r = 79,000$).

The operon is controlled by repression *and* by attenuation. The repressor protein ($M_r = 47,000$) is synthesized constitutively (20 copies per cell) by the *trp R* gene mapping at 100' (Fig. 9-3). Unlike the repressor proteins of the catabolic operons the *trp* repressor can only bind to the operator *in the presence of tryptophan*, the end product of the pathway. Thus the product of the *trp R* gene is more correctly referred to as an aporepressor, since it alone cannot bind to DNA, and tryptophan is the corepressor. The aporepressor cannot bind any amino acid other than tryptophan, and tryptophanyl tRNA^{trp} is also unable to bind.

Control of the operon by the aporepressor–tryptophan complex is simple. When adequate supplies of the amino acid are available tryptophan binds to the aporepressor and the complex binds to the *trp* operator, preventing RNA polymerase from transcribing the gene (Fig. 9-4). When all the tryptophan is consumed, free aporepressor dissociates from the genome, and the operon is derepressed. Thus tryptophan biosynthesis is controlled by a *repressible* operon.

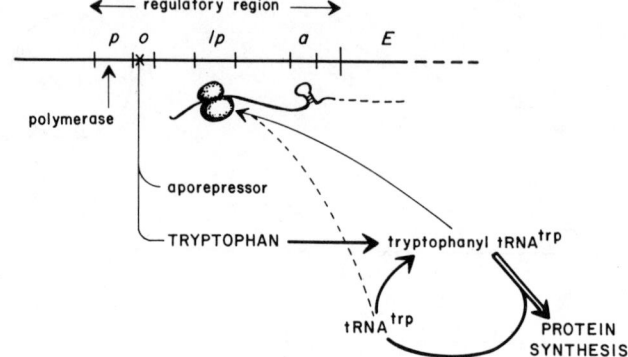

Fig. 9-4 A diagrammatic summary of the regulation of the tryptophan operon showing both the interaction of the ribosome and aminoacyl tRNA with the leader transcript, and the repressor with the operator. *p* and *o* are the promoter and operator, respectively. *lp* and *a* are the sequences of the leader DNA that code for the leader peptide and the attenuator, respectively. *E* is the first structural gene of the operon.

The *trp* operon is also controlled by attenuation, as described in considerable detail in Chapter 2 [see also the review of Yanofsky (5)]. Although repression is the major control element in this operon, attenuation serves to fine-tune the expression of the operon to the metabolic demands of the cell. Moreover, since it is tryptophanyl tRNA^{trp} that controls attenuation and not tryptophan, the expression of the operon can be more precisely attuned to the demand for the amino acid in protein biosynthesis.

The regulatory region of the tryptophan operon is summarized in Fig. 9-4, which shows promoter and operator regions as well as the sites on the leader segment that code for the leader peptide and the attenuator. The expression of the operon is regulated by both the availability of end product in the environment and the demand for the end product in protein synthesis.

The Phenylalanine Operon

Phenylalanine is also a member of the aromatic amino acid family and is derived from chorismic acid (Fig. 9-2). The first two reactions of this pathway are catalyzed by one enzyme, chorismate dismutase-prephenate dehydratase, which is the product of the *phe A* gene mapping at 56' on the *E. coli* genome. This gene is repressible, and the amount of enzyme synthesized is inversely proportional to the concentration of phenylalanine in the cell.

The regulatory region immediately upstream from the gene has been sequenced (Fig. 9-5) and found to contain a classical attenuator arrangement (6). The sequence of the inferred RNA transcript of the leader region is also shown. This sequence is capable of binding a ribosome immediately upstream from a translation start codon (the sequence —UAAGGA— is complementary to the 3' end of the 16S rRNA, Fig. 6-3*b*). A peptide of 15 amino acid residues is

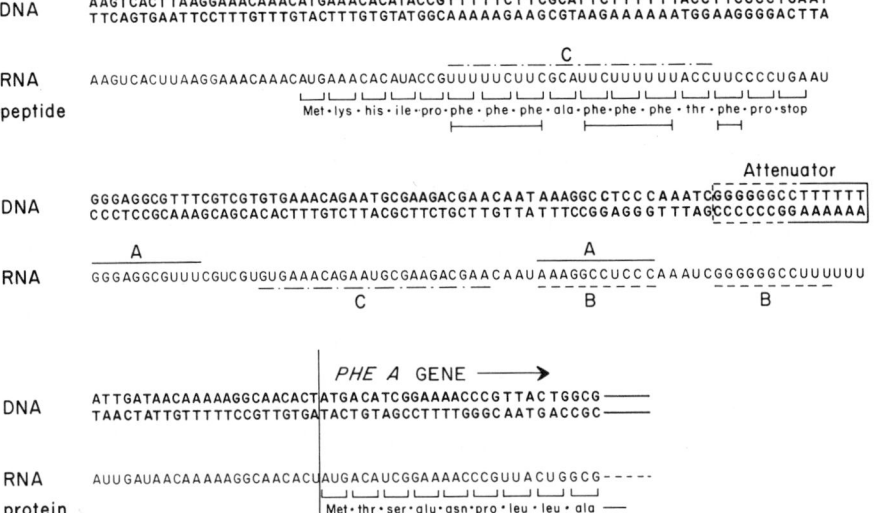

Fig. 9-5 Nucleotide sequence of the 160 base pairs of DNA immediately preceding the start of the *phe A* gene, together with the sequence of an RNA transcript of this sequence and the amino acid sequences of the leader peptide and N-terminal region of the *phe A* gene product. In this and subsequent figures the DNA base-paired sequence commences at the top left with the noncoding (upper) strand running in the 5' → 3' direction. The corresponding nucleotide sequence of the RNA transcript is given underneath the DNA sequence, and the translatable codons are indicated. The RNA transcript sequences A—A, B—B, and C—C are discussed in the text. [Reproduced from Ref. 6 with the permission of Dr. Charles Yanofsky.]

encoded in this segment of the leader, *and 7 of these residues are phenylalanine*, the end product of the biosynthetic pathway.

The leader also has three domains of sequence symmetry; A—A, B—B, and C—C (Fig. 9-5), which correspond very closely to the three domains of the *trp* leader transcript (Fig. 2-11) that are the basis of the attenuation mechanism. Thus the leader transcript is capable of generating terminator, preemptor, and protector base-paired stem-loop structures depending on the behavior of the ribosome translating the leader peptide. There can be no doubt that this ribosome would idle at the *phe·phe·phe* codons in the absence of phenylalanyl tRNA^{phe}.

Regulation of the expression of this gene is directly analogous to that of the *trp* operon by attenuation. Moreover, since no repressor protein has been identified and no operator-constitutive mutants have been isolated, it appears that attenuation is the sole mechanism for the regulation of this gene.

Arginine Biosynthesis

The *arg R* gene codes for an aporepressor protein that controls the expression of the enzymes of arginine biosynthesis. In *E. coli* arginine is synthesized from glutamic acid by a series of reactions involving eight enzymes (Fig. 9-6). Synthesis of all eight enzymes is repressed and derepressed coordinately even though genetic studies have shown that only four of the enzymes are coded by contiguous genes. The other genes are widely distributed on the *E. coli* genome (Fig. 9-6, inset). A total of nine structural genes is involved in coding for the eight enzymes (genes *F* and *I* code for the dissimilar subunits of ornithine transcarbamylase). The product of the *arg R* gene controls the expression of all nine structural genes. The genes are collectively known as the *arg* operons or the *arg regulon* (7), a term sometimes used to describe a number of *coordinately controlled, but widely distributed, structural genes*.

The arginine aporepressor has been isolated and partially purified. It is synthesized constitutively, with about 200 molecules present in each cell. This is some 10–20-fold higher than the concentration of *lac* repressor, presumably because there are seven different operator sites competing for the *arg* aporepressor. Arginine is the corepressor, and control of all the genes in the regulon is by repression. Even though they are controlled coordinately, different amounts of each enzyme are synthesized. This is probably due to intrinsic differences in the efficiencies of the promoters and operators for RNA polymerase and the repressor, respectively.

Recent work (8) has established divergent transcription of the *ECBH* gene cluster. Thus the promoter–operator regions lie between genes *E* and *C*, and transcription of *E* is in the opposite direction to that of *CBH*. At the present

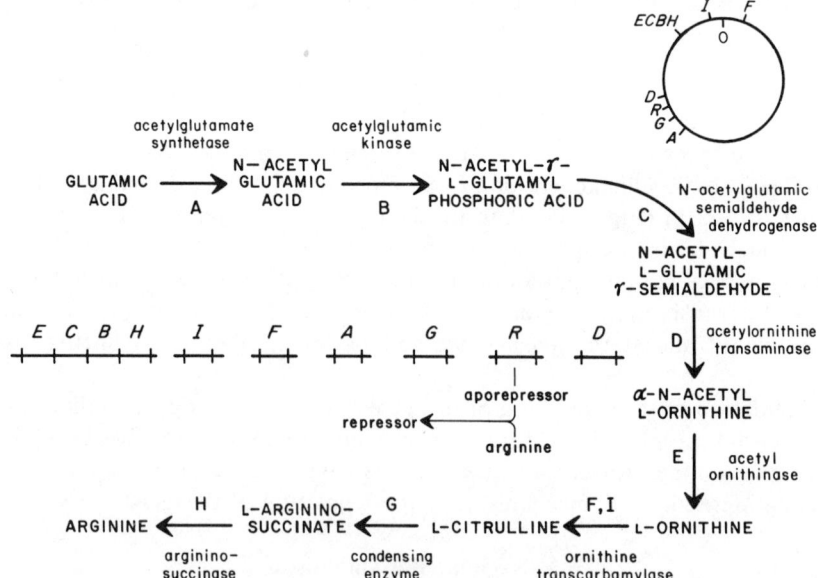

Fig. 9-6 The pathways of arginine biosynthesis from glutamic acid and the organization of the *arg* regulon. The *arg R* gene codes for the aporepressor, and arginine is the corepressor. The map positions of these genes are given in the inset.

time there is no indication that any of these genes are regulated by attenuation, and the *arg* regulon appears to be an example of a biosynthetic pathway regulated solely by repression. The concept of operon control is extended in that it establishes that not all coordinately expressed genes need to be linked together into a single operon.

The Biosynthesis of Threonine

Threonine belongs to the aspartic acid family of amino acids and is synthesized from aspartate by five reactions. The pathway and the organization of the threonine operon are shown in Fig. 9-7 (9). The *thr A* gene is composed of two parts, A_1 and A_2, coding for the bifunctional enzyme aspartokinase-I–homoserine-dehydrogenase I. The *thr B* and *C* genes code for the enzymes responsible for the last two reactions of the pathway. All four genes are contiguous and are controlled by a regulatory region of DNA immediately upstream from the start of the A_1 structural gene.

This region has been studied in detail (10, 11) and found to be another typical example of control by attenuation. There is no evidence for the existence of an aporepressor, and all constitutive mutations map in the attenuator regulatory region. The control region has been sequenced (Fig. 9-8), and transcription is believed to commence slightly upstream from the beginning of this sequence. The leader transcript can code for a peptide of 21 amino acid residues of which 8 are threonine residues. Moreover, these 8 amino acids are clustered within a string of 11 residues. There are also 4 isoleucine codons within this sequence, and this observation offers an explanation for the fact that the threonine operon is repressed in the presence of isoleucine as well as its own end product threonine (isoleucine is synthesized from threonine, Fig. 9-1). The leader transcript also has sequences capable of generating protector, preemptor, and terminator stem-loop structures (A–A, B–B, and C–C in Fig. 9-8).

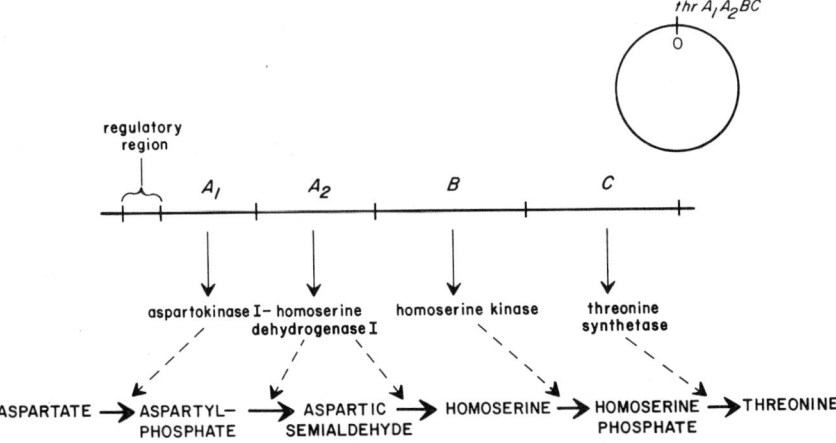

Fig. 9-7 The biosynthesis of threonine from aspartate and the organization of the threonine operon.

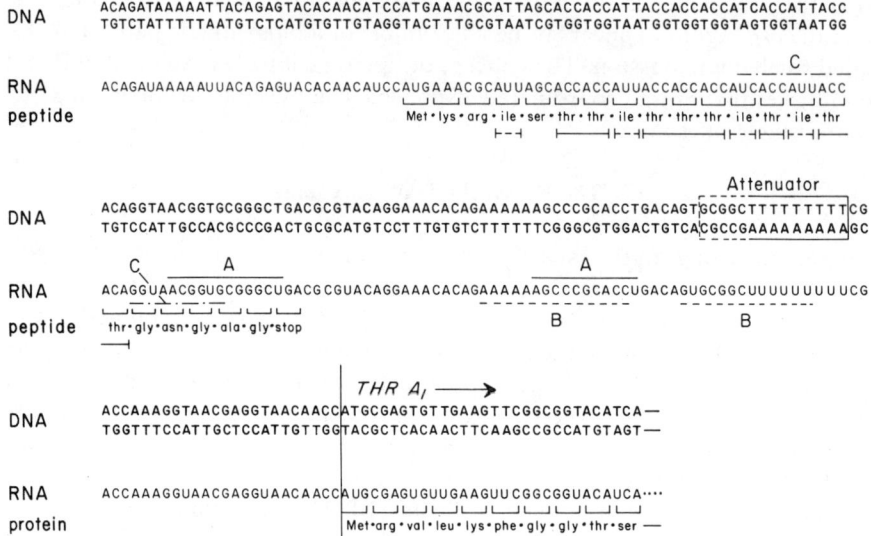

Fig. 9-8 The nucleotide sequence of the regulatory region of the threonine operon, including leader RNA transcript and leader peptide sequences. [Reproduced from Ref. 10 with the permission of Dr. Jeffry Gardner.]

A ribosome will idle during translation of the leader peptide if either threonyl-tRNA^{thr} or isoleucyl-tRNA^{ile} is not present, allowing expression of the operon to make up the shortfall. This is the first example we have encountered in which the expression of an operon is controlled by *two* different amino acid end products. In the past this phenomenon was known as multivalent repression, but now that the mechanism is better understood it should be referred to as *multivalent attenuation*. The mechanism of attenuation is ideally suited to multivalent control because the nucleotide sequence of the leader transcript determines which amino acids are represented by double or multiple codons, and any number of amino acids may be represented. This is particularly well-illustrated in the case of the enzymes of branched-chain amino acid biosynthesis.

Biosynthesis of the Branched-Chain Amino Acids—Leucine, Isoleucine, and Valine

In *E. coli* and several other bacterial species these three amino acids are synthesized from threonine and pyruvate by the reactions shown in Fig. 9-9 (see Refs. 3, 12, and 13 for details). Isoleucine and valine differ by only one methylene group in the side chain, and a parallel series of reactions is used to synthesize them starting with carbon chains of the appropriate length (α-ketobutyrate and pyruvate, respectively). The α-ketoisovalerate derived from pyruvate is either transaminated to valine or converted to leucine.

The enzymes involved in the synthesis of these amino acids are repressible, and their concentrations fluctuate with the availability of leucine, isoleucine, and valine. The enzymes of leucine biosynthesis are coded by four contiguous genes organized into a single operon (Fig. 9-10). Genes *C* and *D* code for

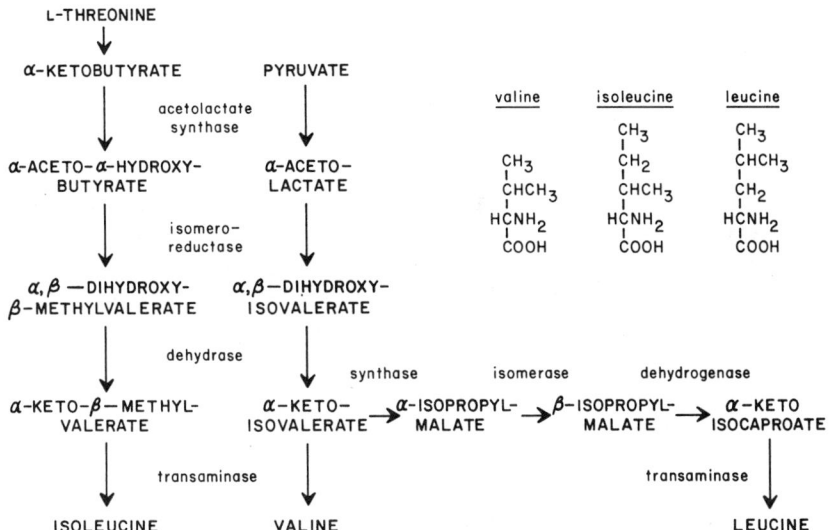

Fig. 9-9 Biosynthesis of the branched-chain amino acids from threonine and pyruvate in *E. coli*.

subunits of the isomerase enzyme, whereas genes *B* and *A* code for the dehydrogenase and synthase, respectively (Figs. 9-9 and 9-10). The final reaction of the pathway is catalyzed by the product of the *ilv E* gene which also transaminates isoleucine and valine. The *leu DCBA* operon is regulated by a control region adjacent to structural gene *A*. The enzymes of isoleucine and valine biosynthesis are also organized into operons, although their precise organization is still not completely understood.

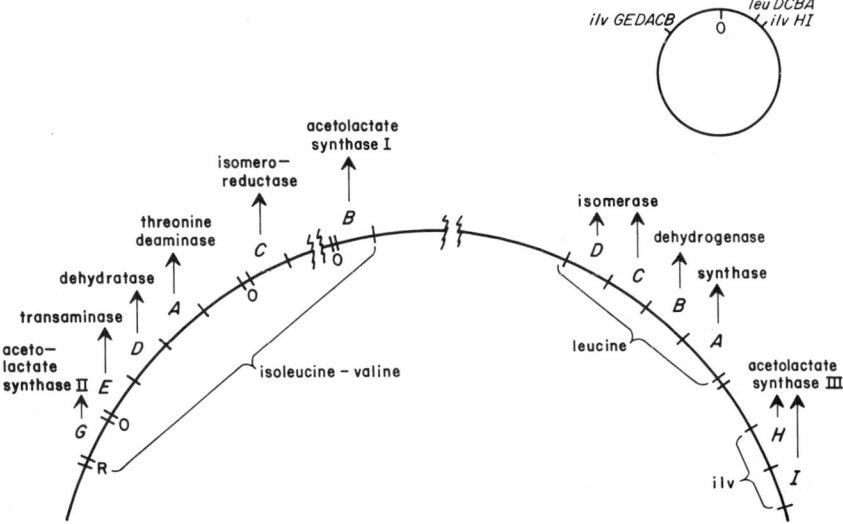

Fig. 9-10 The organization of the *leu* and *ilv* operons of *E. coli*. The gene products catalyze the reactions shown in Fig. 9-9.

192 BIOSYNTHETIC PATHWAYS

A likely arrangement of the *ilv* structural genes in *E. coli* strain K-12, based on the results of several studies (13–18), is given in Fig. 9-10. Acetolactate synthase, the first enzyme of the pathway, exists in three isozymic forms coded by the *ilv G*, *ilv B*, and *ilv H,I* genes. The *ilv G* gene is contiguous with *ilv E, D,* and *A* and forms an operon coding for a polycistronic mRNA. The *C* and *B* genes are *not* connected to *ilv GEDA*, but *are* subject to repression by the end products of the pathway. *ilv H* and *I* are located near the *leu* operon and are also repressible. Moreover, all the structural genes are subject to *multivalent* repression, and their overall expression is the result of a complex series of interactions between all three end products and the genome. Since no repressor molecules have been identified (biochemically or genetically), control appears to be via attenuation. The control regions immediately upstream from the *leu* operon in *S. typhimurium* and the *ilv GEDA* region in *E. coli* have been studied in some detail (17–19).

The nucleotide sequence of the control region of the *leu* operon is shown in Fig. 9-11. A Pribnow-box sequence precedes the putative transcription-start site for the leader transcript. This transcript has the capacity to code for a 28-residue peptide that is somewhat larger than other proposed leader peptides. The presence of four consecutive leucine codons leaves no doubt that a translating ribosome would idle at this point unless leucyl-tRNA^{leu} is present. Three

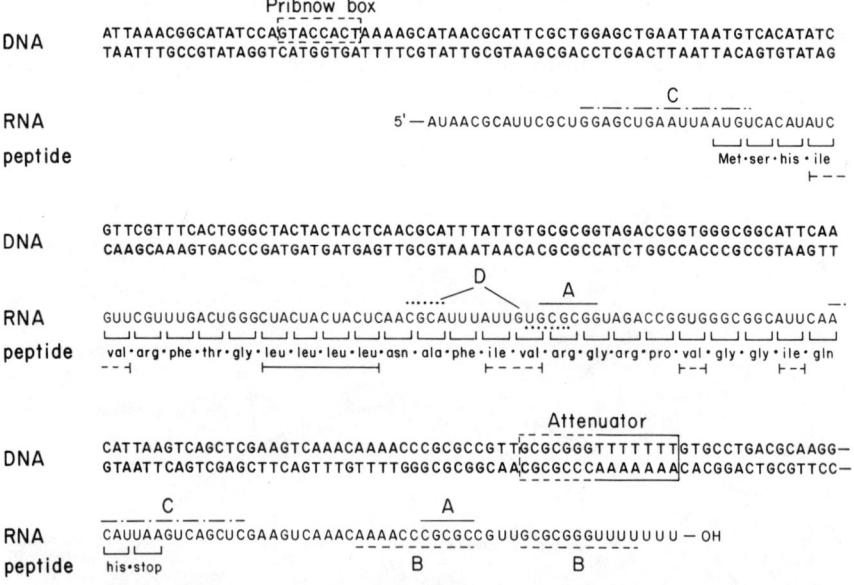

Fig. 9-11 The nucleotide sequence of the regulatory region of the *leucine* operon. The Pribnow box, which is the binding site for RNA polymerase, is shown together with the sequence of the transcript it produces until it reaches the attenuator. The amino acid sequence of a possible leader peptide is also given. [Reproduced from Ref. 19 with the permission of Dr. Robert Gemmill.]

isoleucine and three valine codons are also present, indicating that the availability of these amino acids may affect the expression of the *leu* operon.

The *leu* leader transcript contains nucleotide sequences that are capable of generating preemptor, terminator, and protector secondary structures (19). Moreover, an additional sequence (D–D) is present in a position that could prevent the formation of the preemptor (A–A) and therefore cause attenuation of the operon. Since D–D is formed first, the operon will always be attenuated unless a ribosome is present to disrupt D–D base-pairing, but not A–A base-pairing. This requires very precise alignment of the idling ribosome (since it spans only three codons downstream from the one it idles at) and probably only occurs when it idles at the *leu* codons. Thus a ribosome idling at the first *ile–val* double codon could not derepress the operon. Furthermore, neither could idling at the second *ile–val* double codon cause derepression because it prevents formation of the preemptor. This raises some doubts about the multivalent repression of the *leu* operon.

The sequence of the *ilv GEDA* regulatory region is given in Fig. 9-12. The start of transcription of the leader RNA has been established. Thirty-one nucleotides after the start of transcription lies an AUG codon that could code for a 32-residue polypeptide containing 4 leucine, 6 valine, and 5 isoleucine residues. The leader transcript contains preemptor and terminator sequences (A–A and B–B, respectively) typical of all operons controlled by attenuation. The transcript also contains two sequences, X–X and Y–Y, that either function as protectors or serve to ensure that only a correctly aligned ribosome can derepress the operon. Thus a ribosome must prevent the formation of a Y–Y base-

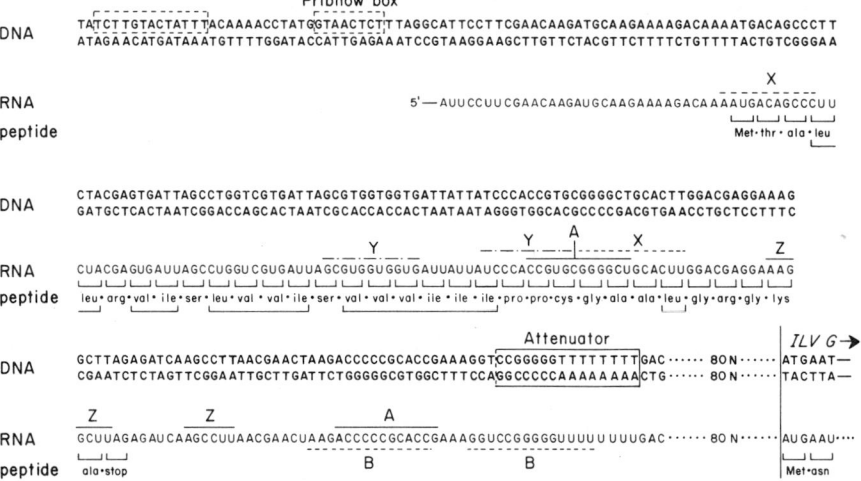

Fig. 9-12 The nucleotide sequence of the regulatory region of the *ilv* operon. The regulation of the expression of the operon by a ribosome translating the coding portion of the leader transcript is described in the text. [Reproduced from Ref. 18 with the permission of Dr. G. W. Hatfield.]

paired stem-loop, but not the preemptor (A–A) in order to derepress the operon, and this can only occur when the ribosome idles at a restricted number of codons as described above. Moreover, the binding of a second ribosome may be required to prevent the formation of an X–X base-paired stem loop.

The extent of expression of this operon is thus likely to depend upon the rates at which one or more ribosomes move along the leader transcript, which is determined by the availability of leucyl, valyl, and isoleucyl tRNAs. Movement of the ribosomes produces dynamic changes in transcript structure, which may result in the formation of the terminator base-paired stem loop. When all amino acyl tRNAs are present the ribosome moves smoothly along the leader peptide until it encounters the base-paired, stem-loop structure Z–Z, which "hides" the stop codon, where it pauses giving sufficient time for the terminator to form and attenuation takes place.

There are many facets of the expression of all the genes of the *ilv* operons that remain unexplained. For example, there is a genetic locus between genes G and E, designated O in Fig. 9-10, that affects the expression of *ilv G*. Thus a mutation at O permits expression of *EDA* but not G, even though transcription is initiated at the *ilv GEDA* regulatory region (*ilv R*) upstream from *ilv G* (20). Also unexplained is the significance of the three isozymic forms of acetolactate synthase. The recent observation (21) that the *ilv B* gene is activated by cAMP is particularly intriguing since it is the first example of a *biosynthetic* enzyme to be induced by cAMP. This is additional evidence for a more general role of cAMP in bacterial metabolism.

Histidine Biosynthesis

The operon coding for the enzymes of histidine biosynthesis is controlled primarily by attenuation in both *E. coli* and *S. typhimurium* (22–26). The nine

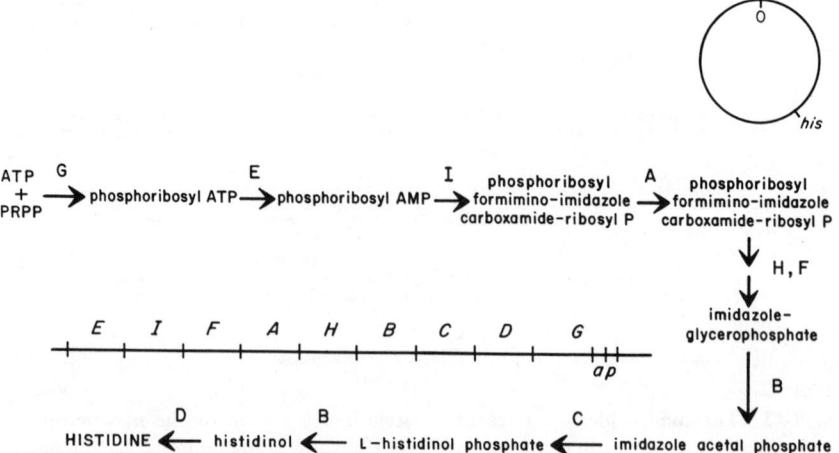

Fig. 9-13 The pathway of histidine biosynthesis and the arrangement of the *his* operon. Note that this operon is transcribed from right to left.

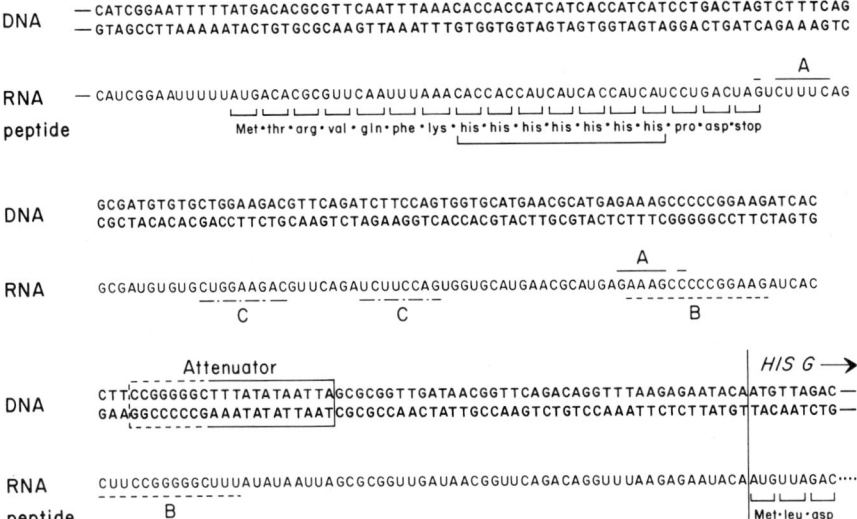

Fig. 9-14 The nucleotide sequence of the regulatory region of the *his* operon, showing typical features of attenuation. Control of the expression of this operon by attenuation is described in the text. [Reproduced from Ref. 23 with the permission of Dr. Wayne Barnes.]

contiguous structural genes are regulated by a single control region adjacent to the *his* G gene (Fig. 9-13). The control region has been sequenced and is presented in Fig. 9-14.

The transcript of the leader DNA sequence could code for a 16-residue peptide that contains *7 consecutive* histidines. All the other sequence and structural features required for attenuation are present in this transcript. Moreover, it has been demonstrated directly (25) that the activity of the operon is controlled by the concentration of histidyl tRNA^{his} as shown in Fig. 9-15. The operon is only fully derepressed when the concentration of aminoacylated tRNA^{his} is very low, and its activity decreases rapidly as the concentration of aminoacylated tRNA^{his} increases. Additional evidence for control by attenua-

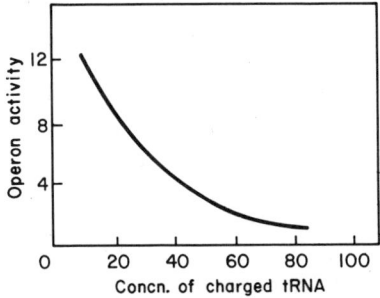

Fig. 9-15 The relationship between the activity of the *his* operon and the concentration of histidyl tRNA^{his}. Redrawn from the data in Ref. 25.

tion is provided by the isolation of a leader RNA transcript attenuated at the predicted site (26). Furthermore, mutants changing the translation-start codon of the leader peptide (AUG) to UAA (a translation-stop codon) seriously affect expression of the operon.

OTHER BIOSYNTHETIC OPERONS

Biotin Biosynthesis

Biotin is one of the few non-amino acids whose biosynthesis has been studied in detail and shown to be under operon control (27–29). The *bio* operon in *E. coli* consists of six structural genes, five of which are known to participate in the conversion of glucose to biotin by way of pimelic acid (Fig. 9-16). The function of the product of the sixth gene (*bio G*) is not known.

Genetic analysis of the operon established that transcription commenced *between* structural genes *A* and *B* (28). This site has now been sequenced (29), as shown in Fig. 9-16a. There are 86 base pairs between the AUG codons of the two genes, and transcription of *bio A* in one direction and *bio BEFGCD* in the opposite direction starts within this region. The transcription-start sites are actually 10 base pairs apart so the transcripts do not overlap, but the promoters do. Indeed, the Pribnow box sequences are only 2 base pairs apart. RNA polymerase binds to this region, and approximately half the molecules transcribe in one direction and half in the opposite direction, producing equal numbers of copies of *bio A* and *bio BEFGCD*.

A palindromic sequence overlaps the promoter sites, and this is probably the binding site for the *bio* repressor. The repressor has not yet been isolated, but the observation that mutants causing constitutive expression of the operon map in this area supports the idea that the operon is regulated by a repressor.

A GENERAL MODEL FOR THE MECHANISM OF ATTENUATION

Role of the DNA Sequence

Operons controlled by attenuation are preceded by a leader DNA sequence 150–300 base pairs in length (Fig. 9-17). This leader DNA starts with a promoter and proceeds to a terminator or attenuator sequence located 20–80 base pairs upstream from the start of the first structural gene. The expression of the operon depends upon, and is regulated by, the interaction of RNA polymerase with these two important leader DNA sequences, the promoter and the attenuator.

In five of the seven operons of amino acid biosynthesis described in this Chapter the binding of RNA polymerase to the promoter is not regulated. Therefore, enzyme molecules will bind frequently to these operons with a frequency determined solely by the affinity between polymerase and the promoter

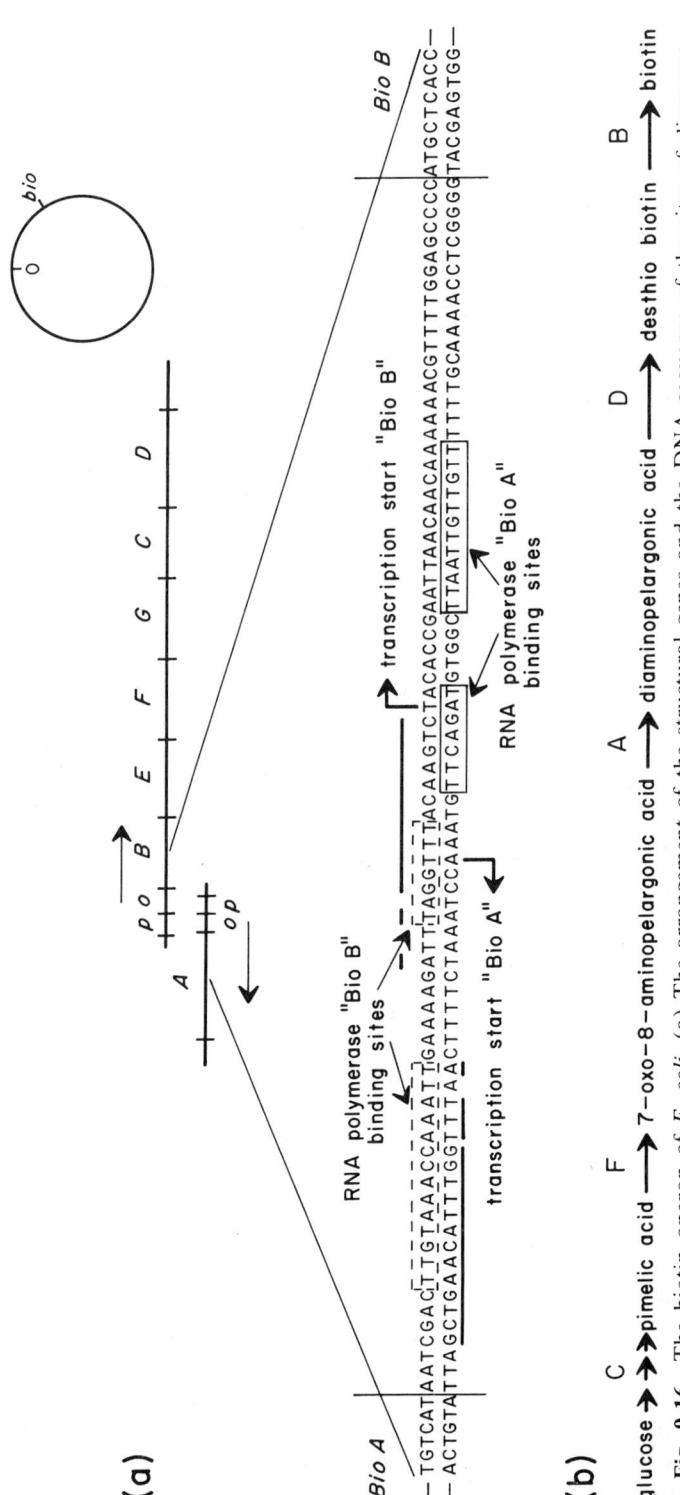

Fig. 9-16 The biotin operon of *E. coli*. (a) The arrangement of the structural genes and the DNA sequence of the sites of divergent transcription between the *A* and *B* genes is shown. *bio BEFGCD* is transcribed on the "upper" strand from left to right, and *bio A* is transcribed on the "lower" strand from right to left. The RNA polymerase binding sites for each gene are shown. The underlined sequences are palindromes and may represent the site of repressor binding. (b) The pathway of biotin biosynthesis.

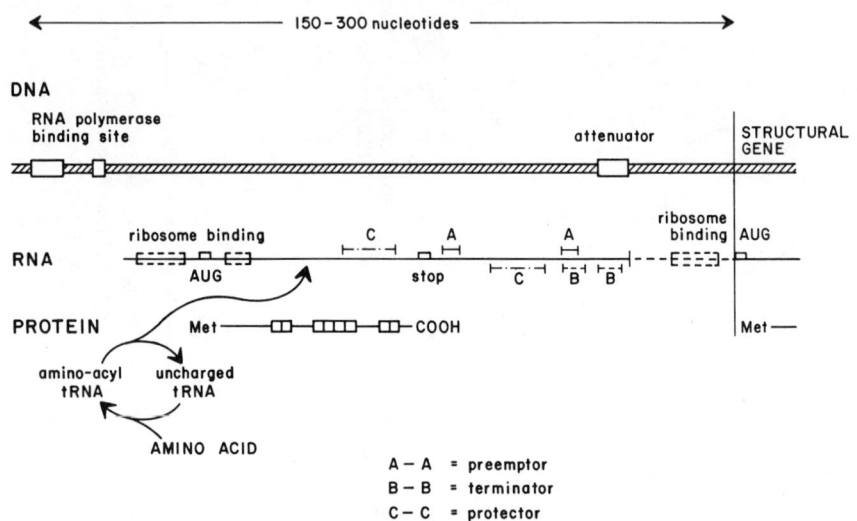

Fig. 9-17 A general model for attenuation showing the important characteristics of the DNA regulatory region, the RNA transcript, and the leader peptide. The roles of each of these components is described in the text.

sequence. In two of the operons (*trp, arg*) a repressor molecule regulates the frequency at which transcription may be initiated. After the initiation of transcription in those operons controlled by attenuation, the RNA polymerase molecule either stops transcription at the attenuator or reads through this sequence and goes on to transcribe the structural genes of the operon. Thus DNA sequences control the locations at which polymerase binds to, or dissociates from, the genome, but RNA molecules modulate the extent of gene expression.

Role of the RNA Transcript of the Leader DNA

RNA polymerase transcribes the leader DNA and produces an RNA transcript with several interesting sequence domains, including two highly conserved sequences. The first sequence is 10–20 nucleotides from the 5′ end; it functions as a ribosome attachment sequence and is followed by an AUG codon. This arrangement is similar to that of most mRNA molecules (Fig. 9-18, cf. Table 6-3). However, unlike most mRNAs the sequence of leader RNAs *past* the AUG codon is also highly conserved and is complementary to 16S rRNA upstream from that normally considered important for mRNA–rRNA interactions. This extensive complementarity permits leader RNA transcripts to compete effectively for the ribosomes that are an absolute requirement for operon expression.

Approximately 100–130 nucleotides past the AUG codon is a second highly conserved sequence (Fig. 9-18) that is also critical for expression of operons regulated by attenuation. This sequence is involved in the formation of the base-paired, stem-loop terminator structure that causes RNA polymerase to stop transcription at the attenuator and dissociate from the genome (Fig. 9-17).

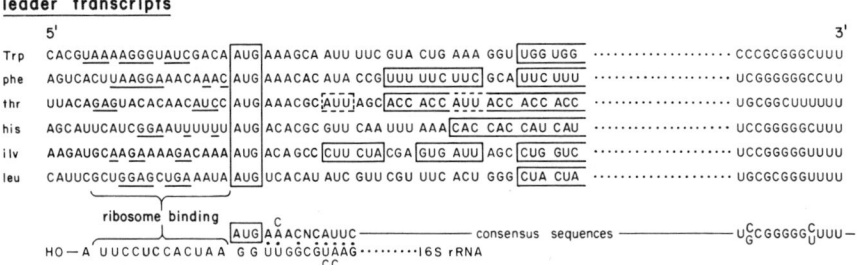

Fig. 9-18 The nucleotide sequences of the 5′ (on the left) and 3′ (on the right) extremes of the leader RNA transcripts of a number of operons. The 5′ sequences are aligned by their AUG codons. A ribosome binding site (complementary to the 3′ end of 16S rRNA (underlined) immediately precedes the AUG codon. Moreover, the transcript sequences immediately past the AUG codons are highly conserved and also complementary to 16S rRNA (the base-pairing of a consensus sequence is shown). The 3′ end of all these transcripts is also highly conserved, and a consensus sequence is given. This sequence participates in the formation of the terminator.

The sequence of the 100–130 nucleotides in-between AUG and the terminator carries sequence information that determines the extent of expression of the operon. There are two important domains. First, commencing with AUG the leader RNA sequence could be translated into a peptide 14–32 residues in length, each of which has two or more adjacent codons for the amino acid end product of the pathway. Second, there are two or more sequences (C–C, A–A in Fig. 9-17) that determine the formation of base-paired stem-loop structures that either stabilize or destabilize the terminator (B–B in Fig. 9-17). The formation of these leader RNA transcript secondary structures is determined by the presence or absence of a ribosome on the coding sequence.

Role of the Ribosome

A ribosome translates the coding portion of the leader transcript in approximately 1 s (0.7–1.6 s, depending on length). During this time RNA polymerase, moving ahead of the ribosome, transcribes C–C, A–A, and then B–B (or the equivalents of the *ilv* operon). Under optimal conditions (i.e., no amino acyl tRNAs limiting) the ribosome moves sufficiently quickly to block, by steric hindrance, the formation of C–C and A–A (protector and preemptor) secondary structures, but cannot prevent the formation of the terminator B–B and the operon is attenuated.

If the concentration of the amino acyl tRNA for the amino acid represented by multiple codons is not optimal the ribosome will slow down, or even pause, during translation. If this occurs it may not reach the portion of the transcript coding for the preemptor (A–A) in time to prevent it from taking up its secondary structure. Formation of the preemptor stops formation of the terminator, polymerase is not destabilized as it passes the attenuator, and it goes on to transcribe the structural genes of the operon.

It is the *rate of movement* of the ribosome along the coding portion of the leader sequence that determines the extent of expression of the operon. The rate of movement is, in turn, determined by the concentration of another RNA molecule, amino acyl tRNA.

Role of the Amino Acyl tRNA Molecule

The amino acyl tRNA pools of bacterial cells turn over with half-lives of fractions of a second and need to be replenished constantly with a fresh supply of amino acids. Because the pools turn over so quickly, any changes in rates of protein synthesis or amino acid biosynthesis produce immediate changes in the concentration of the amino acyl tRNAs, making them ideal candidates for a regulatory role. Since it is the concentration of the amino acyl tRNA that determines the rate of translation of a messenger RNA sequence by a ribosome, the expression of operons controlled by attenuation will be linked directly to the demand for their end product in protein synthesis.

Evidence for the critical role that amino acyl tRNAs play in the mechanism of attenuation has come from studies on mutants affecting the synthesis of the amino acyl tRNA synthetases (30–33). A defect in the production of these enzymes leads to the derepression of those operons of amino acid biosynthesis regulated by attenuation. One such mutant, designated *his T* because it derepressed the histidine operon, attracted more attention when it was found that *leu* and *ilv* were also derepressed in these cells. Moreover, the concentrations of amino acyl tRNAs were normal, indicating that derepression was not caused by a lowered concentration of these molecules. A more thorough genetic analysis revealed that the *his T* locus coded for the enzyme responsible for the modification of the uridine residues in the anticodon loop of the tRNA molecules (Fig. 6-2). The enzyme catalyzes the conversion of uridines into pseudouridines, and only tRNAs modified by this enzyme can participate in the attenuation reaction. The reason for this is unclear since unmodified amino acyl tRNAs can function adequately in the elongation reaction of protein synthesis. This implies that the translation of the leader peptide involves the participation of amino acyl tRNAs in a manner different from their role in the translation of mRNA molecules.

Attenuation and the Regulation of Other Operons

So far only the operons coding for the enzymes of amino acid biosynthesis have been shown to be controlled by attenuation, but there is no reason why this mechanism could not operate, in a modified form, to control the expression of other operons. Such a mechanism would require a leader transcript capable of generating a terminator and a DNA attenuation sequence. Ribosomes and aminoacyl tRNAs would not be necessary. Instead, specific regulatory proteins could bind to the leader RNA transcript and either stabilize or destabilize the terminator. The protein (or even another RNA molecule) could either interact directly with the terminator, as the protein rho appears to do, or it could interact indirectly via preemptor and other secondary structure mechanisms. Indeed, it is conceivable that several proteins could interact (together with their effector metabolites) with the RNA transcript in a manner that integrates the activities of several biochemical pathways.

REGULATION OF THE SYNTHESIS OF THE ENZYMES OF CATABOLISM AND BIOSYNTHESIS: A SUMMARY

Histidine is the one molecule whose biosynthesis and catabolism have been studied in great detail, and it will be used to summarize the regulation of enzyme synthesis in prokaryotes. The components of the regulatory systems for histidine homeostasis are shown in Fig. 9-19. Histidine has three functions in the cell: to provide nitrogen, to provide energy, and to produce proteins. The histidine may be obtained extracellularly, but when unavailable it can be synthesized *de novo*.

When histidine is available in the environment it is taken into the cell, and histidyl tRNA is charged to an extent optimum for protein synthesis. This leads to attenuation of the operon of histidine biosynthesis (*his*). If histidine is present to excess the concentration of uroconate increases to the point where it combines with the *hut* repressor, leading to derepression of the operon for histidine catabolism. The excess histidine is catabolized to produce carbon,

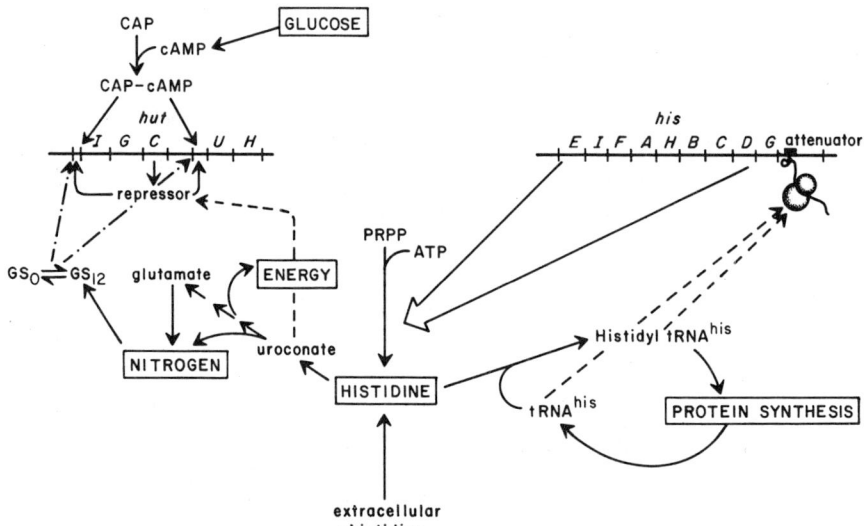

Fig. 9-19 Regulation of the synthesis of the enzymes of histidine metabolism. The roles of the various factors involved in the regulation of the operons of histidine catabolism (*hut*) and biosynthesis (*his*) are described in the text.

energy, and nitrogen provided certain other criteria with regard to the energy status and nitrogen balance are fulfilled.

If glucose, or another more favorable energy source, is being catabolized, *hut* will not be expressed because of a fall in the concentration of CAP–cAMP. However, if there is a demand for nitrogen, the *hut* operon will escape from catabolite repression by virtue of the interaction of unadenylylated glutamine synthetase (GS_0) with the operon (Fig. 9-19).

When the histidine from the environment is exhausted the concentration of intracellular histidine, and consequently the concentrations of uroconate and histidyl tRNA^{his}, falls. This leaves a large concentration of free repressor (it is synthesized autogeneously), leading to immediate repression of *hut* to prevent any further catabolism of the amino acid. Furthermore, the fall in the concentration of histidyl tRNA^{his} stalls the ribosome translating the *his* leader RNA transcript, and the operon of histidine biosynthesis is expressed to produce the enzymes for the *de novo* production of histidine. The rate of *de novo* synthesis is synchronized to the demand for histidyl tRNA^{his} in protein synthesis by the attenuation mechanism.

The two operons are controlled independently, and, surprisingly, neither one of them monitors the concentration of free histidine directly. *The activities of the operons are related to the demand for products of histidine metabolism rather than histidine itself.* Thus histidine is catabolized only if there is a demand for carbon and energy (positive activation of *hut* by CAP–cAMP) or nitrogen (positive activation of *hut* by GS_0). Histidine is only synthesized by *de novo* reactions when the concentration of histidyl tRNA^{his} falls (deattenuation of *his*). The biosynthetic pathway produces histidine for protein synthesis *only* and is not subject to any other controls. Presumably, the concentration of histidine required to optimally charge histidyl tRNA^{his} is much lower than the k_m of histidase for histidine. Moreover, since uroconate, rather than histidine itself, is the inducer, the catabolic operon is not derepressed until there has been a substantial increase in the concentration of histidine.

The relative concentrations of four small molecules—uroconate, cAMP, NH_4^+, and histidyl tRNA^{his}—determine which of the two operons will be expressed and the rate at which the active one is transcribed. The information concerning the concentration of these four compounds is transmitted to the genome by several mechanisms varying considerably in complexity. The concentrations of cAMP and uroconate are monitored by CAP and repressor, respectively, by a relatively simple conformational change that alters their DNA-binding characteristics. The concentration of ammonia is monitored in a somewhat more complex manner involving covalent modifications of the glutamine synthetase cascade, resulting in a change in the DNA-binding ability of glutamine synthetase subunits. Finally, the entire translational apparatus is involved in monitoring the concentration of histidyl tRNA^{his}, leading to a change in the DNA-binding ability of the RNA polymerase molecule transcribing the *his* operon.

REFERENCES

1. O. Maaløe (1979). Regulation of the protein-synthesizing machinery—ribosomes, tRNA, factors and so on. In R. F. Goldberger (Ed.). *Biological Regulation and Development*. Plenum Press, New York, pp. 487–542.
2. S. Zamenhof and H. H. Eichhorn (1967). Study of microbiol evolution through loss of biosynthetic functions: Establishment of 'defective' mutants. *Nature* **216**, 456–458.
3. H. E. Umbarger (1978). Amino acid biosynthesis and its regulation. *Ann. Rev. Biochem.* **47**, 533–606.
4. P. H. Clarke (1979). Regulation of enzyme synthesis in the bacteria: A comparative and evolutionary study. In R. F. Goldberger (Ed.). *Biological Regulation and Development*, Vol. 1. Plenum Press, New York, pp. 109–170.
5. C. Yanofsky (1981). Attenuation in the control of expression of bacterial operons. *Nature* **289**, 751–759.
6. G. Zurawski, K. Brown, D. Killingly, and C. Yanofsky (1978). Nucleotide sequence of the leader region of the phenylalanine operon of *Escherichia coli*. *Proc. Natl. Acad. Sci. USA* **75**, 4271–4275.
7. R. F. Goldberger (1979). Strategies of genetic regulation in prokaryotes. In R. F. Goldberger (Ed.). *Biological Regulation and Development*, Vol. 1. Plenum Press, New York, pp. 1–18.
8. P. H. Pouwels, R. Cunin, and N. Glansdorff (1974). Divergent transcription in the *arg ECBH* cluster of genes in *Escherichia coli* K12. *J. Mol. Biol.* **83**, 421–424.
9. J. Thèze and I. Saint-Girons (1974). Threonine locus of *Escherichia coli* K-12: genetic structure and evidence for an operon. *J. Bacteriol.* **118**, 990–998.
10. J. F. Gardner (1979). Regulation of the threonine operon: tandem threonine and isoleucine codons in the control region and translational control of transcription termination. *Proc. Natl. Acad. Sci. USA* **76**, 1706–1710.
11. M. Katinka, P. Cossort, L. Sibilli, I. Saint-Girons, M. A. Chalvignac, G. Le Bras, G. H. Cohen, and M. Yaniv (1980). Nucleotide sequence of the *thr A* gene of *Escherichia coli*. *Proc. Natl. Acad. Sci. USA* **77**, 5730–5733.
12. M. Iaccarino, J. Guardiola, M. De Felice, and R. Favre (1978). Regulation of isoleucine and valine biosynthesis. *Curr. Topics Cell Reg.* **14**, 29–73.
13. J. M. Calvo, M. Freundlich, and H. E. Umbarger (1969). Regulation of branched-chain amino acid biosynthesis in *Salmonella typhimurium*: isolation of regulatory mutants. *J. Bacteriol.* **97**, 1272–1282.
14. J. M. Smith, D. E. Smolin, and H. E. Umbarger (1976). Polarity and the regulation of the *ilv* gene cluster in *Escherichia coli* strain K-12. *Molec. Gen. Genet.* **148**, 111–124.
15. M. Baez, D. W. Patin, and D. H. Calhoun (1979). Deletion mapping of the *ilv GOEDAC* genes of *Escherichia coli* K-12. *Molec. Gen. Genet.* **169**, 289–297.
16. J. M. Smith, F. J. Smith, and H. E. Umbarger (1979). Mutations affecting the formation of acetohydroxy acid synthase II in *Escherichia coli K-12*. *Molec. Gen. Genet.* **169**, 299–314.
17. F. E. Narguag, C. S. Subrahmanyam, and H. E. Umbarger (1980). Nucleotide sequence of *ilv GEDA* operon attenuator region of *Escherichia coli*. *Proc. Natl. Acad. Sci. USA* **77**, 1823–1827.
18. R. P. Lawther and G. W. Hatfield (1980). Multivalent translational control of transcription termination at attenuator of *ilv GEDA* operon of *Escherichia coli K-12*. *Proc. Natl. Acad. Sci. USA* **77**, 1862–1866.
19. R. M. Gemmill, S. R. Wessler, E. B. Keller, and J. M. Calvo (1979). *leu* operon of *Salmonella typhimurium* is controlled by an attenuation mechanism *Proc. Natl. Acad. Sci. USA* **76**, 4941–4945.

20 D. J. Gayda, T. D. Leathers, J. D. Noti, F. J. Smith, J. M. Smith, C. S. Subrahmanyam, and H. E. Umbarger (1980). Location of the multivalent control site for the *ilv EDA* operon of *Escherichia coli*. *J. Bacteriol.* **142**, 556–567.

21 A. Sutton and M. Freundlich (1980). Regulation by cyclic AMP of the *ilv B*-encoded biosynthetic acetohydroxy acid synthase in *Escherichia coli* K-12. *Molec. Gen. Genet.* **178**, 179–183.

22 S. W. Artz and J. R. Broach (1975). Histidine regulation of *Salmonella typhimurium*: an activator–attenuator model of gene regulation. *Proc. Natl. Acad. Sci. USA* **72**, 3453–3457.

23 W. M. Barnes (1978). DNA sequence from the histidine operon control region: seven histidine codons in a row. *Proc. Natl. Acad. Sci. USA* **75**, 4281–4285.

24 P. P. Di Nocera, F. Blasi, R. Di Lauro, R. Frunzio, and C. B. Bruni (1978). Nucleotide sequence of the attenuator region of the histidine operon of *Escherichia coli* K-12. *Proc. Natl. Acad. Sci. USA* **75**, 4276–4280.

25 J. A. Lewis and B. N. Ames (1972). Histidine regulation in *Salmonella typhimurium*. *J. Mol. Biol.* **66**, 131–142.

26 H. M. Johnston, W. M. Barnes, F. G. Chumley, L. Bossi, and J. R. Roth (1980). Model for regulation of the histidine operon of *Salmonella*. *Proc. Natl. Acad. Sci. USA* **77**, 508–512.

27 B. Rolfe and M. A. Eisenberg (1968). Genetic and biochemical analysis of the biotin loci of *Escherichia coli* K-12. *J. Bacteriol.* **96**, 515–524.

28 A. Guha, Y. Saturen, and W. Szybalski (1971). Divergent orientation of transcription from the biotin locus of *Escherichia coli*. *J. Mol. Biol.* **56**, 53–62.

29 A. Otsuka and J. Abelson (1978). The regulatory region of the biotin operon in *Escherichia coli*. *Nature* **276**, 689–694.

30 C. E. Singer, G. R. Smith, R. Cortese, and B. N. Ames (1972). Mutant tRNA^{his} ineffective in repression and lacking two pseudouridine modifications. *Nature New Biol.* **238**, 72–74.

31 R. Cortese, R. Landsberg, R. A. V. Haar, H. E. Umbarger, and B. N. Ames (1974). Pleiotropy of *his T* mutants blocked in pseudouridine synthesis in tRNA: Leucine and isoleucine–valine operons. *Proc. Natl. Acad. Sci. USA* **71**, 1857–1861.

32 J. E. Brenchley and L. S. Williams (1975). Transfer RNA involvement in the regulation of enzyme synthesis. *Ann. Rev. Microbiol.* **29**, 251–274.

33 R. Cortese (1979). The role of tRNA in Regulation. In R. F. Goldberger (Ed.). *Biological Regulation and Development*, Vol. 1. Plenum Press, New York, pp. 401–443.

SECTION FOUR

Regulation of Enzyme Synthesis in Animal Cells

Animals, like bacteria, have to contend with continual changes in their environment. Bacterial cells are able to survive because they have evolved the capacity to adapt rapidly to these environmental changes. They can proliferate rapidly under favorable conditions, but are also able to survive under adverse conditions. Theirs is a precarious existence, however, since they continuously strive to deplete their environment of all nutrients. Animals have evolved more effective mechanisms to deal with changes in the environment. By having a number of different tissues, each of which is specialized to perform a specific function, the organism can minimize the number of cells exposed to any particular stress. In this section we are concerned with (i) the mechanisms of adaptation in the cells of tissues that do respond to changes in the environment and (ii) the regulation and coordination of these mechanisms so that they contribute to the survival of the whole organism. Consequently, the section is divided into two parts. Part 1 describes the mechanisms of communication between the cells of various tissues. Chapter 10 describes the hormones, and Chapter 11 deals with the components of the intracellular signal transduction systems that convert the information content of hormones into changes in gene expression. In Part 2, Chapter 12, the regulation of the synthesis of adaptable enzymes is discussed, and in Chapter 13 these changes are related to the major environmental stimulus, the light–dark cycle, with a discussion of diurnal rhythms.

PART ONE

Hormones, Second Messengers, and Information Transfer

CHAPTER TEN

Hormones—The Inducers of Enzyme Synthesis

In the higher mammals hormones coordinate the metabolism of diverse tissues so that they function for the benefit of the whole organism. The hormones act as messengers, being secreted (usually into the general circulation) by the cells of one tissue and interacting with specific target tissues. To date, about 50 hormones secreted from 20 different tissues have been identified. These hormones do not act independently of each other, and their secretion is a highly coordinated activity. Indeed, the sole function of some hormones is the control of the secretion of other hormones, and there is a complex hierarchy of hormone interrelationships as illustrated in Fig. 10-1.

The pituitary is responsible for the secretion of several hormones and plays a central role in the coordination of the humoral system. Seven of these hormones are secreted by the cells of the anterior pituitary and two by the cells of the posterior pituitary. All of the hormones secreted by the pituitary are polypeptides, and their release is controlled principally by a number of small peptide hormones released by the hypothalamus (Fig. 10-1). The hypothalamus is at the highest level of the hormone hierarchy. It consists of a number of neurosecretory cells located in the higher centers of the brain that translate changes in neural activity into a change in hormone secretion. When appropriately stimulated the cells secrete minute quantities of *releasing* or *release-inhibiting* hormones into a portal blood vessel that leads directly to the cells of the anterior pituitary (Fig. 10-2). The pituitary itself is also richly innervated, and the release of some hormones is stimulated directly by these nerves. Furthermore, the pituitary is bathed in blood from the general circulation, and some cells are sensitive to specific blood metabolites and hormones. Hence, the secretion of any of the anterior pituitary hormones is the result of a number of different signals.

The target tissues for all of the hormones released by the anterior pituitary (except melanocyte-stimulating hormone, MSH) are specific endocrine tissues each of which, when stimulated, releases a specific nonpolypeptide hormone (Fig. 10-1). Thus the pituitary hormones are collectively referred to as trophic hormones and they are part of the "chain of command" that regulates the

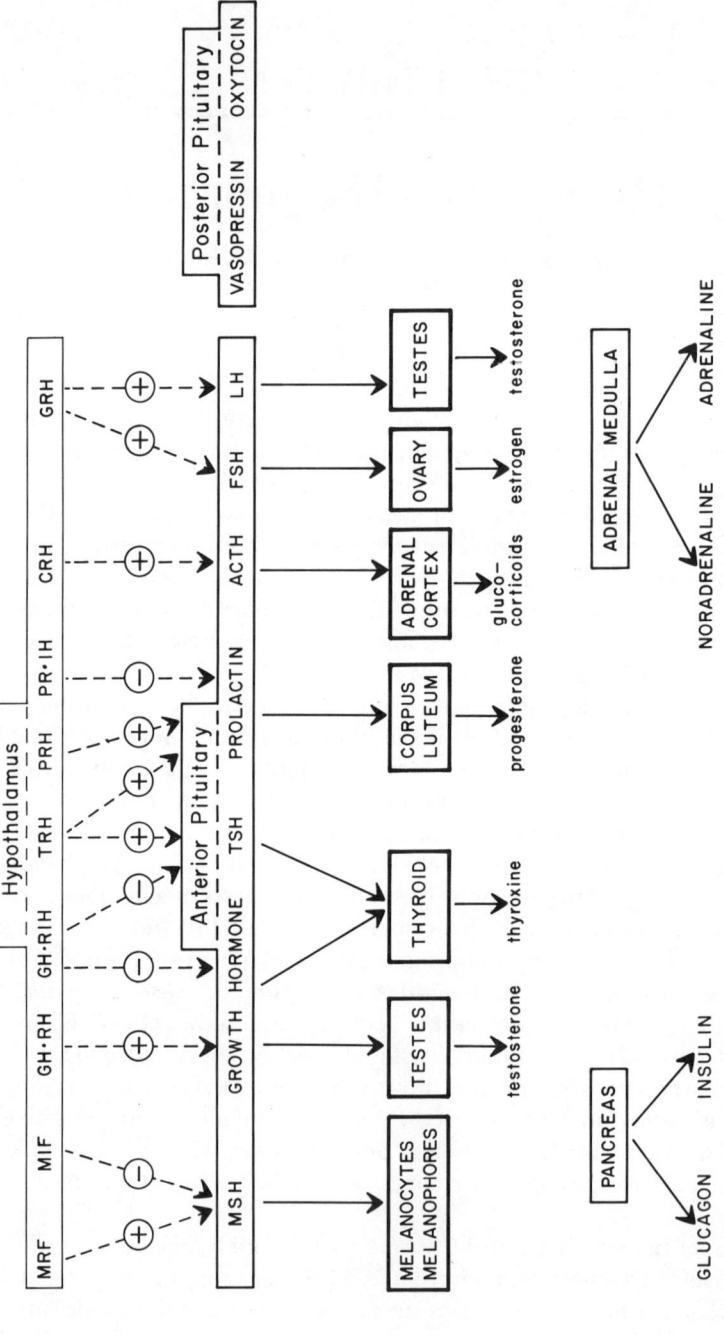

Fig. 10-1 The role of the hypothalamus and the pituitary in the regulation of hormone interrelationships. The hormones secreted by the pancreas and adrenal medulla are also shown. ⊕—release is stimulated, ⊖—release is inhibited.

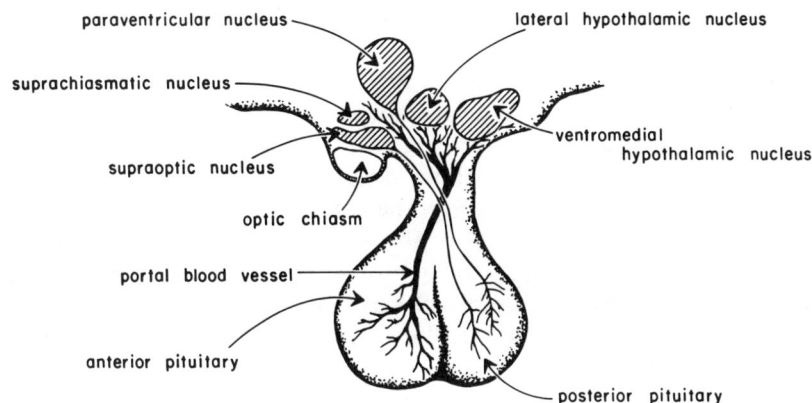

Fig. 10-2 The nuclei of the hypothalamus that are now known to have some influence on the regulation of enzyme synthesis. The venous and neural connections between the hypothalamus and the pituitary are also shown.

secretion of the more physiologically active "metabolic" hormones. Only two of the pituitary trophic hormones, growth hormone and prolactin, have more generalized effects as a result of their interaction with a number of other non-endocrine tissues.

The hormones secreted by the peripheral endocrine glands, in response to the pituitary trophic hormones, are either steroids or iodinated-aromatic peptide derivatives, each of which interacts with a number of target tissues to elicit a biochemical response (Fig. 10-1). For the majority of these hormones this involves an interaction with the genome of their target cells, resulting in a change in the synthesis of specific enzymes or other proteins. Some hormones, for example, progesterone, have very specific effects on a limited number of target tissues; whereas others, for example, thyroxine and the glucocorticoids, interact with many tissues of the body.

Two other endocrine tissues, which secrete important hormones, are also listed in Fig. 10-1. They are the pancreas and the adrenal medulla. The pancreas secretes two polypeptides, insulin and glucagon, whereas the adrenal medulla secretes two amines, adrenaline (epinephrine) and noradrenaline (norepinephrine). In addition to these tissues there are also a number of other endocrine tissues that secrete a variety of hormones [see, e.g., the review of Herman and Taunton (1)], which will not be discussed in detail since we are concerned here with only those hormones involved in the regulation of enzyme synthesis.

All the hormones illustrated in Fig. 10-1 will now be considered in more detail in terms of the regulation of their secretion and the general response that they elicit from target cells. More specific aspects of their action are described in Chapters 11 and 12.

THE PEPTIDES OF THE HYPOTHALAMUS

The hypothalamus is a small, discrete area of the diencephalon located in the forebrain (Fig. 10-2). Neurons terminating in this area are of several types, some of which are capable of releasing small peptide hormones into adjacent capillaries that feed into the portal blood vessels connecting the hypothalamus directly to the pituitary. Since the hypothalamus is such a small tissue and since the amounts of peptide released are also very small, it is only comparatively recently that the structure and function of some of these peptides have been elucidated (for some of these molecules as many as 500,000 pig hypothalami were processed in order to purify milligram quantities). This work has recently been reviewed by Schally (2, 3).

Nine hypothalamic hormones have been isolated so far (Table 10-1), and the amino acid sequence has been determined for four of these. Two of them are tripeptides (MIF and TRH), and in the case of TRH two of the residues are modified: the glutamic acid residue is cyclized to form pyroglutamate, and the C-terminal proline is amidated, giving the structure pyroglutamyl-histidyl-prolylamide.

Pure, or partially pure, preparations of these hormones have been used to demonstrate that their targets are indeed the cells of the anterior pituitary. This was achieved in two ways—by injection into whole animals and by examining their effects on pituitary cells in culture. The mechanism of action of TRH, PRF, GRH, and GH·RH involves binding to target cells, causing an increase in the concentration of cAMP, which then stimulates hormone release (2, 4, 5).

Table 10-1 Oligopeptide Hormones of the Hypothalamus Giving the Number of Amino Acid Residues and Sequence, Where Known

	PEPTIDE	RESIDUES	SEQUENCE
TRH	thyrotropin-releasing hormone	3	(pyro)glu·his·pro(NH$_2$)
MRH	MSH-releasing hormone	—	—
MIF	MSH-release inhibiting hormone	3	pro·leu·gly
PRF	prolactin-releasing hormone	—	—
PIF	prolactin-release inhibiting hormone	—	—
GRH	gonadotropin-releasing hormone	10	(pyro)glu·his·trp·ser·tyr·gly·leu·arg·pro·gly
GH·RF	GH-releasing hormone	—	—
GH·RIH	GH-release inhibiting hormone (somatostatin)	14	ala·gly·cys·lys·asn·phe·phe·trp | | cys·ser·thr·phe·thr·lys
CRF	corticotropin-releasing hormone	—	—

The release-inhibiting hormones, for example, somatostatin (GH·RIH), appear to block this increase in cAMP.

Little else is known about these peptides except that they have very short half-lives [$t_{1/2}$ = 2–10 min (6)] and have very rapid effects on the cells of the pituitary (Fig. 10-1). At the present time the factors regulating the release of the hypothalamic peptides are poorly understood.

HORMONES OF THE PITUITARY

Nine hormones are secreted by the cells of the pituitary gland, all polypeptides (Table 10-2). The pituitary gland is a small ovoid structure attached by a short stalk to the base of the brain (Fig. 10-2). It has a rich blood supply both from the general circulation (into which it secretes hormones) and from the portal vein of the hypothalamus (from which it is controlled). The gland is divided into two parts, the posterior pituitary (neurohypophysis) and the anterior pituitary (adenohypophysis). The anterior pituitary contains populations of different cells, and each type secretes a particular hormone(s). For example, the gonadrotrophs that secrete luteinizing hormone or follicle-stimulating hormone comprise 5% of the total cell mass, whereas the cells specialized in secreting growth hormone and thyroid-stimulating hormone comprise 50–70% of the total cell mass.

In general, the release of hormones from the anterior pituitary is controlled by the peptides secreted by the hypothalamus. These peptides either stimulate or inhibit release, as indicated in Fig. 10-1. However, other factors are also involved. For example, many of the hormones are feedback inhibitors of their own release. Furthermore, the release of some hormones is modulated by the concentration of hormones already present in the blood. Thus the release of follicle-stimulating hormone is stimulated by estradiol, but is inhibited by testosterone. Other examples of these complex interactions can be found in the review of Labrie et al. (4).

Adrenocorticotrophin

Adrenocorticotrophin (ACTH) is a small polypeptide released from the pituitary in response to the secretion of corticotrophin-releasing factor (CRF) by the hypothalamus. The principle target tissue for ACTH is the adrenal cortex, which secretes the glucocorticoids. The synthesis, release, and function of ACTH have recently been reviewed (7–9). The biosynthesis of the molecule is interesting since it appears to be synthesized as a precursor protein (M_r = 30,000) that is processed to yield not only ACTH, but also β-lipotropin, a pituitary glycoprotein, which is then cleaved to form two other hormones, MSH and β-endorphin.

The concentration of CRF in the hypothalamus exhibits a pronounced diurnal rhythm producing large changes in its secretion during each 24-h period.

Table 10-2 The Hormones of the Anterior and Posterior Pituitary

	HORMONE	AMINO ACIDS	M_r	$t_{1/2}$	TARGET TISSUES
Anterior pituitary					
MSH	melanocyte stimulating hormone	13	~1,500	<2 min	melanocytes, melanophores
ACTH	adrenocorticotrophic hormone	39	4,500	2–5 min	adrenal cortex
GH	growth hormone (somatotropin)	188	21,500	20–30 min	thyroid, testes, many tissues
PRL	prolactin	199	23,000	5–30 min	corpus luteum, mammary gland, liver
TSH	thyroid stimulating hormone	—	28,000	~2 h	thyroid gland
FSH	follicle stimulating hormone	—	41,000	~2 h	ovary
LH	luteinising hormone	—	26,000	~2 h	testes
Posterior pituitary					
	oxytocin	9	1,000	—	uterus, mammary gland
	vasopressin	9	1,208	—	blood vessels, kidney

This, in turn, leads to a diurnal rhythm in the release of ACTH by the anterior pituitary and a pronounced rhythm in the concentration of blood glucocorticoids. The significance of these rhythms is discussed in more detail in Chapter 13.

ACTH binds to specific receptors on the surface of the cells of the adrenal cortex. The hormone–receptor complex stimulates adenylate cyclase, which results in an increase in the intracellular concentration of cAMP, followed by the release of glucocorticoids. If the stimulation of the adrenal cells by ACTH is of long duration there is an increase in steroidogenesis and an increase in cortical cell hypertrophy, followed by hyperplasia, to give the cortex a greater capacity of glucocorticoid production.

Thyroid-Stimulating Hormone

TSH is a glycoprotein hormone secreted by the cells of the anterior pituitary. Its primary function is the regulation of the secretion of thyroxine by the thyroid gland (Fig. 10-1). The secretion of TSH is controlled by the smallest of the hypothalamic peptides, the tripeptide thyrotropin-releasing hormone (TRH). TRH stimulates the release of both TSH and prolactin from the anterior pituitary by a mechanism involving the release of preexisting hormone, the synthesis of new hormone, and hyperplasia of the anterior pituitary cells (5).

Once released TSH, like other glycoprotein hormones, is relatively stable in the blood (Table 10-2) and interacts with surface receptors on the cells of the thyroid gland. The release of thyroxine in response to TSH appears to be mediated by cAMP.

Follicle-Stimulating Hormone and Luteinizing Hormone

Follicle-stimulating hormone (in the female) and luteinizing hormone (in the male) are glycoproteins secreted by the gonadotrophic cells of the anterior pituitary in response to the secretion of GRH by the hypothalamus (Fig. 10-1). FSH and LH are relatively stable in the blood ($t_{1/2} \sim 2$ h) and stimulate the ovaries and testis to secrete the steroid sex hormones 17β-estradiol and testosterone, respectively. Figure 10-3 shows the release of luteinizing hormone from the pituitary and testosterone from testis following a single injection of GRH (LH·RH) into an adult male rat, illustrating the magnitude and duration of the responses (10).

The Leydig cells of the testis and the granulosa cells of the ovaries have been shown to possess high-affinity cell-surface receptors for LH and FSH, respectively. Both hormones elicit an increase in cAMP in their respective target cells, and the second messenger stimulates steroid secretion (10–12).

Prolactin

Prolactin is one of the largest polypeptide hormones (see Table 10-2) and appears to carry enough information to direct a variety of responses in a

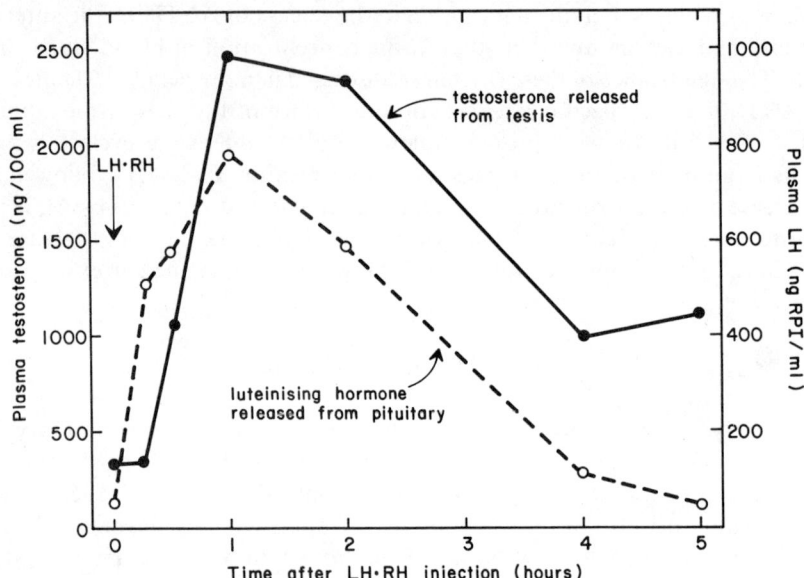

Fig. 10-3 The effect of a single injection of luteinizing hormone-releasing hormone (LH·RH) on the release of luteinizing hormone from the pituitary and its effect on the release of testosterone from the testis. [Reproduced from Fig. 52 of Ref. 10 with permission.]

number of tissues. The hormone has been isolated and purified, and its chemistry and function have been studied in some detail (10, 13, 14) (see also Chapter 15). The release of prolactin from the cells of the anterior pituitary is controlled by both positive (prolactin-releasing hormone, PRH) and negative (prolactin release-inhibiting hormone, PRIH) factors secreted by the hypothalamus. TRH also appears to be able to induce the secretion of prolactin (Fig. 10-1). PRH and PRIH are still poorly characterized, and their mechanism of action is not known.

The half-life of prolactin in the blood is somewhat longer than for most polypeptide hormones (Table 10-2), and it interacts with a number of tissues, including mammary gland, liver, kidney, adrenals, ovaries, and testis. The hormone has been shown to interact with the cells of the corpus luteum in the ovary, where it stimulates the secretion of the steroid progesterone. It also interacts with the mammary gland during pregnancy, where it stimulates the development of the ductal and lobuloalveolar cells and the subsequent production of milk by these cells (see Chapter 15 for details). The action of the hormone on other cells is less certain. Indeed, it is not definitively established that the hormone does have a physiological role in all of these tissues, particularly since prolactin has some affinity for growth hormone receptors.

Prolactin is rapidly internalized into target cells, and its mechanism of action may be somewhat different from other polypeptide hormones, which only bind to cell-surface receptors. Moreover, a second messenger for prolactin action has not been identified (see Chapter 11).

Growth Hormone

Growth hormone is the other large polypeptide hormone secreted by the cells of the anterior pituitary and is synthesized in by far the largest amount of all hormones. Its secretion is under complex control with two hypothalamic hormones (Fig. 10-1), thyroxine, growth hormone itself, and a number of other factors being involved (15, 16). When released, it is relatively stable in the blood (Table 10-2) and has been found to interact with virtually every tissue in the body.

Growth hormone stimulates the cells of the thyroid gland to release thyroxine and the cells of the testis to release testosterone. However, the mechanism of action of the hormone in these cells and in the cells of other tissues is not known. It appears to bind to cell-surface receptors, but no intracellular second messenger has been identified. (cAMP, the second messenger for most polypeptide hormones, has been ruled out.)

Recent work with human growth hormone (17) suggests that the hormone might actually be a *mixture* of polypeptides, each with a specific function, rather than a single molecular species that is responsible for all the actions attributed to growth hormone.

Melanocyte-Stimulating Hormone

Melanocyte-stimulating hormone (MSH) is a small, rapidly turning-over peptide secreted by the anterior pituitary (Fig. 10-1, Table 10-2). Secretion is controlled by positive and negative hypothalamic factors, one of which is a tripeptide (MIF). MSH binds to receptors on melanocytes and melanophores of the skin and stimulates their differentiation, including the synthesis of melanin, the chromatophore that causes darkening of the skin (18).

Oxytocin and Vasopressin

Oxytocin and vasopressin are two well-characterized, small peptide hormones secreted by the cells of the posterior pituitary (19). Oxytocin plays a role in milk secretion by the mammary gland and also induces contractions of the uterine wall during labor. Vasopressin, on the other hand, is involved in changes in the homeostatic mechanisms that regulate blood pressure. In addition, both hormones regulate the rate of water flow and sodium transport in the kidney and bladder.

Cyclic AMP, the second messenger for both hormones, increases following an interaction between the hormones and specific receptors on the surface of target cells. Both hormones bring about short-term, acute responses that do not involve changes in enzyme synthesis.

Hormones of the Hypothalamus and Pituitary: A Summary

Higher eukaryotes have evolved a sophisticated system of neural sensory cells that monitor changes in the external environment. The information from these cells is transmitted to the brain, where some of these signals result in the release of small amounts of hypothalamic peptides into the portal vessel of the pituitary gland. In response, the pituitary releases much larger quantities of a diversified number of hormones into the general circulation. The pituitary trophic hormones, in turn, stimulate a number of distal endocrine glands to release hormones that then elicit specific biochemical changes in target tissues.

This arrangement directs a coordinated series of biochemical changes in a number of tissues in response to a change in the environment. The hypothalamus links together the two important information-carrying systems in the body (the nervous and humoral systems), and the anterior pituitary amplifies and diversifies the activity of the hypothalamus and, in addition, permits endogenous signals to modulate the activity of the humoral system. There is also a cascade effect that is directly analogous to that described in Chapter 7 for certain enzymes.

Only two of the hormones released by the anterior pituitary, growth hormone and prolactin, have biochemical effects on tissues in addition to their action on target endocrine cells. Collectively, the hormones of the anterior pituitary stimulate the secretion of, and usually the synthesis of, either a steroid or thyroxine from endocrine glands located in various parts of the body. The chemistry and function of these hormones will now be described.

THE STEROID HORMONES

Chemistry of the Steroids

Cholesterol, obtained from the diet or synthesized *de novo*, is the precursor of all the steroid hormones. The structures of cholesterol and of some of the more common steroid hormones are shown in Fig. 10-4. These hormones are small lipophilic molecules ($MW = 300-400$) that can easily cross the lipid bilayer of the plasma membrane of virtually every cell in the body. However, they only elicit a response when they are recognized by, and bound to, specific cytoplasmic receptors. Although the structural formulas look quite similar, the side-chain modifications produce quite distinct three-dimensional structures (20) that allow them to be recognized individually by specific protein receptors.

There are two major groups of steroids: the glucocorticoids synthesized by cells of the adrenal cortex and the reproductive hormones synthesized by the testis and ovary.

The Glucocorticoids

The cells of the *zona fasciculata* and the *zona reticularis* of the adrenal cortex sythesize a number of steroid hormones with glucocorticoid activity. Usually a

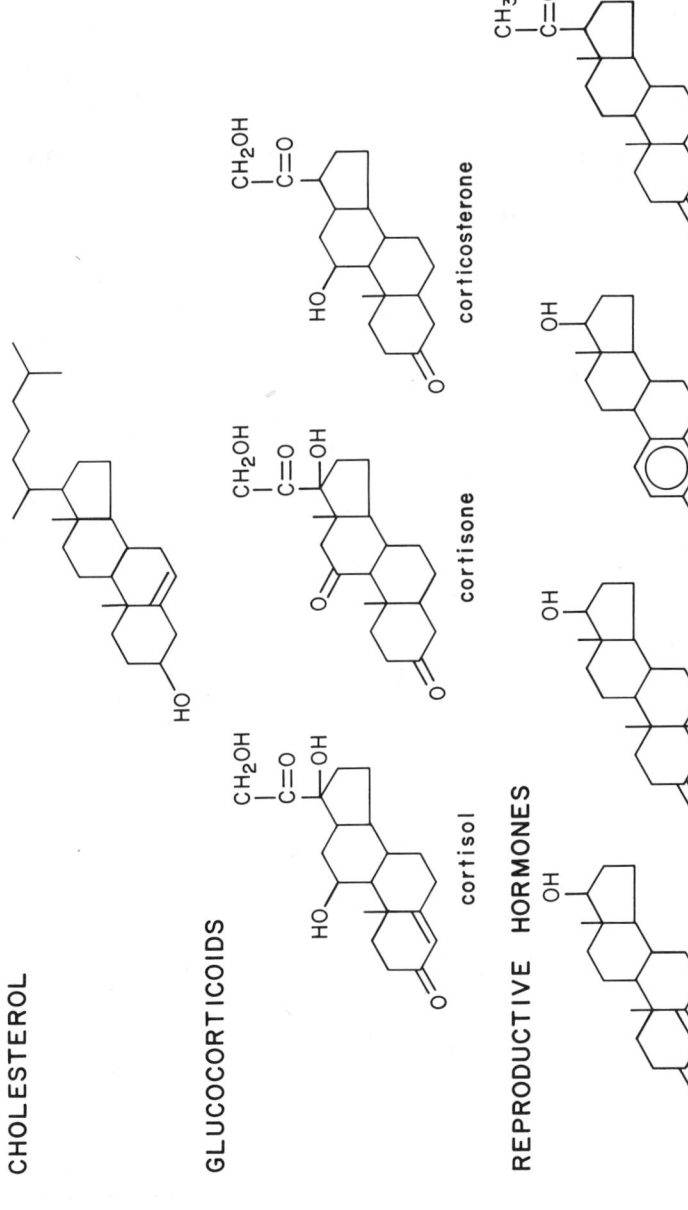

Fig. 10-4 The structures of cholesterol and of the glucocorticoid and reproductive hormones derived from cholesterol.

given species of animal synthesizes and secretes only one principal glucocorticoid hormone. For example, in man the principal glucocorticoid is cortisol, whereas in the rat it is corticosterone (Fig. 10-4), although small amounts of other glucocorticoids such as cortisone and deoxycortisol/deoxycorticosterone may also be released. A number of synthetic glucocorticoids, including dexamethasone, prednisolone, and triamcinolone, are also available and are used therapeutically and for research. These analogues have the advantage of being relatively stable inside the cell, permitting studies of their function to be carried out more easily. The glucocorticoids are the most completely understood (in terms of their structure and function) of all the hormones, and much of this knowledge has recently been reviewed (21–23).

These hormones were first discovered to play a role in the regulation of glucose metabolism, and this is the property for which they are named. The principal role of the hormones is to supply glucose to the heart, brain, and other vital centers during conditions of prolonged stress. This is achieved by the stimulation of muscle and peripheral-tissue protein catabolism to generate gluconeogenic precursors that are converted to glucose and glycogen by the liver (Fig. 10-5). There are simultaneous increases in liver amino acid catabolism and urea production to remove the excess nitrogen. Thus the glucocorticoids have both anabolic and catabolic effects. The effects of the hormones may last from a few hours to several days, and during this time there are changes in the concen-

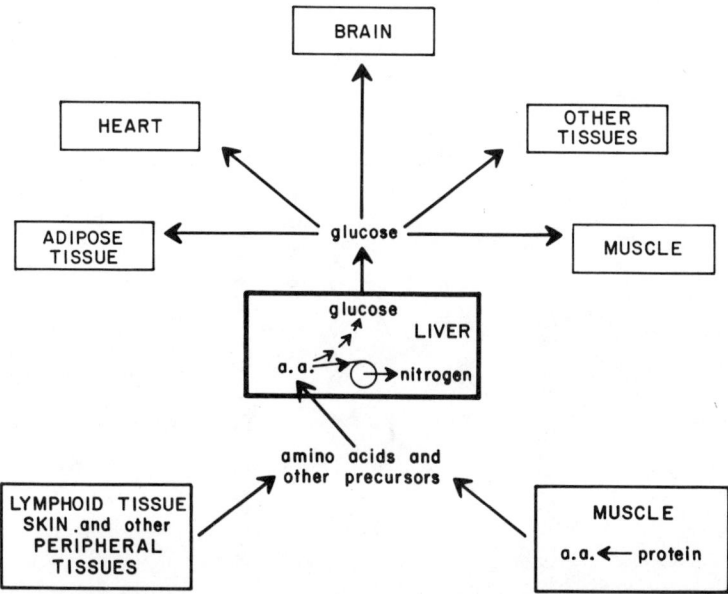

Fig. 10-5 The effects of glucocorticoids on the flow of carbon (and nitrogen) between body tissues during stress. The hormone mobilizes carbon from muscle and peripheral tissues, and these precursors are converted into glucose in the liver and sent to vital organs.

tration of a number of enzymes. This role of glucocorticoids in the regulation of enzyme synthesis in the liver is discussed in more detail in Chapter 12.

The half-life of steroid hormones in the blood ($t_{1/2} \sim 90$ min) is somewhat longer than that of the polypeptide hormones. Moreover, there is a pronounced diurnal rhythm of plasma glucocorticoid levels, with levels being higher during the period without food. The physiological significance of these rhythms is discussed at length in Chapter 13.

Androgens and Estrogens

The two major reproductive hormones are testosterone, produced by the testis in the male, and the estrogens, principally 17β-estradiol, produced by the ovaries in the female. Both hormones are responsible for the development of male and female characteristics at various stages of maturation and in adult animals. For example, the hormones are secreted shortly before and during puberty and promote the maturation of a number of tissues (Fig. 10-6). In addition, the estrogens play an important role in the estrous cycle and during pregnancy. A third steroid hormone, progesterone, is synthesized by the cells of the corpus luteum in the ovary and also participates in the regulation of events during estrus, pregnancy, and lactation (Fig. 10-6).

The reproductive hormones regulate long-term developmental changes in their target cells, and their postnatal effects on the accessory reproductive tissues have been well-studied as model systems for terminal differentiation. Indeed, much of our current knowledge on the regulation of gene expression in eukaryotes has come from studies on these model systems (see Chapter 15).

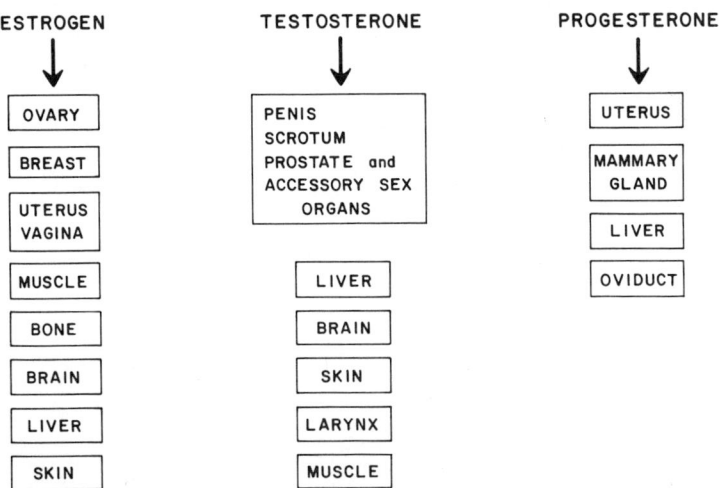

Fig. 10-6 The principal target tissues for the reproductive hormones.

THYROID HORMONES

The Chemistry of T_3 and T_4

The hormone thyroxine (T_4) is secreted by the follicular cells of the thyroid gland in response to changes in the activity of the hypothalamus–anterior-pituitary regulatory system (Fig. 10-1). Thyroxine is synthesized from tyrosine residues that are, initially, an integral part of the protein thyroglobulin that forms a matrix upon which synthesis takes place. The residues are iodinated to form diiodotyrosyl residues, which are conjugated and released from the protein matrix as thyroxine. The structures of thyroxine and of a closely related form of the hormone, triiodothyronine (T_3), are shown in Fig. 10-7. About 90% of the hormone is synthesized as T_4, but it is monodeiodinated in target cells to generate T_3, the physiologically active form of the hormone.

Secretion and Function of Thyroid Hormone

The secretion of thyroxine is under complex control. Both growth hormone and thyroid-stimulating hormone stimulate secretion. The concentrations of these two hormones are, in turn, regulated by a total of four hypothalamic peptides (Fig. 10-1). Moreover, since thyroxine also stimulates the synthesis and release of growth hormone by the pituitary, the release of the two appears to be highly coordinated. The availabiliy of dietary iodine also affects the concentration of thyroxine.

The thyroid hormones are small molecules ($MW = 650–777$) and appear to be able to rapidly penetrate the plasma membrane of most cells. Once inside the cell the hormone is rapidly translocated to the nucleus, where it evokes changes in gene expression in a number of tissues. The hormone plays a major role in the development of mammals and amphibians as well as contributing to the maintenance of the differentiated state of several tissues in the adult, including liver, adipose tissue, heart, brain, and the pituitary gland. The hormone induces both

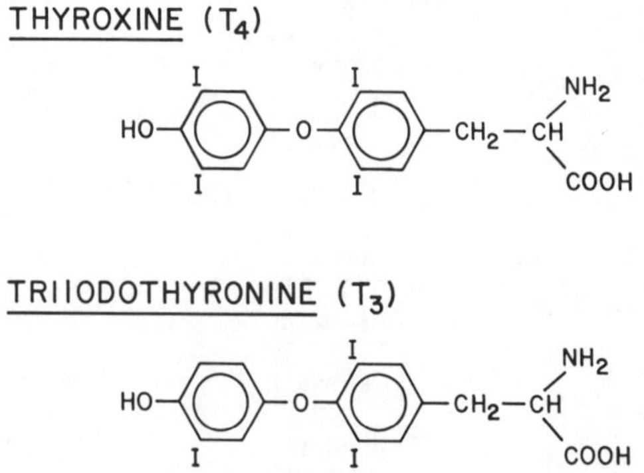

Fig. 10-7 The structures of thyroxine and triiodothyronine.

short-term (hours) and long-term (days) changes in target tissues; a number of these effects have recently been reviewed (24–26) and are discussed in Chapters 12 and 15.

HORMONES NOT DIRECTLY UNDER THE CONTROL OF THE HYPOTHALAMIC–PITUITARY AXIS

The secretion of growth hormone, prolactin, thyroxine, and all the steroid hormones is under the control of the hypothalamus and the pituitary gland. These hormones usually initiate major changes in gene expression, often resulting in changes in the state of differentiation of the target tissues. In most cases several tissues are affected simultaneously.

In addition to these hormones there are a number of others that are responsible for the organization and coordination of the metabolism of various tissues to meet more short-term changes in the environment. Two of these, vasopressin and oxytocin, have already been discussed. Four other hormones also fall into this category: the polypeptides insulin and glucagon, and the catecholamines adrenaline (epinephrine) and noradrenaline (norepinephrine). The properties of these hormones are summarized in Table 10-3.

Insulin and Glucagon

Insulin and glucagon are both secreted by cells within the islets of Langerhans in the pancreas. Insulin is secreted by the beta cells and glucagon by the alpha cells. The release of each hormone is controlled by the concentration of blood glucose; when it rises insulin is released, if it falls glucagon is released. The ratio of insulin to glucagon in the blood appears to determine the overall pattern of energy metabolism in various tissues of the body.

Table 10-3 Hormones Not Directly under the Control of the Hypothalamic–Pituitary Axis

HORMONE	TISSUE	M_r	STRUCTURE	TARGET TISSUES
INSULIN	β-cells of pancreas	6,000	polypeptide (51 residues)	liver, muscle, peripheral tissues
GLUCAGON	α-cells of pancreas	3,674	polypeptide (29 residues)	liver, muscle, peripheral tissues
ADRENALINE	adrenal medulla	183	HO–C$_6$H$_3$(OH)–CH(OH)·CH$_2$·NH$_2$	liver and adipose tissue
NORADRENALINE	adrenal medulla	169	HO–C$_6$H$_3$(OH)–CH(OH)·CH$_2$NH·CH$_3$	liver and adipose tissue

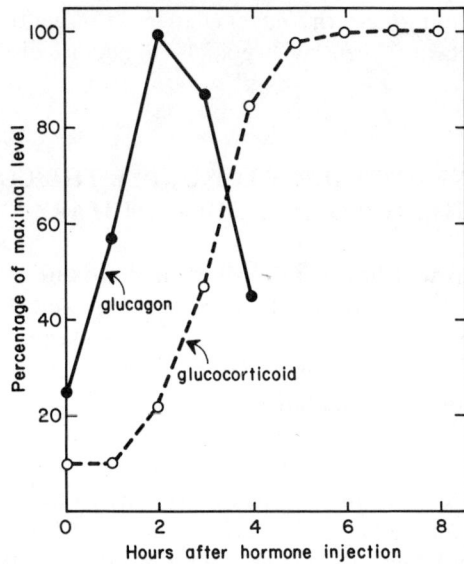

Fig. 10-8 Temporal differences in the induction of tyrosine aminotransferase in the liver by polypeptide and steroid hormones. See also Fig. 12-5.

Many of the responses by target tissues to these hormones are of an acute, short-term nature that help redress imbalances in the body's energy metabolism, for example, the breakdown of liver glycogen in response to a fall in the insulin:glucagon ratio, or an increase in lipogenesis in the adipose tissue in response to an increase in the ratio. These effects are mediated by covalent modification of regulatory enzymes as described in Chapter 7 (see Fig. 7-2). However, both hormones are also capable of inducing changes in enzyme synthesis to bring about longer-term changes in target cell metabolism, although they are usually of shorter duration than those induced by the glucocorticoids. For example, glucagon induces the synthesis of the hepatic, amino acid-catabolizing enzyme tyrosine aminotransferase, an enzyme also induced by glucocorticoids (Fig. 10-8). The enzyme responds immediately to glucagon, reaches a peak at 2 h, and rapidly declines in activity. In contrast, there is a delay of 2 h in the response of the enzyme to glucocorticoid, followed by a slower increase. However, once achieved the elevated enzyme level is maintained for much longer.

Adrenaline and Noradrenaline

Whereas the glucocorticoids (secreted by the adrenal cortex) coordinate the actions of tissues in response to long-term stresses, the catecholamines, adrenaline and noradrenaline (secreted by the adrenal medulla), elicit more acute responses to stress. They are small amines (Table 10-3), and they interact with specific receptors of liver and adipose tissue, where they produce a rapid mobilization of carbohydrate reserves.

These two hormones are not considered to play a major role in the regulation of enzyme synthesis. They do, however, exert their acute effects via changes in the same intracellular second messengers that other hormones used to mediate changes in gene expression. It is possible, therefore, that they may play some modulatory role under certain circumstances.

REFERENCES

1. R. H. Herman and O. D. Taunton (1980). The mechanism of action of hormones. In R. H. Herman, R. M. Cohn, and P. D. McNamara (Eds.). *Principles of Metabolic Control In Mammalian Systems*. Plenum Press, New York, pp. 535-620.
2. A. V. Schally, D. H. Coy, and C. A. Meyers (1978). Hypothalamic regulatory hormones. *Ann. Rev. Biochem.* 47, 89-128.
3. A. V. Schally, D. H. Coy, C. A. Meyers, and A. J. Kastin (1979). Hypothalamic peptide hormones: Basic and clinical studies. In C. H. Li (Ed.). *Hormonal Proteins and Peptides*, Vol. 7. Academic Press, New York, pp. 1-54.
4. F. Labrie, L. Lagacé, M. Beaulieu, L. Ferland, A. De Léan, J. Drouin, P. Borgeat, P. A. Kelly, L. Cusan, A. Dupont, A. Lemay, T. Antakly, G. H. Pelletier, and N. Borden (1979). Mechanisms of action of hypothalamic and peripheral hormones in the anterior pituitary gland. In C. H. Li (Ed.). *Hormonal Proteins and Peptides*, Vol. 7. Academic Press, New York, pp. 206-277.
5. T. F. J. Martin and A. H. Tashjian, Jr. (1977). Cell culture studies of thyrotropin-releasing hormone action. In G. Litwack (Ed.). *Biochemical Actions of Hormones*, Vol. 4. Academic Press, New York, pp. 269-312.
6. H. P. J. Bennett and C. McMartin (1978). Peptide hormones and their analogues: distribution, clearance from the circulation, and inactivation *in vivo*. *Pharmacol. Rev.* 30, 247-292.
7. G. N. Gill (1979). ACTH regulation of the adrenal cortex. In G. N. Gill (Ed.). *Pharmacology of Adrenal Cortical Hormones*. Pergamon Press, Oxford, pp. 35-65.
8. A. Brodish (1979). Control of ACTH secretion by corticotropin-releasing factor(s). *Vit. Horm.* 37, 111-152.
9. D. L. Krieger, A. S. Listta, M. J. Brownstein, and E. A. Zimmerman (1980). ACTH, β-lipotropin and related peptides in brain, pituitary and blood. *Rec. Prog. Horm. Res.* 36, 277-344.
10. M. L. Dufau and K. J. Catt (1978). Gonadotropin receptors and regulation of steroidogenesis in the testis and ovary. *Vit. Horm.* 36, 461-592.
11. J. H. Dorrington and D. T. Armstrong (1979). Effects of FSH on gonadal functions. *Rec. Prog. Horm. Res.* 35, 301-342.
12. K. J. Catt, J. P. Harwood, R. N. Clayton, T. F. Davies, V. Chan, M. Katikineni, K. Wozu, and M. L. Dufau (1980). Regulation of peptide hormone receptors and gonadal steroidogenesis. *Rec. Prog. Horm. Res.* 36, 557-622.
13. C. H. Li (1980). The Chemistry of Prolactin. In. C. H. Li (Ed.). *Hormonal Proteins and Peptides*, Vol. 8. Academic Press, New York, pp. 1-36.
14. J. J. Elias (1980). The role of prolactin in normal mammary gland growth and function. In C. H. Li (Ed.). *Hormonal Proteins and Peptides*, Vol. 8. Academic Press, New York, pp. 37-74.
15. E. E. Müller (1979). The control of somatotropin secretion. In C. H. Li (Ed.). *Hormonal Proteins and Peptides*, Vol. 7. Academic Press, New York, pp. 123-204.

16 A. Negro-Vilar, S. R. Ojeda, and S. M. McCann (1980). Hypothalamic control of LHRH and somatostatin: Role of central neurotransmitters and intracellular messengers. In G. Litwack (Ed.). *Biochemical Actions of Hormones*, Vol. 7. Academic Press, New York, pp. 245–285.
17 U. J. Lewis, R. N. R. Singh, G. F. Tutwiles, M. B. Sigel, E. F. Vanderlaan, and W. P. Vanderlaan (1980). Human growth hormone: A complex of proteins. *Rec. Prog. Horm. Res.* **36**, 477–508.
18 S.-T. Chen, H. Wahn, W. A. Turner, J. D. Taylor, and T. T. Chen (1974). MSH, cyclic AMP and Melanocyte differentiation. *Rec. Prog. Horm. Res.* **30**, 319–345.
19 M. S. Soloff and A. F. Pearlmutter (1979). Biochemical actions of neurohypophysial hormones and neurophysin. In G. Litwack (Ed.). *Biochemical Actions of Hormones*, Vol. 6. Academic Press, New York, pp. 265–333.
20 J. P. Glusker (1979). Structural aspects of steroid hormones and carcinogenic polycyclic aromatic hydrocarbons. In G. Litwack (Ed.). *Biochemical Actions of Hormones*, Vol. 6. Academic Press, New York, pp. 121–204.
21 J. D. Baxter and G. G. Rousseau (Eds.) (1979). *Glucocorticoid hormone action*. Springer-Verlag, Berlin.
22 J. D. Baxter (1979). Glucocorticoid hormone action. In G. N. Gill (Ed.). *Pharmacology of Adrenal Cortical Hormones*. Pergamon Press, Oxford, pp. 67–121.
23 A. K. Roy and J. H. Clark (Eds.) (1980). *Gene Regulation by Steroid Hormones*. Springer-Verlag, New York.
24 J. H. Oppenheimer (1979). Thyroid hormone action at the cellular level. *Science* **203**, 971–979.
25 N. L. Eberhardt, J. W. Apriletti, and J. D. Baxter (1980). The molecular biology of thyroid hormone action. In G. Litwack (Ed.). *Biochemical Actions of Hormones*, Vol. 7, Academic Press, New York, pp. 311–394.
26 J. D. Baxter, N. L. Eberhardt, J. W. Apriletti, L. K. Johnson, R. D. Ivarie, B. S. Schachter, J. A. Morris, P. H. Seeburg, H. M. Goodman, K. R. Latham, J. R. Polansky, J. A. Martial, and T. Cobley (1979). Thyroid hormone receptors and responses. *Rec. Prog. Horm. Res.* **35**, 97–153.

CHAPTER ELEVEN

Transfer of Information from Hormones to the Nucleus

Hormones responsible for the regulation of enzyme synthesis must transmit their information to the nucleus of target cells and stimulate a change in gene expression. There are three general mechanisms of intracellular information transfer, each characterized by the extent to which the hormone itself can penetrate the plasma membrane and reach the nucleus. These are illustrated in Fig. 11-1. The actual changes of gene expression are mediated by proteins that are capable of recognizing both the hormone (or a second messenger) and a specific location on the genome (see Chapters 3 and 4).

The thyroid hormones, thyroxine (T_4) and triiodothyronine (T_3), are the only hormones known to enter the nucleus in a free form. Steroid hormones also enter the nucleus, but only after an initial interaction with specific proteins in the cytoplasm. Most polypeptide hormones and the catecholamines, on the other hand, are believed not to even enter the cell. They interact with specific receptor proteins on the surface of the cell, and other molecules are required to transmit the hormone's information from the plasma membrane to the nucleus. The mechanisms of information transfer for the thyroid hormones and for the steroids are fairly well-understood. However, for some polypeptide hormones, our knowledge is far from complete, particularly for those that also have acute actions in the cytoplasm or use more than one kind of second messenger. Each of the three mechanisms of information transfer will now be discussed in more detail.

THYROID HORMONES TRANSMIT INFORMATION DIRECTLY TO THE NUCLEUS

Thyroxine, together with small amounts of T_3, released from the thyroid gland is transported in the blood attached to one of three proteins—thyroxine-binding α-globulin, thyroid-hormone-binding prealbumin, and serum albumin. The properties of these proteins are summarized in Table 11-1.

The hormones appear to be able to cross the plasma membrane of target cells quite easily, although it is not yet resolved if this is due solely to passive

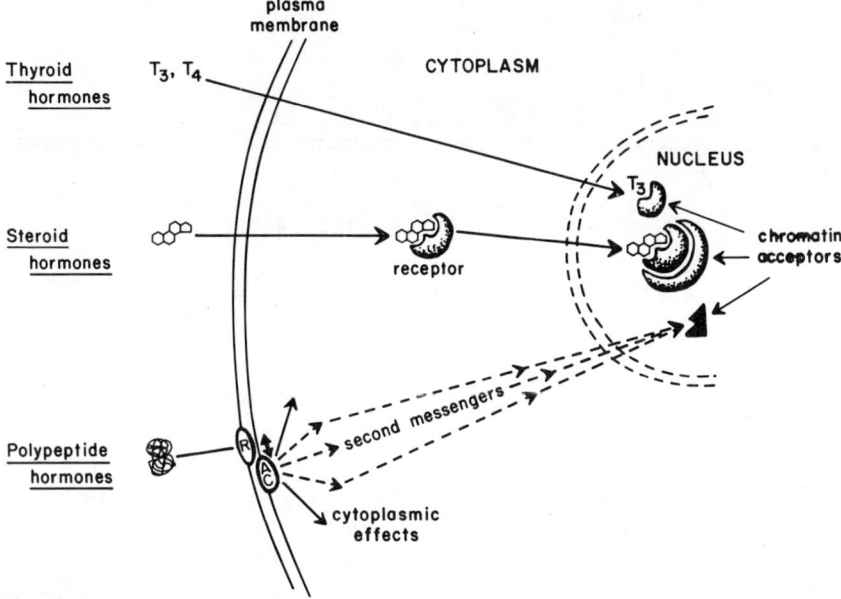

Fig. 11-1 The three general pathways for information flow from extracellular hormones to the nucleus. The pathways are described in detail in the text.

diffusion or whether carrier proteins are involved (1). Once inside the cell the hormones bind to several cytoplasmic proteins, but this binding is not of sufficiently high affinity for any of them to be considered specific receptors. The function, if any, of these binding proteins is not known, although one of them is likely to be the enzyme responsible for the monodeiodination of T_4 to form T_3, the biologically active form of the hormone.

Table 11-1 Thyroid-Hormone-Binding Proteins in the Serum and in the Cell

PROTEIN	M_r	LOCATION	FUNCTION	AFFINITY CONSTANT
THYROXINE-BINDING α-GLOBULIN	54,000	serum	transports 70% of hormone	$2.5 \times 10^9 \, M^{-1}(T_4)$
THYROID HORMONE-BINDING PROTEIN	55,000	serum	transports 10% of hormone	$1.1 \times 10^8 \, M^{-1}(T_4)$
SERUM ALBUMIN	67,000	serum	transports 20% of hormone	—
INTRACELLULAR PROTEINS	—	cytoplasm	mostly unknown	—
NUCLEAR RECEPTOR	50,000	nucleus	interacts with genome	$5.0 \times 10^{11} \, M^{-1}(T_3)$

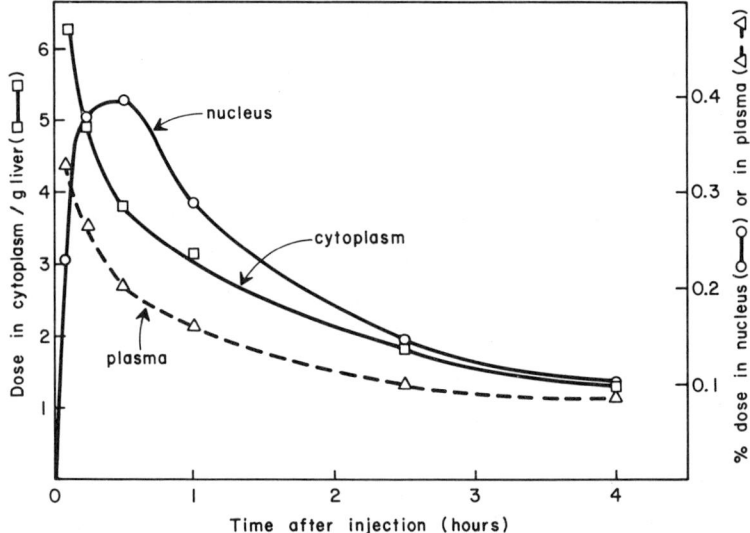

Fig. 11-2 The concentration of thyroid hormone in the plasma, cytoplasm, and nucleus at various times after the injection of radioactively labeled hormone into a rat. [Reproduced from Fig. 1 of J. H. Oppenheimer (1979). *Science* **202**, 971–979. Copyright 1979 by the American Association for the Advancement of Science.]

T_3 enters the nucleus in a free form (i.e., it is not bound to a cytoplasmic protein); typical kinetics of accumulation are shown in Fig. 11-2. Following injection, the hormone is rapidly cleared from the blood into the cells of target tissues, such as liver, but it does not accumulate in the cytoplasm. It is rapidly transported to the nucleus, and a maximum concentration is reached within 1 h. The nuclear concentration of the hormone subsequently declines with a half-life of about 15 min.

Nuclear Receptors

Evidence for the existence of specific protein receptors for thyroid hormones in the nucleus was first obtained in 1972, and since then the receptor protein has been isolated from a number of tissues (2–4) and characterized. It is a nonhistone protein (M_r = 50,000), and there are approximately 5000 molecules per nucleus in the cells of responsive tissues (Tables 11-1 and 11-2), but virtually none in tissues such as spleen and testis that do not respond to thyroid hormones. The receptor turns over relatively rapidly [$t_{1/2}$ = 3–4 h (5)] and has a much higher affinity for T_3 than for T_4.

The number of receptor molecules in the nucleus is very small even in the most active tissues (1 receptor per 2000 nucleosomes), and its location in chromatin has not yet been firmly established. Nuclease digestion studies, designed to establish a relationship between the receptor and transcriptionally active genes,

Table 11-2 The Number of Receptors for Triiodothyronine (T_3) in Various Tissues[a]

TISSUE	T_3 RECEPTORS/NUCLEUS
Anterior pituitary	5847
Liver	4515
Kidney	3923
Heart	2960
Brain	1998
Lung	1120
Spleen	133
Testis	17

[a] Reproduced from Table 1 of Oppenheimer (1979). *Science* **202**, 971–979. Copyright 1979 by the American Association for the Advancement of Science.

have produced conflicting results [discussed in reviews (2–4)]. Some reports indicate that 70% of the receptor molecules are released when only 10% of the DNA (i.e., the active sequences) is digested (5), whereas other reports suggest that the receptors are evenly distributed between active and inactive DNA sequences. Moreover, there are indications that the receptor is associated with higher orders of chromatin structure (the 20–30-nm fiber), implying that the genes activated by thyroid hormones reside at this structural level.

Receptor molecules bind strongly to both single- and double-stranded DNA, but not to RNA. However, it is unlikely that the receptors bind to *specific regulatory DNA sequences* because bacterial-cell DNA is as effective as eukaryotic-cell DNA in *in vitro* binding experiments. There is also evidence that the receptor molecules are in contact with histones, since high-affinity binding of T_3 to receptor molecules *in vitro* is only observed in the presence of histones, particularly the histones of the nucleosome core. It appears, therefore, that the receptor molecules are associated with nucleosomes of the 20–30-nm fiber and may make contact with both DNA and histones.

Mechanism of Transcriptional Activation

The concentration of cellular mRNA is lower in the livers of hypothyroid animals than in normal (euthyroid) animals, and sequence complexity studies have shown this to be due to a decrease in the synthesis of all mRNA species rather than a specific subset. Thus, the hormone influences the total transcriptional activity of the nucleus in addition to its ability to induce a small number of specific proteins such as malic enzyme and α_{2u}-globulin.

The generalized increase in transcriptional activity in the presence of the hormone appears to be due to activation of RNA polymerases I (rRNA) and II (mRNA). The activities of both polymerases slowly rise over a period of days in thyroxine-treated hypothyroid animals, in parallel with the general increase in template availability.

THYROID HORMONES

The mechanism by which the hormone brings about *specific* changes in transcription is proving to be much more difficult to elucidate. Evidence that the binding of T_3 to nuclear receptors leads to a change in chromatin structure has come from the results of "rifampicin-challenge" studies (6). This assay was originally devised as a measure of the number of sites of transcription initiation on chromatin of eukaryotic cells *using E. coli RNA polymerase* as a probe. Briefly, the technique involves an initial reaction between chromatin and *E. coli* RNA polymerase during which each transcription-start site (equivalent to the number of genes being transcribed) accepts one polymerase molecule. Transcription is initiated by adding nucleotides in the presence of rifampicin, a drug that prevents any more polymerase molecules from binding to the transcription-start sites. Transcription is allowed to continue, and each RNA chain synthesized is equivalent to the number of start sites. The number of chains is measured, and this is taken as the number of active genes in that particular sample of chromatin.

Experiments carried out on chromatin from control T_3-treated animals show that the hormone does increase the number of transcription-start sites (Fig. 11-3). However, when the amount of RNA synthesized is converted into the number of genes activated by the hormone, the data indicate that the hormone increases the number of active genes from 1 million to 1.5 million! These numbers are clearly impossible, since the maximum number of genes in a cell is unlikely to exceed 50,000 (Chapter 3). Therefore, the bacterial polymerase must

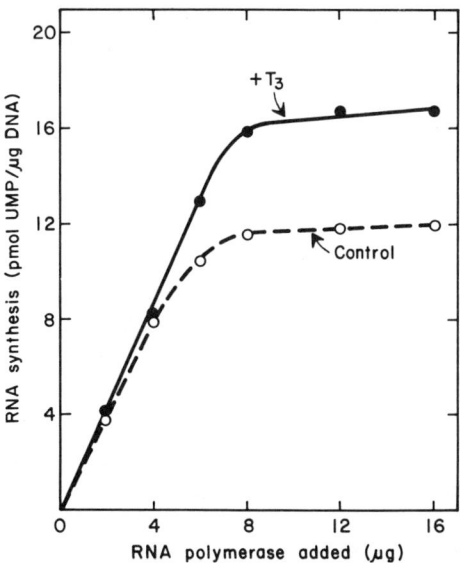

Fig. 11-3 The stimulation of *E. coli* RNA polymerase binding to chromatin by triiodothyronine (T_3). This is the basis of the rifampicin-challenge assay as described in the text. [Reproduced from Fig. 16 of Baxter et al. (4) with permission.]

start transcription at sites other than those recognized by the eukaryotic RNA polymerase, and the number or chains synthesized is not a reflection of the number of active genes. However, the assay is useful in that it indicates *a change in chromatin structure* in the presence of T_3, exposing more sites for bacterial polymerase binding. In this sense, the bacterial enzyme is a probe of chromatin structure rather than a specific measure of transcriptional activity.

The number of extra polymerase binding sites exposed (~500,000) is far more than the number of thyroid receptors (~5,000 molecules), indicating that either the hormonal signal is greatly amplified or the binding of hormone causes each receptor molecule to expose approximately 100 sites. Amplification could occur via covalent modification of chromatin proteins if the hormone–receptor complex were to activate a protein kinase or acetylase because each molecule could modify a large number of proteins. An increase in nuclear protein kinase activity following T_3 injection has been reported (7). Thyroid hormones also affect the concentration of at least two other nuclear proteins that may be involved in maintaining chromatin structure (8). It appears, therefore, that even though the total number of genes activated at T_3 is quite small, quite complex and extensive changes in chromatin structure are required for their expression.

STEROID HORMONES TRANSMIT INFORMATION VIA A CYTOPLASMIC RECEPTOR

Steroids readily penetrate the plasma membrane [in some cells protein carriers may also exist (1)] and bind to a number of cytoplasmic proteins, one of which is a high-affinity steroid *receptor*. Each receptor has a high affinity for only one steroid hormone, and when the hormone binds it is *activated* and rapidly translocated to the nucleus. In the nucleus, the hormone–receptor complex (HRC) interacts with a specific nuclear *acceptor* protein, causing a change in chromatin structure that leads to an alteration in gene expression. This information-transfer process can be conveniently broken down into a number of stages to facilitate discussion: interaction of the hormone with its receptor, activation of the receptor and its translocation to the nucleus, binding to the acceptor protein, and, finally, changes in chromatin structure and gene expression.

Cytoplasmic Receptors

The receptors for estrogen (9–11), testosterone (12–14), progesterone (15–17), and glucocorticoids (18–21) are currently being studied in great detail; some of their properties are summarized in Table 11-3. Receptors are usually isolated as aggregates in buffers of low ionic strength (sedimentation coefficients of 7–12S), and these are dissociated into individual entities ($S_{20,w}$ = 4–6S, M_r = 70,000–120,000) by increasing the salt concentration to weaken ionic interactions. Each receptor molecule binds one molecule of its specific steroid with high affinity

Table 11-3 Characteristics of the Steroid Hormone Receptors[a]

HORMONE	TISSUE	SIZE OF RECEPTOR ($S_{20,w}$)	SIZE OF RECEPTOR (M_r)	AGGREGATE ($S_{20,w}$)	K_D (M)	ACTIVATION ($S_{20,w}$)	RECEPTORS/CELL	REF.
Estrogen	uterus	4	70–80,000	8	0.8×10^{-9}	$3.9 \rightarrow 5.4$	20,000	10, 11
DHT	prostate	3–5	90,000	7–12	$3-4 \times 10^{-10}$	$3.8 \rightarrow 2.9$	—	13, 14
DHT	kidney	4.2	90,000	8	1×10^{-9}	—	—	12
Glucocorticoid	liver	4	102,000	7	3×10^{-9}	—	2,000–10,000	19
Progesterone	oviduct	6	A = 79,000 B = 115,000	6–8	5×10^{-9}	—	35,000	15, 16

[a] $S_{20,\omega}$ is the sedimentalian coefficient (a measure of size), and K_D is the dissociation constant (a measure of affinity). The references are at the end of the chapter.

($K_D \sim 10^{-9}$ M). Other steroid hormones, or hormone analogues, may also bind, but with a lower affinity, which may or may not be sufficient to trigger a biochemical response.

The number of receptors per cell varies from tissue to tissue and hormone to hormone, but is usually in the range 10,000–35,000. The presence of receptors enables target tissues to concentrate the hormone [the steroid hormones penetrate most cells but cannot be accumulated by cells without receptors (21)]. With the exception of the progesterone receptor the structure of these cytoplasmic receptors has not been studied in detail. The progesterone receptor exists as two subunits (see Chapter 4 and Refs. 16 and 17), and each one binds a molecule of hormone. The activated AB dimer is translocated to the nucleus. Other receptors do not appear to have dissimilar subunits, although they may exist in interconvertible forms with similar chromatin and DNA-binding characteristics as the progesterone subunits (22).

Receptor Activation and Translocation to the Nucleus

In unstimulated cells the cytoplasmic steroid receptor molecules do not enter the nucleus. However, upon binding hormone, the receptors undergo a conformational change that greatly increases their affinity for the nucleus, and 80–90% of them translocate into this organelle (typical kinetics are shown in Fig. 11-4). This conformational change is referred to as *receptor activation* [see Milgrom (23) for a recent review]. The phenomenon of activation has proved

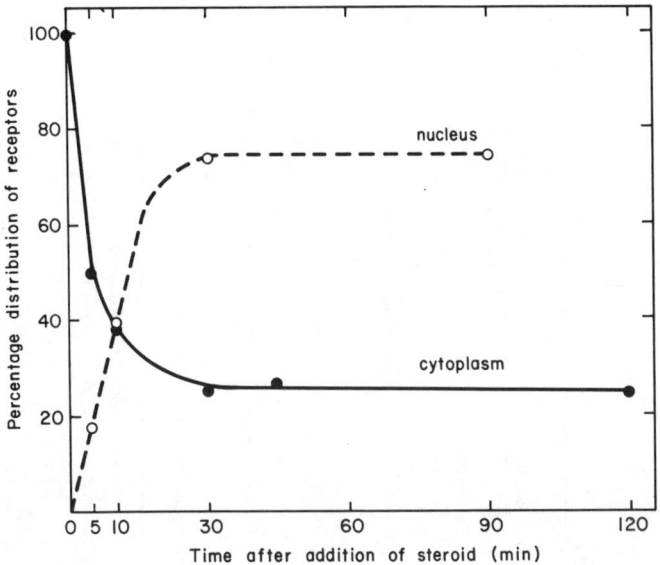

Fig. 11-4 The kinetics of glucocorticoid hormone uptake and transport into the nucleus of hepatoma tissue culture cells. [Reproduced from Fig. 2 of Higgins et al. (24) with permission.]

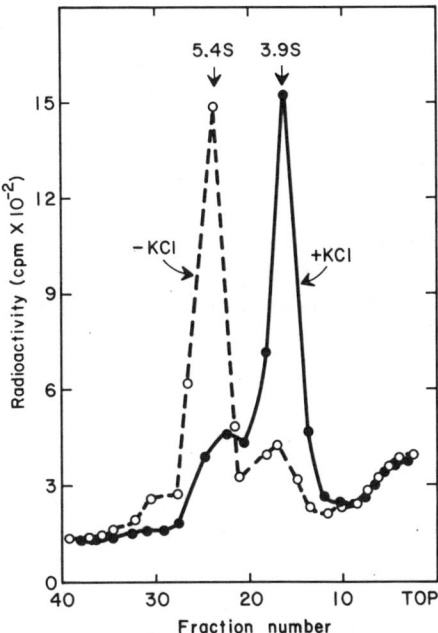

Fig. 11-5 Salt-induced conformational changes in the estrogen receptor. The removal of salt causes a dimerization of the receptor that has a greater sedimentation coefficient in the ultracentrifuge. [Reproduced from Fig. 1 of Notides (11) with permission.]

difficult to study for two reasons. First, the concentration of receptors in any given tissue is usually very low; and, second, it is difficult to assay the activation process directly.

Kinetic analyses of steroid–receptor interactions (23) indicate that the reaction is zero order, suggesting that the steroid induces a conformational change in the receptor, and no other molecules are involved. In the case of the estrogen receptor, activation may also be accompanied by a dimerization of receptors, and the dimer is translocated to the nucleus. Figure 11-5 shows sedimentation profiles of the unactivated 3.9S receptor and the activated 5.4S dimer (in this experiment activation was mimicked by lowering the KCl concentration). The dihydrotestosterone receptor also undergoes a change in sedimentation characteristics upon activation (Table 11-3), and this probably reflects the conformational change induced by steroid binding.

The nature of the conformational change that gives the receptor an increased affinity for a particular organelle is intriguing. Moreover, it is an important factor in the mechanism of information transfer. Unfortunately, virtually nothing is known about the physical property changes that accompany activation. It has been established, however, that the molecules are acidic and possess high axial ratios (15–20 in the case of the progesterone receptor). A high axial ratio (the ratio of length to width of the molecule) is compatible with the molecule's function as a translocator of information from cytoplasm to nucleus.

Finally, it is noteworthy that the hormone–receptor interaction *does not amplify* the hormonal signal, since one steroid molecule activates only one steroid–receptor complex. In fact, if two steroid molecules bind to one progesterone receptor dimer, as proposed (16, 17), the signal is actually deamplified.

Nuclear Acceptor Sites

Translocation of cytoplasmic HRCs into the nucleus is completed approximately 30 min after hormone injection (Fig. 11-4), and although changes in nuclear activity can be seen within 1 h, new mRNAs are often not seen in the cytoplasm for about 2 h (see Fig. 12-5). During this time the HRCs locate their correct binding sites on the genome and direct the transcription of specific DNA sequences. The nature of the binding of HRC to the genome is considered in this section (see Refs. 13, 19, and 24–34 for more details).

Although HRCs have a high affinity for DNA ($K_D = 10^{-10}M$), it is unlikely that DNA is the primary site of HRC-binding for two reasons. First, the affinity of binding is not high enough for it to be sequence-specific; and, second, HRCs bind to DNA regardless of source, including bacterial DNA. Furthermore, HRC binding is not specific for DNA, since RNA molecules can compete quite effectively for HRC (28, 29).

Two approaches have been taken to try to find the location of HRC binding sites in the nucleus—the more conventional technique of salt extraction and the more recent technique of nuclease digestion with nucleases specific for active gene sequences. Although these studies are far from complete and have often produced conflicting results, some general conclusions can be drawn. There appear to be two classes of binding sites in the nuclei of steroid responsive cells (prostate, uterus, liver, etc.), and these can be distinguished by the ease with which they are extracted from the nucleus. One class is liberated by 0.4 M NaCl or brief incubations with nucleases, such as micrococcal nuclease or DNase-I (26, 32–34). The other class is much more resistant to salt and is not released by nucleases. This latter class can only be extracted with high concentrations of chaotropic agents such as guanidine hydrochloride (27).

The "easily-extractable" sites appear to reside in a nonhistone protein fraction that is thought to contain one or more protein acceptor molecules to which the HRC binds (histones do not bind HRC). The protein acceptor molecule(s) has not yet been purified and characterized. The binding sites that are resistant to extraction are either tightly bound to DNA or are associated with the nuclear matrix (27, 34). A partial purification of the tightly bound, progesterone–HRC-acceptor protein from chicken oviduct nuclei has been achieved (27). It is a small protein ($M_r = 12,000–16,000$) and only shows high-affinity HRC-binding when it is associated with DNA.

During this study of the binding of progesterone–receptor complexes to oviduct nuclei it was noticed that removal of the nonhistone protein (NHP) fraction with salt actually increased the number of HRC-acceptor binding sites in the residual chromatin. Thus the NHP must *mask* a number of sites, render-

Table 11-4 The Masking of Nuclear Progesterone Acceptors in Various Tissues[a]

TISSUE	ACCEPTORS/NUCLEUS	% MASKED
OVIDUCT nuclei *in vivo*	10,600	58
OVIDUCT chromatin	7,441	71
OVIDUCT chromatin minus histone	9,278	58
OVIDUCT chromatin minus histone minus NHP	25,290	0
SPLEEN chromatin	48	100
SPLEEN chromatin minus histone minus NHP	24,156	0
ERYTHROCYTE chromatin	416	100
ERYTHROCYTE chromatin minus histone minus NHP	15,018	0

[a] The number of binding sites for the hormone was measured in either nuclei or chromatin in the presence or absence of histone and nonhistone proteins (NHP). [Data reproduced from Spelsberg et al. (1979), in *Ontogeny of Receptors and Reproductive Hormone Action* (Ref. 27), with the permission of Raven Press, New York.]

ing them inaccessible to the HRC. In oviduct nuclei as many as 58% of the acceptors are inaccessible (Table 11-4 and Ref. 27). In isolated chromatin 71% are inaccessible, and this is reduced only to 58% if histones are removed, but to 0% if nonhistone proteins are removed. When this study was extended to other tissues, particularly those that are not responsive to progesterone (such as spleen cells and erythrocytes), Spelsberg et al. (27) found that unresponsive cells contained as many acceptor sites as oviduct cells (~20,000), but all of them were masked (Table 11-4).

Since it is the presence of *cytoplasmic* receptors that determines a tissue's responsiveness, the presence of nuclear acceptors in cells lacking receptors is puzzling. Furthermore, if there is no cytoplasmic receptor there should be no need to mask the nuclear acceptor sites. These observations have several important implications because they establish a *repressive* role for some NHP and imply that during differentiation the cells of many tissues develop acceptor activity that is later masked by specific NHP in unresponsive cells.

The Mechanism of Transcriptional Activation

It is well-established that steroid hormones effect specific changes in gene expression. For example, glucocorticoids induce the synthesis of tryptophan oxygenase and tyrosine aminotransferase (TAT) mRNAs in liver, and estrogen induces the synthesis of ovalbumin, conalbumin, and other mRNAs in chick oviduct (these and other examples of steroid hormone action are discussed in Chapters 12–15). The mechanisms by which steroid hormones achieve an increase in gene transcription are discussed in this section.

The magnitude of the response of a cell to a steroid hormone is determined by the fractional occupancy of the nuclear acceptor sites, as shown in Fig. 11-6a

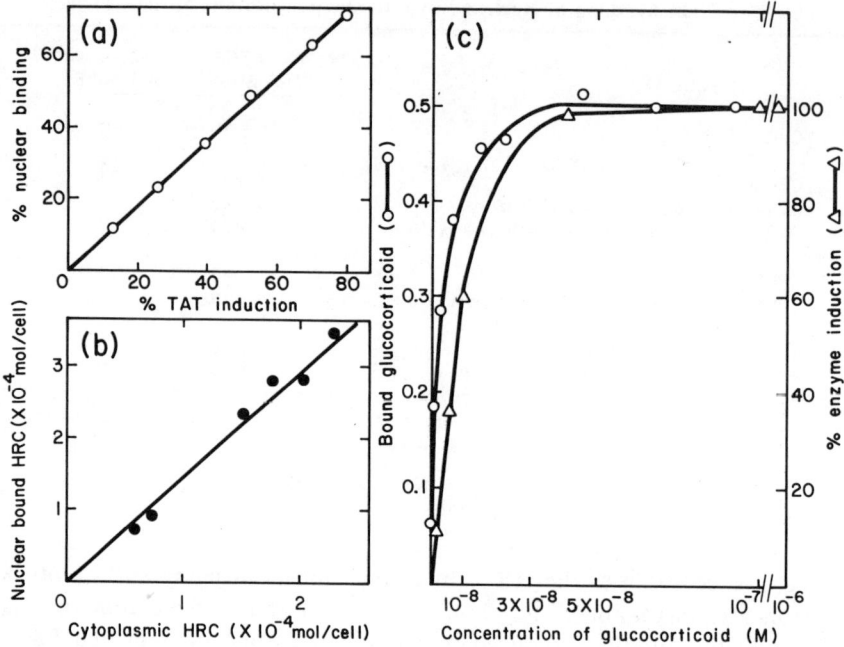

Fig. 11-6 The relationship between receptor occupancy and increase in enzyme synthesis. (a) A direct correlation between the extent of liver tyrosine aminotransferase induction and the percentage of nuclear acceptor sites binding hormone; (b) a direct relationship between the concentration of cytoplasmic hormone receptor complexes and the number of occupied nuclear acceptor sites; (c) the relationship between hormone concentration and extent of enzyme induction. [Data in (a) reproduced from Fig. 8 of Johnson et al. (25), with permission. Data in (b) and (c) reproduced from Figs. 2 and 10 of Baxter and Ivarie (19), with permission.]

for the induction of TAT by glucocorticoids. The fractional occupancy of nuclear acceptor sites is, in turn, dependent upon the concentration of cytoplasmic HRCs (Fig. 11-6b; see also Chapter 15 for a discussion of more complex kinetics of interactions between estrogen HRCs and nuclear binding sites in oviduct cells). The concentration of cytoplasmic HRCs is a function of both the concentration of hormone and the concentration of receptors, the extent of induction being determined, initially, by the concentration of hormone, but eventually the number of receptors becomes limiting (Fig. 11-6c). Thus the *degree* of gene expression is determined at the cytoplasmic level rather than at the nuclear level.

The net result of the interaction between HRC and nuclear acceptor molecules is an appreciable change in chromatin structure. This has been observed in the electron microscope (35) and is supported by the results of rifampicin challenge assays (18, 19, 21, 25). The results of a typical rifampicin challenge assay are shown in Fig. 11-7 for the action of dexamethasone (a glucocorticoid analogue, Chapter 10) on liver. Within 1 h after hormone injection there is a

50% increase in *E. coli* RNA polymerase binding to isolated chromatin, a result that is remarkably similar to that obtained with thyroid hormone (cf. Fig. 11-3). Similar changes in chromatin structure have been obtained for a number of tissues in response to estrogen, progesterone, and testosterone, indicating that it is a common feature of steroid hormone action. The effect of estrogen on endometrial nuclei has been studied microscopically (35), and a dispersion of condensed chromatin was observed within 1 h after hormone injection. It is likely that this dispersion of chromatin is responsible for the increase in *E. coli* RNA polymerase binding sites. The amount of chromatin that becomes dispersed (~25% of the total) is far more than that required to code for the small number of proteins each steroid hormone is known to induce. Therefore, the action of steroids is likely to be a two-stage process. The first stage disperses a large amount of DNA (probably by "fastening" it to the matrix) that contains all the sequences to be transcribed, and the second stage involves the location of the specific DNA sequences that are to be transcribed.

Steroid hormones also increase the activity of RNA polymerase II molecules as shown in Fig. 11-8 for the action of estradiol on cells of the uterus (31). The polymerase response is characteristically biphasic (Fig. 11-8). There is a peak of activity within half an hour, and polymerase activity returns to basal levels by 1 h. In the continued presence of hormone there is a second, sustained increase in enzyme activity that is not observed if hormone is withdrawn. Thus the continued presence of hormone is required for a maximal response. The biphasic nature of the activity profile may reflect a redistribution of HRC molecules between the different pools of acceptor (32).

There are many aspects of the mechanism of transcriptional activation by steroid hormones yet to be explained. The lack of knowledge in this area re-

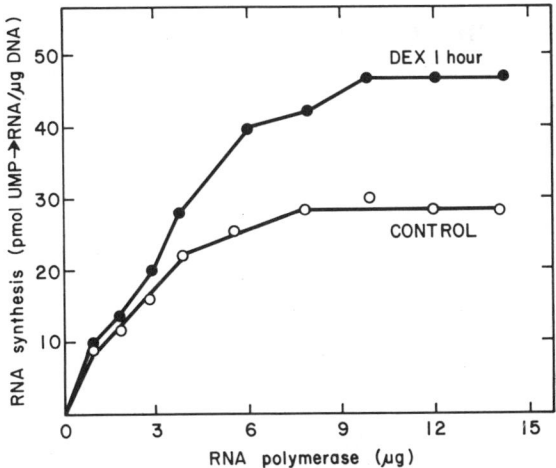

Fig. 11-7 Rifampicin challenge assay carried out to show the changes in chromatin structure induced by the glucocorticoid hormone analogue dexamethasone. [Data reproduced from Fig. 10 of Johnson et al. (25), with permission.]

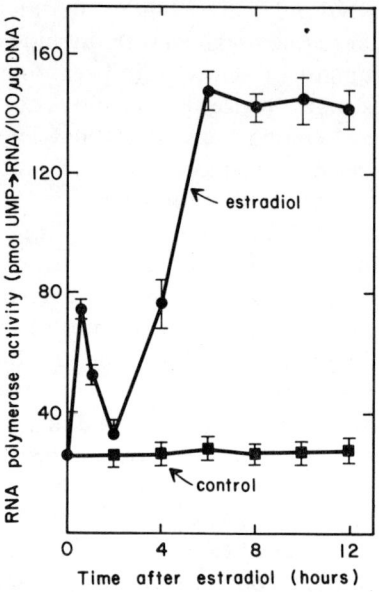

Fig. 11-8 Biphasic changes in the concentration of RNA polymerase II activity in uterus triggered by estradiol. [Reproduced from Fig. 1 of Borthwick and Smellie (31) with the permission of the Biochemical Society, London.]

flects our general lack of understanding of gene expression in eukaryotes. In fact, the study of steroids has contributed greatly to what little we do actually know (see Chapters 3 and 4) and should contribute much more in the future.

POLYPEPTIDE HORMONES TRANSMIT INFORMATION TO THE PLASMA MEMBRANE

The polypeptide hormones are a diverse group of 40 or so molecules ranging in size from tripeptides to large polypeptides. Despite such diversity of structure, and an even greater diversity in their effects, virtually all of these hormones are believed to share one common feature in their mechanism of action. That is, they transmit their information only as far as the plasma membrane of target cells. The reason for this is not immediately obvious, but may be related to the fact that, unlike thyroid and steroid hormones, which only appear to exert effects on the nucleus, the polypeptide hormones also affect general biochemical processes in the cytoplasm and other cell compartments. In fact, some polypeptide hormones may not effect changes in gene expression at all. In general, polypeptide hormones are responsible for coordinating very short-term changes in metabolism (minutes to hours) rather than being involved in initiating long-term, complex programs of growth and differentiation. Another feature of the actions of polypeptide hormones is their interaction with each other during the stimulation of target cells, and the plasma membrane is an excellent site to integrate and multiplex these interactions.

Plasma-Membrane Receptors

Every polypeptide hormone that has been examined has been found to have a plasma-membrane receptor (Table 11-5), and the binding characteristics of these receptor proteins indicate that they are capable of binding hormones at their physiological concentrations (10^{-9}–10^{-13} M). Although a large number of receptors are known to exist, based on hormone-binding data, only a few have been isolated and studied biochemically. The insulin receptor from liver cells and fat cells, the gonadotrophic hormone (LH and FSH) receptors of testis and ovary, and the prolactin receptor from a number of tissues have been isolated (36–38). All are large polypeptides (M_r = 200,000–300,000), and the number of receptors per cell varies considerably, from 15,000 to 20,000 LH receptors on the cell surface of testicular Leydig cells to 250,000–300,000 insulin receptors on liver and fat-cell plasma membranes.

A characteristic property of plasma-membrane receptors is the low percentage occupancy required to generate a maximal biological response. For example, LH and FSH trigger maximal responses when only 1% of the target cell surface receptors are occupied by hormone (Table 11-5). Since there are only

Table 11-5 Polypeptide Hormones and General Characteristics of Their Receptors and Secondary Messengers[a]

HORMONE	TARGET TISSUE	SURFACE RECEPTOR	DOWN REGULATION	↑cAMP	% OCCUPANCY
LH	testis	+	+	+	1.0
FSH	testis (immature)	+	+	+	1.0
FSH	ovary	+	+	+	
ACTH	adrenal	+	+	+	1→10
GLUCAGON	liver, adipose	+		+	
OXYTOCIN	kidney, bladder	+		+	
VASOPRESSIN	bladder	+		+	
PTH	kidney	+		+	
CALCITONIN	kidney	+		+	
TSH	thyroid	+		+	
LHRH	pituitary	+		+	
TRH	pituitary	+	+	+	3→5/20→30
GH	pituitary	+	+	−	
INSULIN	liver	+	+	−	
PROLACTIN	testis	+	↑	−	
PROLACTIN	mammary gland	+	↑	−	

[a] ↑cAMP means the hormone triggers an increase in intracellular cAMP concentration. The % occupancy is a measure of the fraction of receptors that need to be occupied to trigger a full response. +, positive; −, negative for each characteristic.

15,000–20,000 receptors, this means that a full response can be triggered by only 150–200 molecules of hormone. Other hormones behave similarly, including ACTH, which can trigger steroidogenesis with only 1–10% occupancy (39), and luteinizing hormone-releasing hormone, which triggers short-term changes when 3–5% of receptors are occupied and longer-term changes when 20–30% are occupied (40). It is not clear what biological function the 70–99% of "spare" receptors have, unless either they ensure that at very low concentrations enough receptors are occupied for the hormone to function, or they serve as a pool of receptors to replace those internalized during the process of hormone degradation.

Since the hormones bind only to surface receptors, information transfer to the nucleus (and other cell compartments) must be mediated by *intracellular molecules*. Moreover, if as few as 200 molecules of hormone can trigger a biological response some mechanism for *amplifying* the signal must also exist. The discovery of cAMP and the subsequent discovery that it is synthesized on the intracellular side of the plasma membrane were major breakthroughs in this area. A small molecule that can be synthesized rapidly in large quantities is an ideal second messenger. This concept of the second messenger was developed initially by Sutherland (41) but has now been refined and diversified to include other molecules such as calcium ions and cyclic GMP (3′,5′-cyclic guanosine monophosphate). the second-messenger role of cAMP will be discussed first.

Cyclic AMP as an Intracellular Second Messenger

cAMP was discovered as a heat-stable intermediate in the glucagon stimulation of liver glycogen breakdown. This intermediate was capable of activating glycogen phosphorylase in a crude cell extract, and when the location of the enzyme responsible for its synthesis, adenylate cyclase (Fig. 11-9), was found to be at the plasma membrane cAMP was proposed to be the *intracellular* second messenger for an *extracellular* hormonal stimulation. This work and the ideas that led to the development of the second messenger concept are summarized in Sutherland's Nobel lecture (41) and in a subsequent monograph (42).

The advent of sensitive techniques for the measurement of changes in intracellular cAMP led to a bewildering amount of activity in this field, and the cyclic nucleotide was postulated to be involved in the regulation of a vast number of biochemical processes. It was only after the discovery of a cAMP-activated protein kinase (43) that a rational explanation for the role of cAMP in information transfer could be advanced. Thus, it is now believed that cAMP exerts all its actions via this kinase (in most cells there are actually two cAMP-dependent protein kinase isozymes, see Chapter 7) and an essential feature of information transfer is the covalent modification, by phosphorylation, of specific target proteins. This mechanism of information transfer also amplifies the original signal (see Fig. 7-7a).

Although cAMP is involved in the activation and inhibition of many diverse processes, the role of cAMP is identical in all of them. Thus it is an *entirely*

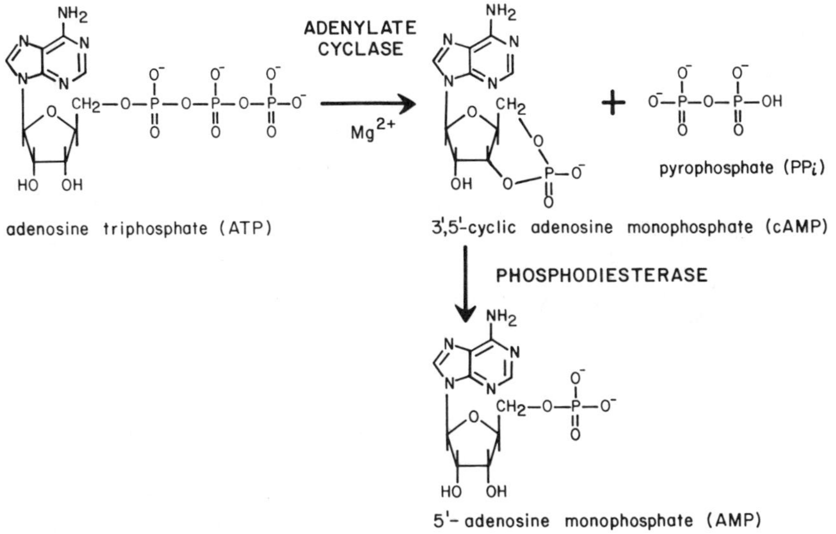

Fig. 11-9 The synthesis of cAMP by adenylate cyclase and its degradation by phosphodiesterase.

nonspecific transducer of a hormonal signal into a change in intracellular protein kinase activity. The *specificity* of the response is determined by (i) the hormone–receptor interaction and (ii) the presence of specific proteins that are targets for the activated cAMP-dependent protein kinase. The spectrum of phosphoproteins is a characteristic of each cell and is determined during the course of differentiation.

Virtually all of the oligopeptide and polypeptide hormones that have been examined use cAMP as their second messenger (see Table 11-5; a more complete list is given in Table 6 of Ref. 44). There are, however, three notable exceptions—growth hormone, insulin, and prolactin. These hormones do not use cAMP as a second messenger. All three hormones have multiple effects on several tissues, rather than a single effect on one highly specialized target tissue, and at the present time the nature of their second messengers, if any, is not understood.

Information transfer via cAMP is very versatile; it can mediate control over processes lasting for only a few seconds as well as over processes taking many hours to reach completion. Typical examples are shown in Fig. 11-10. The cyclic nucleotide mediates the development of contractile force *and* the supply of energy for muscle contraction, a process that is complete in a matter of seconds (Fig. 11-10a), as well as the biosynthesis of testosterone by cells of the testis following gonadotrophic stimulation, a process occurring over a period of 20 h or more (Fig. 11-10b).

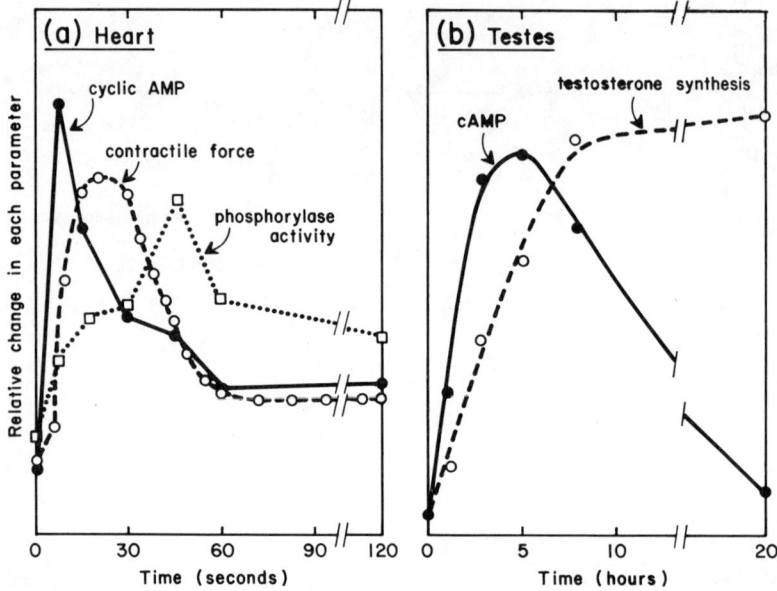

Fig. 11-10 Temporal differences in cellular responses to hormones in which cAMP acts as second messenger. (*a*) The response of heart cells to adrenaline, showing the increase in cAMP concentration, contractile force, and phosphorylase activity; (*b*) the increase in cAMP and testosterone biosynthesis in cells of the testes stimulated by gonadotrophic hormones. Note the differences in time scales. [Data in (*a*) reproduced from Fig. 6 of Sutherland (41) with permission, © The Nobel Foundation 1972. Data in (*b*) reproduced from Fig. 4 of Catt et al. (45) with permission.]

Activation of Plasma-Membrane Adenylate Cyclase

An understanding of the interaction between hormones, receptors, and adenylate cyclase requires some knowledge of plasma-membrane structure. Most membrane proteins are suspended in the fluid, lipid bilayer and have a considerable freedom for lateral movement (Fig. 11-11). Some cells have surface receptors for only one polypeptide hormone, whereas other cells may have several different kinds of receptors all of which are free to move and to interact with each other. Adenylate cyclase, the enzyme responsible for cAMP biosynthesis, is located on the inner portion of the plasma membrane with its active site facing into the cytoplasm. In the absence of hormone, receptors do not make contact with the cyclase, and it remains inactive. In the presence of hormone, the hormone–receptor complexes move laterally and make contact with and activate the enzyme.

There is evidence that receptor movement, in the lipid bilayer, is directed by the submembranous structures such as microfilaments and microtubules. These structures form a layer on the inside of the plasma membrane and interact with membrane proteins. The fact that drugs such as colchicine and cytochalasin B, which disrupt the formation of microtubules and microfilaments, respectively, interfere with information transfer from polypeptide hormones confirms the importance of these structures. Moreover, these structures also mediate the integration of hormonal signals since the binding of one hormone often restricts the movement of the receptors for other hormones to prevent them from triggering a response in the cell. The role of membrane structure and other factors in the regulation of adenylate cyclase has recently been reviewed (45–53).

Adenylate cyclase is a large enzyme (M_r = 160,000–200,000) consisting of a catalytic subunit and a small (M_r = 42,000) regulatory subunit. The catalytic subunit is essentially inactive unless four other components are present: hormone, receptor, regulatory subunit, and guanosine triphosphate (GTP). The requirement for the hormone–receptor complex is obvious, but the requirement for GTP and the role played by the regulatory subunit are still not certain.

The presence of GTP is an absolute requirement for enzyme activity, and the following model has been proposed for its role. Following interaction between a hormone-occupied receptor and the regulatory subunit, GTP is believed to bind to the regulatory subunit and promote its association with the catalytic subunit to produce an active enzyme. The active enzyme continues to synthesize cAMP (Fig. 11-11) until the GTP molecule is hydrolyzed to GDP by a GTPase that is an intrinsic part of the regulatory subunit. The hydrolysis of GTP causes dissociation of the active enzyme complex. Support for this model comes from observations that GTP analogues that are resistant to hydrolysis, such as Gpp(NH)p, are powerful activators of adenylate cyclase. Moreover, cholera toxin, which catalyzes the ADP-ribosylation of the regulatory subunit, is also an activator, since covalently modified regulatory subunits do not have GTPase activity.

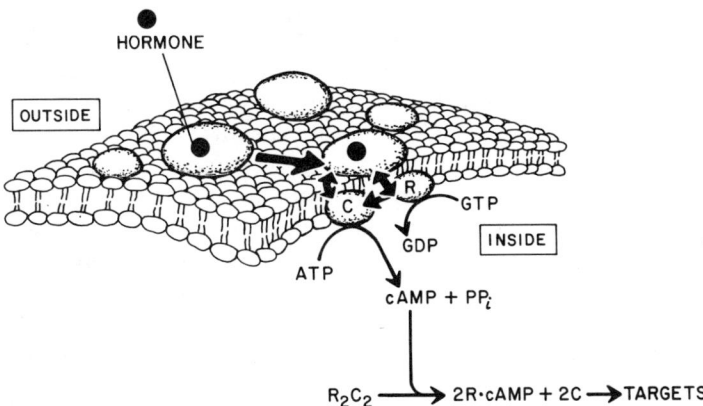

Fig. 11-11 A diagrammatic representation of the events at the cell surface when a polypeptide hormone binds to its receptor and triggers an increase in the concentration of cAMP inside the cell. cAMP causes the activation and dissociation of cAMP-dependent protein kinase (R_2C_2).

This mechanism for the activation of adenylate cyclase permits an *extracellular*, information-carrying molecule to increase the *intracellular* concentration of cAMP *without entering the cell*. Moreover, this is the only mechanism for the activation of adenylate cyclase and cAMP *can only be synthesized in response to extracellular effectors*. There are no known examples of an increase in cAMP synthesis being generated by a solely intracellular event. However, the *magnitude* of the response of cAMP to an extracellular signal can be modified by intracellular factors. For example, changes in the GTPase activity of the regulatory subunit could alter the activity of adenylate cyclase. Furthermore, activation or inhibition of phosphodiesterase, the enzyme that catalyzes the hydrolysis of cAMP (Fig. 11-9), is another important site for intracellular regulators to modulate the signals originating extracellularly. In addition, factors that alter the degree of polymerization of the submembranous cytoskeletal network also affect the extent of the cell's response to polypeptide hormones.

cAMP, Protein Kinases, and the Transfer of Information to the Nucleus

Most of the effects of polypeptide hormones on their target cell(s) do not require changes of enzyme synthesis. The hormones achieve their desired effect via covalent modification of existing enzymes by the catalytic subunit of cAMP-dependent protein kinase. For example, glucagon effects the switch from glucose storage to glucose production by the phosphorylation of the key enzymes of the two opposing pathways (see Fig. 7-2). A number of polypeptide hormones do actually change the rate of synthesis of enzymes (and other proteins), and some of these effects are mediated at the transcriptional level. For those polypeptide hormones using cAMP as second messenger the change in transcription is likely to occur via the phosphorylation of the nuclear proteins by a cAMP-dependent protein kinase.

Most cells contain two types of cAMP-dependent protein kinase (designated types I and II in order of their elution from ion-exchange resins, Fig. 11-12). The proportions of the two vary considerably from tissue to tissue, and even in the same tissue of different species. The enzymes have been isolated and purified, and their properties have been studied in detail (54–56). Both enzymes are tetramers built up from two regulatory and two catalytic subunits (R_2C_2). The regulatory subunit is different in type I and type II kinases, but the catalytic subunits are identical. Since the enzymes are activated by the binding of cAMP to the regulatory subunits causing the dissociation of active catalytic subunits (Fig. 11-11), it is generally difficult to assess which isozyme is being activated by a particular hormone.

The enzymes appear to be predominantly cytoplasmic and were initially reported to be absent from the nucleus. This would require the translocation of catalytic subunits (or a protein phosphorylated by catalytic subunits) from the cytoplasm to the nucleus to cause the change in gene expression. Evidence has been presented that this can, indeed, take place (55). However, since it has now been established that nuclei actually contain *both* types of cAMP-dependent

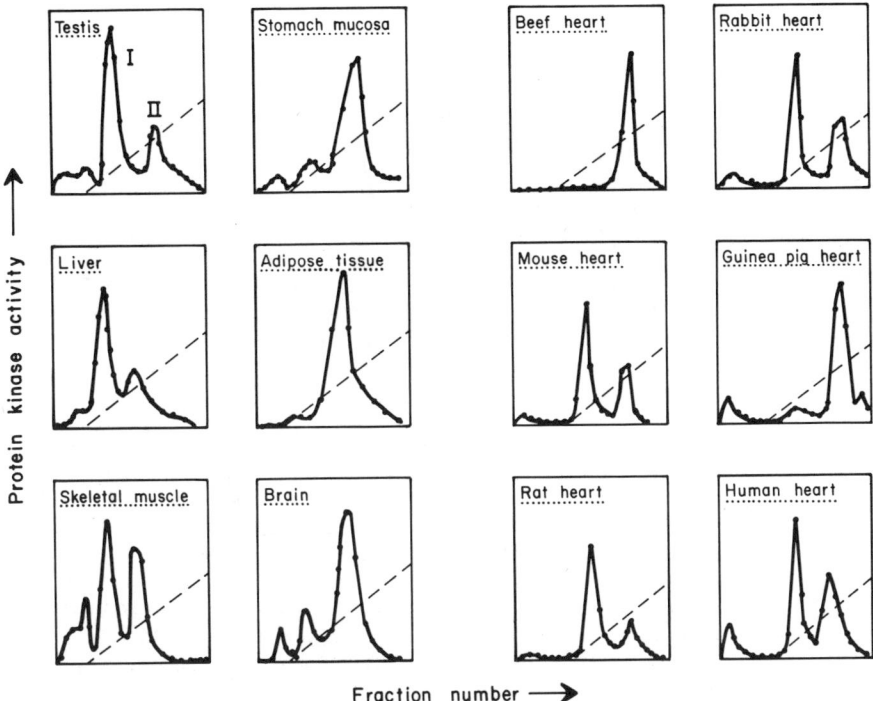

Fig. 11-12 Chromatographic profiles of protein kinase isoenzymes in various tissues of the rat and in heart tissue from various species. The chromatographic material is DEAE-cellulose, and the dashed line indicates the salt gradient used to elute the enzymes. The type I enzyme elutes first, followed by the type II enzyme. [Data reproduced from Figs. 5 and 15 of Walsh and Cooper (55), with permission.]

protein kinase (57), it is not clear whether additional C subunits are translocated during hormonal stimulation, or whether cAMP enters the nucleus and stimulates the activation of intranuclear kinase molecules (assuming that free cAMP can exist in the cell long enough to reach the nucleus). Either way, the net result is the release of active catalytic subunits inside the nucleus.

Phosphorylation and the Mechanism of Transcriptional Activation

There is little doubt that covalent modifications, including phosphorylation, play a major role in gene activation in response to hormonal stimuli (see Chapter 4 for the mechanisms involved). For example a direct correlation has been established between the phosphorylation of the serine residue at position 37 of H1 histone and the extent of induction of tyrosine aminotransferase in liver by glucagon (or cAMP), as shown in Fig. 11-13 (58, 59). Since the phosphorylation of H1 has been implicated in the conversion of the 20–30-nm chromatin fiber to

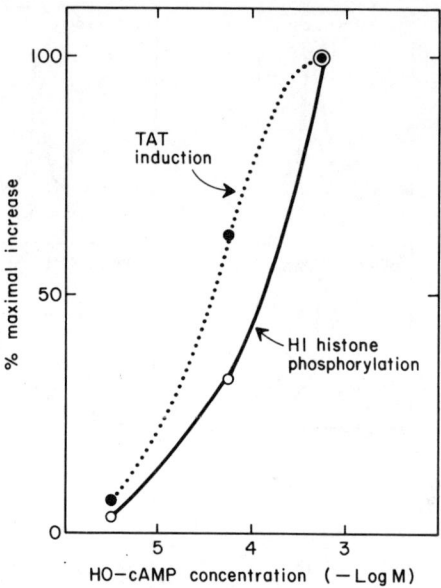

Fig. 11-13 The relationship between the phosphorylation of H1 histone and the extent of tyrosine aminotransferase (TAT) induction by a cAMP analogue. M = molarity. [Reproduced from Fig. 2 of W. D. Wicks et al. (1975). Possible participation of protein kinase in enzyme induction *J. Cyclic Nucl. Res.* **1**, 49–58, with the permission of Raven Press, New York.]

the 10-nm fiber (Fig. 4-9), the gene for tyrosine aminotransferase may reside at this higher structural level when not being expressed.

Protein kinase-mediated activation of transcription is not preceded by the massive changes in chromatin structure that precede transcriptional activation induced by steroids and thyroid hormone. The reason for this difference is not clear. Indeed, we have very little knowledge of the way in which protein kinases mediate changes in gene expression.

Some Polypeptide Hormones Do Not Use cAMP as Second Messenger

The three large polypeptide hormones—insulin, prolactin, and growth hormone—interact with target cells via cell-surface receptors, but do not alter the concentration of cAMP (Table 11-5). Thus, either they enter the cell (intact or as physiologically active fragments) or they use a different second messenger. Very little is known about information transfer from prolactin and growth hormone. These hormones are particularly difficult to study because they participate in multihormone processes in which individual pieces of information cannot usually be assigned to a particular hormone. Insulin, on the other hand, has been studied in detail and reviewed recently (60).

There is evidence (60) that insulin is rapidly internalized and binds to various subcellular organelles. However, it is still not clear whether this insulin is biologically active or whether it is in the process of degradation. It is never found in the nucleus, and any effects of the hormone on gene activation must, therefore, be mediated by a second messenger. Such a second messenger has not been found. Most recently, insulin has been suggested to mediate its effects via a considerable number of different mechanisms including inhibition of adenylate cyclase, activation of phosphodiesterase, activation of phosphoprotein phosphatase, activation of guanylate cyclase (see next section), inhibition of Ca^{2+} release (see Fig. 11-14), inhibition of protein degradation, and specific protease activation. Thus, either the hormone interacts with different cells in different ways or there is some common factor in all these mechanisms that has not yet been discovered. It is noteworthy that most of the proposed mechanisms involve an interaction of insulin with proteins responsible for the generation of second messengers for other hormones.

Other Possible Hormone Second Messengers—cGMP

In addition to cAMP most cells have low concentrations of another cyclic nucleotide cGMP (guanosine 3',5'-cyclic monophosphate), and a role for this cyclic nucleotide in information transfer has been predicted. A plasma membrane-associated guanylate cyclase is present in these cells, although appreciable amounts may be in the cytoplasm and in other organelles including the nucleus. cGMP-activated protein kinases are also present in cells, providing all the necessary components for an information-transfer system. However, despite considerable effort there has been no clear demonstration of an involvement of cGMP in information transfer in response to hormones at physiological cGMP levels, although it has been reported that some neurotransmitters, notably acetylcholine, mediate their effects through cGMP (see the references in Ref. 44).

The cGMP-dependent protein kinase (M_r = 160,000) has been isolated from a number of tissues and found to consist of two dissimilar subunits, each of which binds cGMP. However, the enzyme does not dissociate into regulatory and catalytic subunits upon activation (61). A heat-stable, acidic protein located in the cytoplasm is a powerful activator of the enzyme. This modulator protein appears to exist in an aggregated form (M_r = 180,000) but dissociates into active subunits (M_r = 34,000). Since the spectrum of proteins phosphorylated by cGMP-activated kinase is similar to that of cAMP-dependent kinases (61), and since the concentration of cAMP-dependent kinases is some 10-fold higher, the physiological role of the cGMP information transfer system is not clear (62).

The fact that cGMP-dependent protein kinase activity has been found in the nucleus (63), coupled with the recent demonstration of the phosphorylation of the serine residue at position 37 of histone H1 by this enzyme (64), suggests a function in gene activation, but this remains to be demonstrated.

Other Possible Hormone Second Messengers—Calcium Ions

The inorganic divalent cations calcium and magnesium participate in numerous biochemical reactions and processes. For example, virtually every enzyme involved in the transfer of phosphate, and certain other inorganic groups, require magnesium ions in the vicinity of the active site, and many other enzymes require calcium ions. In addition to their function as catalysts, calcium ions appear to *modulate* the rate of certain reactions. This is a somewhat surprising observation since calcium ions, like magnesium ions, have no intrinsic information content. However, upon closer examination there are considerable differences between calcium and magnesium. Thus the cytoplasmic calcium concentration is maintained at 10^{-7}–10^{-8} M in contrast to a magnesium concentration of 1–5 $\times$ 10^{-3} M. An elaborate system of pumps is required to keep the calcium level so low because the extracellular concentration (10^{-3} M) is some 10,000 times greater. In addition, *every* eukaryotic cell contains high-affinity nonenzyme, calcium-binding proteins (no equivalent magnesium-binding proteins have been found). These proteins undergo conformational changes in the presence of calcium ions to forms that specifically activate a number of cellular processes. These special aspects of calcium metabolism form the basis of a system for signal transduction with calcium ions as second messengers (65).

The number of reactions purported to be controlled by calcium ions is legion, and our discussion here will be confined to calcium ions fluxes in response to hormones (and neurotransmitters) acting at the plasma membrane. To understand these changes some knowledge of the distribution of calcium ions in the cell is necessary. More than 90% of the ion is sequestered inside intracellular organelles, principally mitochondria and the endoplasmic reticulum. The membrane of these vesicles contain calcium-dependent ATPases, which actively pump calcium into these stores to maintain a low cytoplasmic concentration (calcium ions are also pumped out of the cell by plasma-membrane ATPases). Some hormones and neurotransmitters binding to surface receptors appear to open calcium channels within the membrane and allow calcium ions to flow into the cell. Other effectors are able to stimulate the flow of calcium ions into the cytoplasm from the intracellular stores. Both mechanisms result in an increase in the cytoplasmic free-calcium concentration.

The flow of calcium ions across the plasma membrane in response to polypeptide hormones such as vasopressin and α-adrenergic neurotransmitters (α- and β-adrenergic receptors are described in the section on catecholamines in this chapter) is usually accompanied by a turnover of membrane phospholipids, especially phosphatidyl inositol (Fig. 11-14*a*). The membrane phospholipids, particularly phosphatidic acid, the product of phosphatidyl inositol turnover, have a high affinity for calcium ions and may actually function as carriers (66–68).

Hormones such as glucagon, the catecholamines, and neurotransmitters binding to β-adrenergic receptors all stimulate adenylate cyclase and increase the intracellular concentration of cAMP. The cyclic nucleotide stimulates the re-

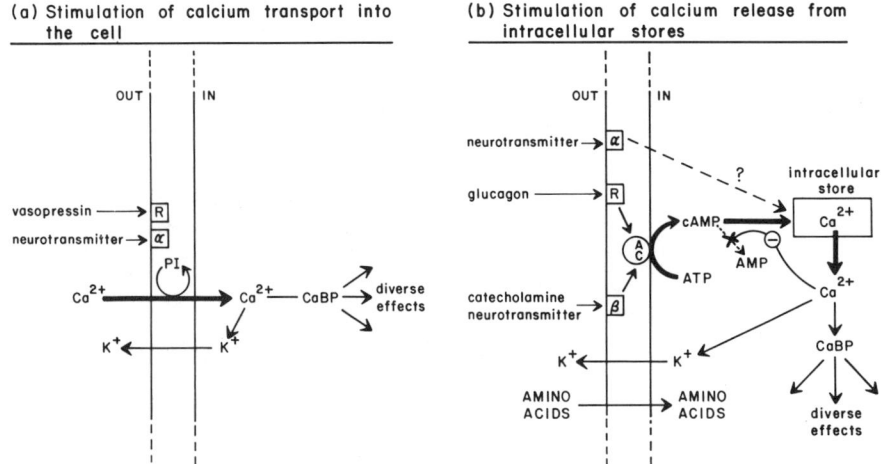

Fig. 11-14 The regulation of calcium ion fluxes by hormones. The fluxes either take place across the plasma membrane (*a*) or across membranes of intracellular stores (*b*). Certain hormones and α-adrenergic neurotransmitters stimulate calcium influx into the cell by stimulating plasma membrane phosphatidyl inositol (PI) turnover (*a*). Other hormones and β-adrenergic neurotransmitters increase adenylate cyclase activity, and the increased concentration of cAMP stimulates intracellular calcium release. CaBP = calcium-binding protein (e.g., calmodulin).

lease of calcium ions from intracellular stores (Fig. 11-14*b*). Neurotransmitters binding to α-adrenergic receptors may also stimulate calcium release from intracellular stores by an unknown mechanism. Both modes of calcium flux (Fig. 11-14*a*, 11-14*b*) are usually accompanied by an efflux of potassium ions from the cell or an uptake of amino acids or both.

Of the calcium-binding proteins that are sensitive to these changes in calcium concentration the ubiquitously distributed (among eukaryotes) calmodulin has received the most attention. The discovery, characterization, and properties of calmodulin have been recently reviewed (69–72). It is a small (M_r = 17,000), heat-stable, acidic protein, and each molecule can bind four calcium ions ($K_D \sim 10^{-6}\ M$). The binding of calcium ions induces a large conformational change in the molecule (it is not known if the molecule has different, characteristic shapes depending on the number of sites occupied). This conformational change increases the affinity of the calcium–calmodulin complex for a number of intracellular targets, virtually all of which are activated in its presence.

A number of proteins, many of them protein kinases, have been shown to be activated by calcium–calmodulin (Table 11-6). Of these, the most characterized are the proteins in smooth muscle and nerve cells, where the actions of an influx of calcium, or calcium release from intracellular stores, on cell activity are fairly well established (see below). Calcium–calmodulin also activates adenylate cyclase in some tissues as well as phosphodiesterase. These actions, together with the activation of the Ca^{2+}-Mg^{2+} ATPase, establish a close relationship between

Table 11-6 Proteins That Are Modulated by Calcium–Calmodulin

Protein kinases and other proteins activated by calcium–calmodulin

1. Myosin light chain kinase
2. Phosphorylase (–synthetase) kinase
3. Sarcoplasmic reticulum kinase
4. Neuronal / brain kinases
5. Rat liver nuclear kinase
6. Ca^{2+} dependent protease–activated kinase
7. Adenylate cyclase in brain, liver
8. Phosphodiesterase
9. Ca^{2+} Mg^{2+} ATPase
10. Phospholipase A_2
11. Microtubules

cAMP levels and calcium fluxes to the point where it is often difficult to separate the second-messenger role of each one. The activation of phospholipase A_2 links calcium–calmodulin to another potential messenger system, the prostaglandins (see next section). Calcium–calmodulin also binds to microtubules and under appropriate conditions appears to dissociate them. However, this process requires ~10^{-4} M calcium compared with 10^{-6}–10^{-7} M for the activation of other proteins listed in Table 11-6.

The ability of calcium–calmodulin to activate a nuclear protein kinase (57) provides this second-messenger system with the ability to effect changes in gene expression. However, such changes have not yet been demonstrated. Moreover, there are some aspects of calcium ion metabolism that are inconsistent with such a role. For example, calcium fluxes in cells are *transient*, lasting from milliseconds to minutes but seldom longer. In muscle cells this flux is sufficiently long-lived to activate myosin light-chain kinase, which phosphorylates myosin to develop muscle tension, and to simultaneously activate phosphorylase kinase to stimulate enough glycogen breakdown to supply energy for the contraction. Similarly, in some nerve cells the calcium transient causes the release of neurotransmitter by stimulating the stimulus–secretory coupling system. Both of these are short-lived processes, and calcium–calmodulin only activates its targets during the time the calcium levels are elevated. When the calcium levels fall calmodulin is no longer an activator.

It has also been shown that, in all cells, calcium ions from the cytoplasm are rapidly sequestered into intracellular stores (73). Therefore, calcium ions cannot travel great distances inside the cell and ions entering through plasma-membrane channels are unlikely to be able to reach the nucleus to stimulate calcium–calmodulin activation of protein kinases. Thus, unless some as yet undiscovered system exists for the conversion of the transient calcium signal into one of longer duration, a role for this second-messenger system in the regulation of enzyme synthesis by polypeptide hormones remains to be established. The presence in some cells of a protein kinase that is proteolytically activated (i.e., irreversibly activated) by a calcium–calmodulin-dependent protease could be a component of such a system. At the present time calcium ions appear to regulate only those processes occurring close to the membranes across which the ion fluxes occur.

Prostaglandins: Hormones or Second Messengers, or What?

The prostaglandins are a group of polyunsaturated, hydroxy-fatty acids derived from arachidonic acid (see Refs. 74 and 75 for summaries of their structure and function). They have been found in almost every tissue of the body and are synthesized by a membrane-associated (plasma membrane and endoplasmic reticulum) enzyme complex. The arachidonic acid for prostaglandin biosynthesis is stored in the membrane phospholipids, and membrane phospholipid turnover is a prerequisite to prostaglandin biosynthesis. This turnover is catalyzed by the calcium–calmodulin-activated phospholipase A_2.

The prostaglandins have characteristically short half-lives ranging from seconds to minutes. Indeed, the turnover of some of them, particularly the endoperoxides and thromboxanes, is so rapid that they were originally considered to be only reaction intermediates and not biologically active molecules. However, they are now known to be powerful activators of such processes as platelet aggregation and smooth muscle contraction.

The observation that some of the prostaglandin and related metabolites, particularly the E series of prostaglandins, can mimic the actions of certain hormones led to the speculation that they may be intracellular second messengers. However, the diverse nature of their sites of biosynthesis and changes in concentration that appear to be independent of hormonal stimulation have cast some doubt on this idea. Moreover, the prostaglandins are able to modulate both calcium ion fluxes and the concentration of cAMP. Thus, although there is no real direct evidence for prostaglandins as specific second messengers, what appears to be emerging is the existence of a highly integrated intracellular signal-transduction system. Exogenous stimuli, such as hormones, interact with this network in a variety of ways through cell surface receptor-mediated interactions. The net result of this interaction may be subtle changes in one or more of these intracellular molecules all of which have some capacity to transmit and amplify information.

Prostaglandins are associated with many reactions of the cardiovascular system and consequently are present in the blood. Moreover, many cells, including liver, adipose tissue, adrenal, thyroid, and corporaluteal cells, possess cell-surface receptors for prostaglandins. The E series of prostaglandins is capable of activating plasma-membrane adenylate cyclase and raising the intracellular concentration of cAMP. However, as yet, there is no consensus for a hormonal role of these molecules or any involvement in the regulation of enzyme synthesis.

CATECHOLAMINES AND NEUROTRANSMITTERS ALSO ONLY TRANSMIT INFORMATION TO THE CELL MEMBRANE

Even though they are small molecules (Table 10-3) the catecholamines, adrenaline and noradrenaline, bind to cell-surface receptors and elicit their effects via second messengers. Initially, two distinct receptors for catecholamine binding were identified and designated α- and β-adrenergic receptors (reviewed in Refs. 76–78). Subsequent work has revealed subclasses of each of these main classes, and these are designated α_1, α_2, β_1, and β_2. Many cells possess all four types of receptors, whereas other have a restricted number. In rat liver, for example, more than 80% of the α-receptors are α_1, whereas in the uterus they are 60% α_2. The physiological significance of these differences is not known.

As shown in Fig. 11-14a the binding of a hormone or neurotransmitter to α-receptors leads to an increased turnover of phosphatidyl inositol and an influx of calcium ions into the cell (they may also stimulate calcium release from intracellular stores, Fig. 11-14b). On the other hand, binding to β-receptors activates adenylate cyclase via an interaction with the GTP-binding regulatory subunit (Fig. 11-14b) followed by the release of calcium from intracellular stores. In general, the α-adrenergic receptors have a greater affinity for adrenaline, whereas the β-receptors have a greater affinity for noradrenaline.

The catecholamines usually stimulate very acute changes in their target cells that do not require changes in enzyme synthesis. However, prolonged exposure of cells to catecholamines (usually at pharmacological doses) can lead to changes in the synthesis of specific enzymes. For example, noradrenaline induces the synthesis of lactate dehydrogenase in cultured brain (C6-glial) cells, an effect that appears to be mediated by cAMP (see Fig. 12-13 for the induction of this enzyme by isoproterenol, a catecholamine analogue).

REGULATION OF RECEPTOR CONCENTRATION AND THE PHENOMENON OF DOWN REGULATION

The most important factor determining the ability of cells to respond to a hormone is the presence of receptors. Moreover, the concentration of receptors often defines the magnitude of the cell's response. It is not surprising, therefore, that cells carefully regulate receptor concentrations. With the exception of prolactin (45, 79), hormones do not usually induce the synthesis of their own receptors, although they are frequently involved in the synthesis of receptors for other hormones. For example, follicle-stimulating hormone and progesterone induce the synthesis of luteinizing hormone receptors in the testis (37), whereas estrogen induces the synthesis of progesterone receptors in the uterus and prolactin receptors in the liver (prolactin induces the synthesis of its own receptors in liver and testis).

Examples of hormonal control of receptor concentration are given in Fig. 11-15. Estradiol induces a seven- to eightfold increase in uterine progesterone receptor concentration within 24 h (Fig. 11-15a), and the concentration of receptors subsequently falls to basal levels over the next few days. However, if progesterone is injected there is a marked *decrease* in the concentration of its own receptors (80). A loss of receptors for a particular hormone *in the presence of that hormone* is a phenomena common to all types of hormones and is usually referred to as receptor "down regulation" (see Table 11-5). *Thus cells become*

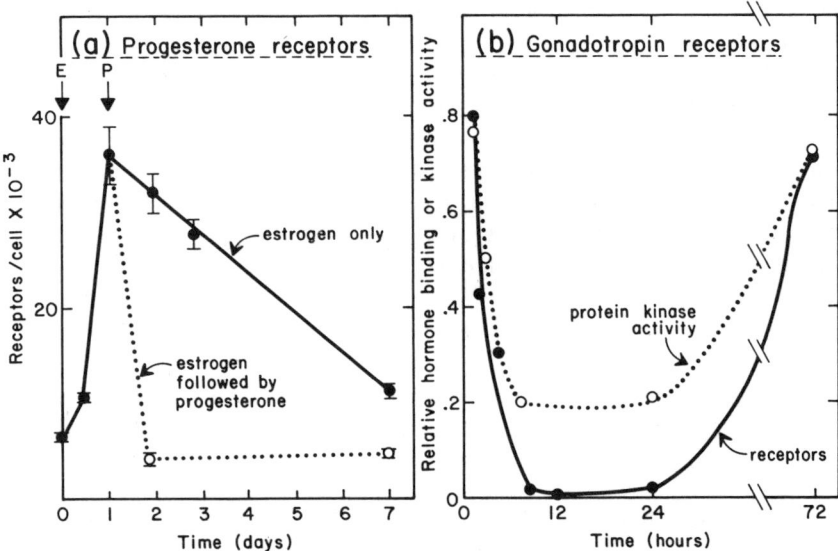

Fig. 11-15 Examples of hormone–receptor down regulation. The down regulation of uterine progesterone receptors by progesterone is shown in (a), and the down regulation of ovarian gonadotropin (reproductive hormone) receptors in the presence of hormone is shown in (b). In the latter cells there is also a decline in intracellular protein kinase levels. [Data in (a) are reproduced from Fig. 4 of Milgrom (80) with permission, and in (b) from Fig. 2 of Sen, Aghar, and Menon (81) with permission of th American Society of Biological Chemists, Inc.]

less sensitive to a hormone as the duration of exposure increases. In extreme cases this can lead to an almost total loss of receptors. For example, there is a total loss of gonadotropin (follicle-stimulating hormone) receptors from rat ovarian plasma membrane following injection of the hormone (Fig. 11-15 and Ref. 81). This loss of receptors is also accompanied by a decrease in the concentration of protein kinase, indicating that there is a loss of all of the components of the information-transfer system. Down regulation has now been studied in many cells (reviewed in Ref. 82) and shown to be a common physiological response by which cells regulate their hormone responsiveness.

Once a hormone binds to its receptor it invariably triggers a response, and receptor down regulation is the mechanism by which cells avoid being overstimulated. This is particularly important for hormones that induce a complex series of reactions in a cell that may take several days to complete. Although down regulation usually occurs with homologous hormones, it may also occur in other situations. For example, partial hepatectomy (surgical removal of two-thirds of the liver) of rats stimulates the remaining cells to proliferate. Proliferate activation causes a major change in gene expression and mRNA synthesis (83), and this is accompanied by a loss of glucocorticoid receptors from the cytoplasm (84–85) and thyroid hormone receptors from the nucleus (86). Presumably, this action prevents hormone-induced structural rearrangements of chromatin that may be incompatible with proliferation. The fate of receptors removed from the cell surface, cytoplasm, or nucleus is not known, although there are indications that they are removed to an inaccessible cell compartment rather than being degraded (87).

Changes in the concentration of hormone receptors during experimental protocols designed to assess the influence of hormones on enzyme synthesis have often been ignored. However, receptor concentration is one of the most important factors in determining the responsiveness of a cell, and more attention needs to be paid to these changes.

REFERENCES

1. G. S. Rao (1981). Mode of entry of steroid and thyroid hormones into cells. *Mol. Cell. Endocrinol.* **21**, 97-108.
2. N. L. Eberhardt, J. W. Apriletti, and J. D. Baxter (1980). The molecular biology of thyroid hormone action. In G. Litwack (Ed.). *Biochemical Actions of Hormones*, Vol. 7. Academic Press, New York, pp. 311-394.
3. J. H. Oppenheimer (1979). Thyroid hormone action at the cellular level. *Science* **203**, 971-979.
4. J. D. Baxter, N. L. Eberhardt, J. W. Apriletti, L. K. Johnson, R. D. Ivarie, B. S. Schachter, J. A. Morris, P. H. Seeburg, H. M. Goodman, K. R. Latham, J. R. Polansky, J. A. Martial, and T. Cobley (1979). Thyroid hormone receptors and responses. *Rec. Prog. Horm. Res.* **35**, 97-153.
5. H. H. Samuels, F. Stanley, J. Casanova, and T. C. Shao (1980). Thyroid hormone nuclear receptor levels are influenced by the acetylation of chromatin-associated proteins. *J. Biol. Chem.* **255**, 2499-2508.
6. B. W. O'Malley, H. C. Towle, and R. J. Schwartz (1977). Regulation of gene expression in eucaryotes. *Ann. Rev. Genetics* **11**, 239-275.
7. J. Kruh and L. Tichonicky (1976). Effect of triiodothyronine on rat liver chromatin protein kinase. *Eur. J. Biochem.* **62**, 109-115.
8. C. P. Barsano, A. H. Coleoni, and L. J. DeGroot (1980). Description and partial characterization of thyroid hormone-specific and thyroid hormone-dependent rat liver nuclear proteins. *Endocrinology* **106**, 1475-1488.
9. J. H. Clark, E. J. Peek, Jr., J. W. Hardin, and H. Eriksson (1978). The biology and pharmacology of estrogen receptor binding: relationships to uterine growth. In B. W. O'Malley and L. Birnbaumer (Eds.). *Receptors and Hormone Action*, Vol. 2. Academic Press, New York, pp. 1-31.
10. J. H. Clark, S. Upchurch, B. Markeverich, H. Eriksson, and J. W. Hardin (1978). Estrogen receptor heterogeneity and uterotropic response. In A. K. Roy and J. H. Clark (Eds.). *Gene Regulation by Steroid Hormones*. Springer-Verlag, New York, pp. 89-105.
11. A. C. Notides (1978). Conformational forms of the estrogen receptor. In B. W. O'Malley and L. Birnbaumer (Eds.). *Receptors and Hormone Action*, Vol. 2. Academic Press, New York, pp. 33-61.
12. C. W. Bardin, L. P. Bullock, N. C. Mills, Y.-C. Lin, and S. T. Jacob (1978). The role of receptors in the anabolic actions of androgens. In B. W. O'Malley and L. Birnbaumer (Eds.). *Receptors and Hormone Action*, Vol. 2. Academic Press, New York, pp. 83-103.
13. J. L. Tymoczko, T. Liang, and S. Liao (1978). Androgen receptor interaction in target cells: biochemical evaluation. In B. W. O'Malley and L. Birnbaumer (Eds.). *Receptors and Hormone Action*, Vol. 2. Academic Press, New York, pp. 121-156.

REFERENCES

14 E. M. Wilson, O. A. Lea, and F. S. French (1978). Androgen-binding proteins of the male rat reproductive tract. In B. W. O'Malley and L. Birnbaumer (Eds.). *Receptors and Hormone Action*, Vol. 2. Academic Press, New York, pp. 491–531.

15 W. W. Leavitt, T. J. Chen, Y. S. Do, B. D. Carlton, and T. C. Allen (1978). Biology of progesterone receptors. In B. W. O'Malley and L. Birnbaumer (Eds.). *Receptors and Hormone Action*, Vol. 2. Academic Press, New York, pp. 157–188.

16 W. T. Schrader and B. W. O'Malley (1978). Molecular structure and analysis of progesterone receptors. In B. W. O'Malley and L. Birnbaumer (Eds.). *Receptors and Hormone Action*, Vol. 2. Academic Press, New York, pp. 189–224.

17 W. T. Schrader, Y. Seleznev, W. V. Vedeckis, and B. W. O'Malley (1978). Steroid receptor subunit structure. In A. K. Roy and J. H. Clark (Eds.). *Gene Regulation by Steroid Hormones*. Springer-Verlag, New York, pp. 78–88.

18 P. Feigelson, L. Ramanarayanan-Murphy, and P. D. Colman (1978). Studies on the cytoplasmic glucocorticoid receptor and its nuclear interaction in mediating induction of tryptophan oxygenase messenger RNA in liver and hepatoma. In B. W. O'Malley and L. Birnbaumer (Eds.). *Receptors and Hormone Action*, Vol. 2. Academic Press, New York, pp. 226–249.

19 J. D. Baxter and R. D. Ivarie (1978). Regulation of gene expression by glucocorticoid hormones: studies of receptors and responses in cultured cells. In B. W. O'Malley and L. Birnbaumer (Eds.). *Receptors and Hormone Action*, Vol. 2. Academic Press, New York, pp. 251–296.

20 G. G. Rousseau and J. D. Baxter (1979). Glucocorticoid receptors. In J. D. Baxter and G. G. Rousseau (Eds.). *Glucocorticoid Hormone Action*. Springer-Verlag, Berlin, pp. 49–77.

21 C. D. Green (1980). The regulation of gene expression by steroid hormones in animal cells. In M. J. Clemens (Ed.). *Biochemistry of Cellular Regulation.* Vol. I. *Gene Expression.* CRC Press, Boca Raton, Florida, pp. 59–83.

22 M. Gschwendt (1980). The general validity of the subunit model of the progesterone receptor from chick oviduct appears questionable. *Mol. Cell. Endocrinol.* **19**, 57–67.

23 E. Milgrom (1981). Activation of steroid-receptor complexes. In G. Litwack (Ed.). *Biochemical Actions of Hormones*, Vol. 8. Academic Press, New York, pp. 465–492.

24 S. J. Higgins, J. D. Baxter, and G. G. Rousseau (1979). Nuclear binding of glucocorticoid receptors. In J. D. Baxter and G. G. Rousseau (Eds.). *Glucocorticoid Hormone Action*. Springer-Verlag, Berlin, pp. 136–160.

25 L. K. Johnson, S. K. Nordeen, J. L. Roberts, and J. D. Baxter (1978). Studies on the mechanism of glucocorticoid hormone action. In A. K. Roy and J. H. Clark (Eds.). *Gene Regulation by Steroid Hormones*. Springer-Verlag, New York, pp. 151–187.

26 S. S. Simons, Jr. (1978). Factors influencing association of glucocorticoid receptor–steroid complexes with nuclei, chromatin, and DNA: Interpretation of binding data. In J. D. Baxter and G. G. Rousseau (Eds.). *Glucocorticoid Hormone Action*. Springer-Verlag, Berlin, pp. 161–187.

27 T. C. Spelsberg, C. Thrall, G. Martin-Dani, R. A. Webster, and P. A. Boyd (1980). Steroid receptor interaction with chromatin. In T. H. Hamilton, J. H. Clark and W. A. Sadler (Eds.). *Ontogeny of Receptors and Reproductive Hormone Action*. Raven Press, New York, pp. 31–63.

28 M. Feldman, J. Kallos, and V. P. Hollander (1981). RNA inhibits estrogen receptor binding to DNA. *J. Biol. Chem.* **256**, 1145–1148.

29 S. Liao, S. Smythe, J. L. Tymoczko, G. P. Rossimi, C. Chen and R. A. Hiipakka (1980). RNA-dependent release of androgen– and steroid–receptor complexes from DNA. *J. Biol. Chem.* **255**, 5545–5551.

30 R. E. Leake (1981). Problems associated with dose response in steroid-hormone activation of structural genes. *Mol. Cell. Endocrinol.* **21**, 1–13.

31 N. M. Borthwick and R. M. S. Smellie (1975). The effects of oestradiol-17β on the ribonucleic acid polymerases of immature rabbit uterus. *Biochem. J.* **147**, 91–101.

32 N. Defer, L. Tichonicky, B. Paris, A. Kitzis, and J. Kruh (1981). Glucocorticoid hormones are successively present in two sites with different accessibilities to nucleases in chromatin from HTC cells. *Biochem. Biophys. Res. Commun.* **98**, 169–175.
33 M. B. Senior and F. R. Frankel (1978). Evidence for two kinds of chromatin binding sites for the estradiol–receptor complex. *Cell* **14**, 857–863.
34 D. R. Schoenberg and J. H. Clark (1981). Nuclear association states of rat uterine oestrogen receptors as probed by nuclease digestion. *Biochem. J.* **196**, 423–432.
35 P. Vic, M. Garcia, C. Humeau, and H. Rochefort (1980). Early effects of estrogen on chromatin ultrastructure in endometrial nuclei. *Mol. Cell. Endocrinol.* **19**, 79–92.
36 B. H. Ginsberg (1979). The insulin receptor: properties and regulation. In G. Litwack (Ed.). *Biochemical Action of Hormones*, Vol. 4. Academic Press, New York, pp. 313–349.
37 M. L. Dufau and K. J. Catt (1978). Gonadotropin receptors and regulation of steroidogenesis in the testis and ovary. *Vit. and Horm.* **36**, 461–592.
38 H. Nagasawa, S. Sakai, and M. R. Bannerjee (1979). Prolactin Receptor. *Life Sciences* **24**, 193–208.
39 G. N. Gill (1978). ACTH regulation of the adrenal cortex. In G. N. Gill (Ed.). *Pharmacology and Adrenal Corticol Hormones*. Pergamon Press, Oxford, pp. 35–65.
40 M. Jutisz, A. Berault, J. Debeljuk, B. Kedelhué, and M. Théoleyre (1979). Gonadoliberin. In C. H. Li (Ed.). *Hormonal Proteins and Peptides*. Academic Press, New York, pp. 55–122.
41 E. W. Sutherland (1972). Studies on the mechanism of hormone action. *Science* **177**, 401–408.
42 G. A. Robinson, R. W. Butcher, and E. W. Sutherland (1971). *Cyclic AMP*. Academic Press, New York.
43 D. A. Walsh, J. P. Perkins, and E. G. Krebs (1968). An adenosine 3′,5′-monophosphate-dependent protein kinase from rabbit skeletal muscle. *J. Biol. Chem.* **243**, 3763–3765.
44 R. H. Herman and O. D. Taunton (1980). The mechanism of action of hormones. In R. H. Herman, R. M. Cohn, and P. D. McNamara (Eds.). *Principles of Metabolic Control in Mammalian Systems*. Plenum Press, New York, pp. 535–630.
45 K. J. Catt, J. P. Harwood, R. N. Clayton, T. F. Davies, V. Chan, M. Katikineni, K. Nozu, and M. L. Dufau (1980). Regulation of peptide hormone receptors and gonadal steroidogenesis. *Rec. Prog. Horm. Res.* **36**, 557–622.
46 M. Rodbell (1978). The role of nucleotide regulatory components in the coupling of hormone receptors and adenylate cyclase. In G. Falco and R. Paoletti (Eds.). *Molecular Biology and Pharmacology of Cyclic Nucleotides*. Elsevier/North-Holland, Amsterdam, pp. 1–11.
47 J. N. Fain (1979). Hormones, membranes and cyclic nucleotides. *Receptors and Recognition (Series A)* **6**, 1–61.
48 J. Abramowitz, R. Iyengar, and L. Birnbaumer (1979). Guanyl nucleotide regulation of hormonally-responsive adenylyl cyclases. *Mol. Cell. Endocrinol.* **16**, 129–146.
49 K. H. Jakobs (1979). Inhibition of adenylate cyclase by hormones and neurotransmitters. *Mol. Cell. Endocrinol.* **16**, 147–156.
50 A. Levitzki and E. J. M. Helmreich (1979). Hormone–receptor–adenylate cyclase interactions. *FEBS Lett.* **101**, 213–219.
51 E. M. Ross and A. G. Gilman (1980). Biochemical properties of hormone-sensitive adenylate cyclase. *Ann. Rev. Biochem.* **49**, 533–564.
52 G. M. Edelman (1976). Surface modulation in cell recognition and cell growth. *Science* **192**, 218–226.
53 L. E. Limbird (1981). Activation and attenuation of adenylate cyclase. *Biochem. J.* **195**, 1–13.
54 F. Hofman (1978). Interaction of subunits of cAMP-dependent protein kinase. In G. Folco and R. Paoletti (Eds.). *Molecular Biology and Pharmacology of Cyclic Nucleotides*. Elsevier/North-Holland, Amsterdam, pp. 129–140.

REFERENCES

55 D. A. Walsh and R. H. Cooper (1979). The physiological regulation and function of cAMP-dependent protein kinases. In G. Litwack (Ed.). *Biochemical Actions of Hormones*, Vol. 6. Academic Press, New York, pp. 1-75.

56 G. M. Carlson, P. J. Bechtel, and D. J. Graves (1979). Chemical and regulatory properties of phosphorylase kinase and cyclic AMP-dependent protein kinase. *Adv. Enzymol.* **50**, 41-115.

57 M. Sikorska, J. P. MacManus, P. R. Walker, and J. F. Whitfield (1980). The protein kinases of rat liver nuclei. *Biochem. Biophys. Res. Commun.* **93**, 1196-1203.

58 W. D. Wicks, J. Koontz, and K. Wagner (1975). Possible participation of protein kinase in enzyme induction. *J. Cyclic Nuc. Res.* **1**, 49-58.

59 J. Wimalasena, B. H. Leichtling, E. J. Lewis, T. A. Langan, and W. D. Wicks (1980). Coordinate regulation of adenylate cyclase, protein kinase, and specific enzyme synthesis by cholera toxin in hormonally unresponsive hepatoma cells. *Arch. Biochem. Biophys.* **205**, 595-605.

60 I. D. Goldfine (1981). Effects of insulin on intracellular functions. In G. Litwack (Ed.). *Biochemical Actions of Hormones*, Vol. 8. Academic Press, New York, pp. 273-305.

61 J. D. Corbin and T. M. Lincoln (1978). Comparison of cAMP and cGMP-dependent protein kinases. *Adv. Cyclic Nuc. Res.* **9**, 159-170.

62 F. Murad, W. P. Arnold, C. K. Mittal, and J. M. Braughler (1979). Properties and regulation of guanylate cyclase and some proposed functions for cyclic GMP. *Adv. Cyclic Nuc. Res.* **11**, 175-204.

63 A. Linnala-Kankkunen and P. H. Mäenpää (1979). A cyclic GMP-dependent histone kinase bound to liver nucleoli. *Biochim. Biophys. Acta* **587**, 324-332.

64 C. E. Zeilig, T. A. Langan, and D. B. Glass (1981). Sites on histone H1 selectivity phosphorylated by guanosine 3':5'-monophosphate-dependent protein kinase. *J. Biol. Chem.* **256**, 994-1001.

65 R. H. Kretsinger (1979). The informational role of calcium in the cytosol. *Adv. Cyclic. Nuc. Res.* **11**, 1-26.

66 R. H. Michell (1975). Inositol phospholipids and cell surface receptor function. *Biochim. Biophys. Acta* **415**, 81-174.

67 D. E. Green, M. Fry, and G. A. Blondin (1980). Phospholipids as the molecular instruments of ion and solute transport in biological membranes. *Proc. Natl. Acad. Sci. USA* **77**, 257-261.

68 J. W. Putney, Jr., S. J. Weiss, C. M. Van de Walle, and R. A. Heddas (1980). Is phosphatidic acid a calcium ionophore under neurohormonal control? *Nature* **284**, 345-347.

69 D. J. Wolff and C. O. Brostron (1979). Properties and functions of the calcium-dependent regulator protein. *Adv. Cyclic Nuc. Res.* **11**, 27-88.

70 C. B. Klee, T. H. Crouch, and P. G. Richman (1980). Calmodulin. *Ann. Rev. Biochem.* **49**, 489-515.

71 A. R. Means and J. R. Dedman (1980). Calmodulin—An intracellular calcium receptor. *Nature* **285**, 73-77.

72 W. Y. Cheung (1980). Calmodulin plays a pivotal role in cellular regulation. *Science* **207**, 19-27.

73 B. Rose and W. R. Lowenstein (1975). Calcium ion distribution in cytoplasm visualized by aequorin diffusion in cytosol restricted by energized sequestering. *Science* **190**, 1204-1206.

74 B. Samuelson, M. Goldyne, E. Granström, M. Hamberg, S. Hammarström, and C. Malmsten (1978). Prostaglandins and thromboxanes. *Ann. Rev. Biochem.* **47**, 997-1079.

75 U. Zor and S. A. Lamprecht (1979). Mechanism of prostaglandin action in endocrine cells. In G. Litwack (Ed.). *Biochemical Actions of Hormones*, Vol. 4. Academic Press, New York, pp. 85-133.

76 J. H. Exton (1979). Mechanisms involved in effects of catecholamines on liver carbohydrate metabolism. *Biochem. Pharmacol.* **28**, 2237-2240.

77 R. J. Lefkowitz and B. B. Hoffman (1980). Adrenergic receptors. *Adv. Cyclic Nuc. Res.* **12**, 37-47.

78 A. Levitzki (1981). The β-adrenergic receptor and its mode of coupling to adenylate cyclase. *CRC Critical Review Biochem.* **10**, 81–112.

79 R. J. Barkey, J. Shani, M. Lahav, T. Amit, and M. B. H. Yondim (1981). Effect of prolactin and prostaglandins on the stimulation of prolactin binding sites in the male rat liver. *Mol. Cell. Endocrinol.* **21**, 129–138.

80 E. Milgrom (1978). Progesterone-binding proteins in plasma and the reproductive tract. In B. W. O'Malley and L. Birnbaumer (Eds.). *Receptors and Hormone Action*, Vol. 2. Academic Press, New York, pp. 473–490.

81 K. K. Sen, S. Azhar, and K. M. J. Menon (1979). Receptor-mediated gonadotropin action in the ovary. *J. Biol. Chem.* **254**, 5664–5671.

82 K. J. Catt, J. P. Harwood, G. Aguilera, and M. L. Dufau (1979). Hormonal regulation of peptide receptors and target cell responses. *Nature* **280**, 109–116.

83 P. R. Walker and J. F. Whitfield (1981). Regulation of the prereplicative changes in the synthesis and transport of messenger and ribosomal RNA in regenerating livers of normal and hypocalcemic rats. *J. Cell. Physiol.* **108**, 427–437.

84 F. Isoharshi, K. Tsukanaka, M. Terada, Y. Nakanishi, H. Fukushima, and Y. Sakamoto (1979). Alteration in binding of dexamethasone to glucocorticoid receptors in regenerating liver after partial hepatectomy. *Cancer Res.* **39**, 5132–5135.

85 J. N. Loeb and W. Rosner (1979). Fall in hepatic cytosol glucocorticoid receptor induced by stress and partial hepatectomy: evidence for separate mechanism. *Endocrinol.* **104**, 1003–1006.

86 W. H. Dillmann, H. L. Schwartz, and J. H. Oppenheimer (1978). Selective alterations in hepatic enzyme response after reduction of nuclear triiodothyronine receptor sites by partial hepatectomy and starvation. *Biochem. Biophys. Res. Commun.* **80**, 259–266.

87 M. Krupp and D. M. Lane (1981). On the mechanism of ligand-induced down regulation of insulin receptor levels in the liver cell. *J. Biol. Chem.* **256**, 1689–1694.

PART TWO

Physiological Control of Enzyme Adaptation

CHAPTER TWELVE

Induced Enzymes

The rates of synthesis of enzymes usually change in response to changes in the extracellular environment. In bacterial cells these changes in enzyme synthesis are very pronounced, with several-hundred-fold increases in concentration commonly being observed. In animal cells the situation is quite different because, although the whole organism may be exposed to marked changes in the environment, mechanisms exist to minimize the number of cells required to respond. For example, tissues such as liver and adipose tissue do adapt to nutritional changes in the environment in order to supply optimal concentrations of metabolites to tissues such as muscle and brain. Consequently, the enzymes of muscle and brain are expressed constitutively. In fact, the vast majority of enzymes in multicellular organisms are expressed constitutively, and even in liver only about 20 enzymes respond to nutritional and other environmental changes.

There are additional differences in adaptation between bacteria and eukaryotic cells. For example, in bacterial cells *all* the enzymes of a pathway are either expressed or repressed; when repressed there may be only a few molecules of each enzyme per cell. Eukaryotic cells, on the other hand, maintain substantial basal concentrations of all enzymes at a cost in energy that would be intolerable for a bacterial cell. Furthermore, adaptation is both temporally and quantitatively quite different. Thus even the most rapidly induced enzymes in eukaryotic cells take 2–4 h to reach peak activities that are usually only 10–20-fold higher than the basal levels. Moreover, only enzymes at the beginning of a metabolic pathway undergo adaptation in eukaryotic cells.

Most research has been focused on the adaptive enzymes in the liver of the higher mammals since reproducible changes in concentration can be easily induced by changes in diet. This chapter focuses on the regulation of the synthesis of these enzymes. No attempt will be made to document exhaustively all the changes in enzyme activities that have been reported in the literature. Two criteria have been used to select the enzymes that will be discussed. First, only those enzymes for which an actual change in enzyme protein concentration during adaptation has been directly demonstrated will be described. Second, emphasis will be placed on those enzymes for which the changes in the concentration of their mRNA have been determined, since this provides the most insight into the regulatory mechanisms that form the basis of adaptation.

Figure 12-1 outlines the general pathways of carbon flow in liver and shows some of the enzymes that will be discussed. For the sake of convenience these enzymes are broken down into groups, the two major groups being the enzymes of the pathways involved in the synthesis and storage of energy reserves (principally glycolysis and lipogenesis) and the pathways of glucose production from noncarbohydrate precursors (gluconeogenesis and amino acid catabolism). The specific enzymes that will be considered are listed in Table 12-1. These are the

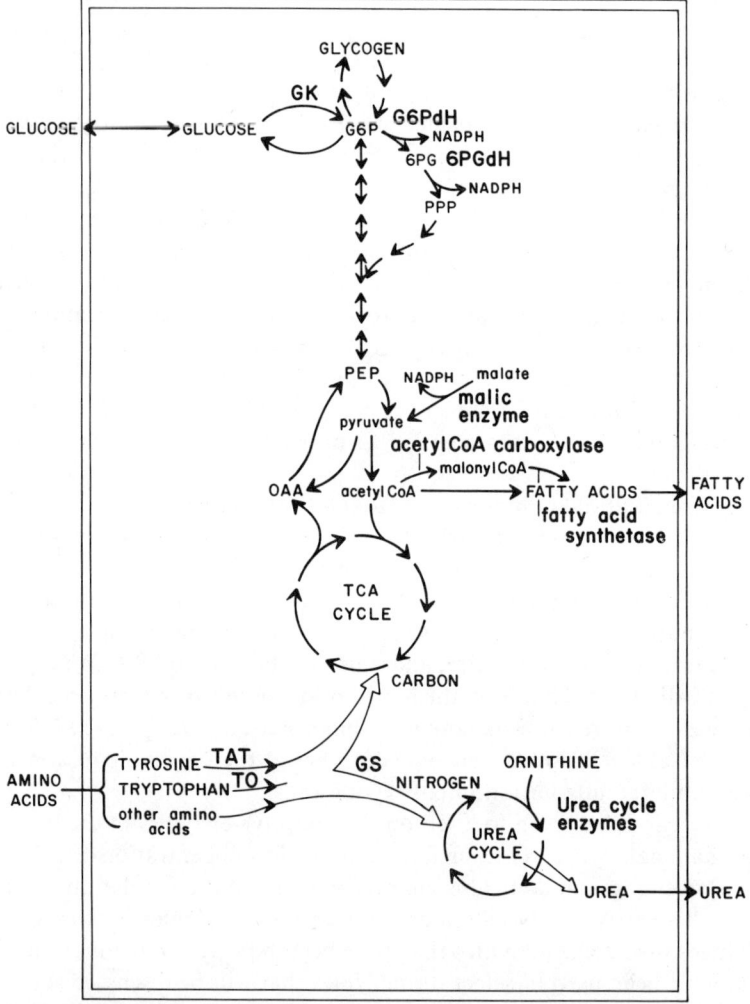

Fig. 12-1 The pathways of carbon flow between amino acids, carbohydrates, and lipid showing the reactions catalyzed by the adaptive enzymes that are discussed in the text. GK = glucokinase, G6PdH and 6PGdH = glucose 6-phosphate and 6-phosphogluconate dehydrogenases, respectively (PPP = pentose phosphate pathway), TAT = tyrosine aminotransferase, TO = tryptophan oxygenase, and GS = glutamine synthetase.

Table 12-1 Adaptive Enzymes of Carbon Storage and Glucose Production

Enzymes of Carbon Storage	Enzymes of Glucose Production
GLUCOKINASE	PHOSPHOENOLPYRUVATE CARBOXYKINASE
GLUCOSE 6-PHOSPHATE DEHYDROGENASE	TYROSINE AMINO TRANSFERASE
6-PHOSPHOGLUCONATE DEHYDROGENASE	TRYPTOPHAN OXYGENASE
FATTY ACID SYNTHETASE	AMINO ACID TRANSFERASES
MALIC ENZYME	UREA CYCLE ENZYMES
ACETYL CoA CARBOXYLASE	GLUTAMINE SYNTHETASE

only enzymes that are inducible, and in general they share certain common characteristics:

(a) Catalyzes the rate-limiting reaction of the pathway
(b) Is often the first enzyme in a reaction sequence
(c) Catalyzes reactions involved in opposing pathways
(d) Is not usually subject to covalent modification
(e) Exists in isozymic forms

The other enzymes in the pathways are synthesized constitutively, although small changes in concentration during chronic, long-term periods of adaptation do occur (this is likely because of changes in *total* mRNA or protein synthesis rather than specific enzyme induction). Many of these constitutive enzymes are controlled by *covalent modifications* (usually phosphorylation, see Fig. 7-2) that can alter their activities to match the changes in flux through the pathway that are evoked by changes in the *concentration* of the rate-limiting adaptive enzymes. These covalent modifications are often regulated by the same hormones that regulate the synthesis of the adaptive enzymes.

An enzyme that is inducible in the cells of one tissue may not be inducible in the cells of another tissue. The specificity of the response may be controlled in a number of ways. Perhaps the simplest is by the presence or absence of receptors for the hormone. Alternatively, the rate of synthesis or the inducibility of the enzyme may be controlled by the structure of the genome, which may create different *transcriptons* for the same gene in different tissues (as discussed for α-amylase in Chapter 5). In many instances, however, separate gene products appear to be involved. Thus, there are several proteins capable of catalyzing the phosphorylation of glucose, a reaction that is essential for the survival of all cells. These proteins are collectively referred to as the isoenzymes of hexokinase, and only one of them is inducible, the high-K_m glucokinase that is found only in liver parenchymal cells. In this way, a reaction that occurs in all cells can be selectively induced in one cell type, even by a hormone that reacts with many other tissues. The genetic and molecular characteristics of isoenzymes will not be discussed in detail here, but excellent books are available on this subject (1, 2).

Hormones are the signals that carry information concerning changes in the external environment and are responsible for inducing the synthesis of adaptive enzymes in their target cells. Before discussing the induction of specific enzymes, or enzyme groups, some consideration will be given to changes in hormone concentration in response to changes in the environment.

HORMONAL BASIS OF NUTRITIONAL ADAPTATION

Insulin, Glucagon, and the Concentration of Hepatic cAMP

Changes in the nutritional status of an organism are reflected in changes in the concentrations of glucose and amino acids in the blood, and these metabolites signal changes in the secretory activity of the α and β cells of the pancreas. The net result is a shift in the ratio of insulin to glucagon in the blood, and, if the nutritional status is severe enough, there may also be changes in the secretion of glucocorticoids.

During starvation the concentration of glucagon in the blood increases, whereas the insulin levels remain unaltered (Fig. 12-2a and Refs. 3 and 4). Refeeding a starved animal with a high-carbohydrate diet causes both a sharp increase in the concentration of insulin and a decrease in the concentration of glucagon (Fig. 12-2b). In both cases the concentration of cAMP in the liver parallels the change in glucagon. However, such a direct correlation is not always observed, as illustrated in Fig. 12-2c for animals refed a high-protein, low-carbohydrate diet. However, in this and all other conditions there is a good negative correlation between the concentration of cAMP and the *ratio* of insulin to glucagon in the blood (Fig. 12-2d).

Changes in Other Hormones

The concentrations of thyroid hormones (T_4 and T_3) and thyroid-stimulating hormone (released from the pituitary; Table 10-1) decrease during periods of starvation and refeeding (5, 6). Moreover, there is a fall in the concentration of nuclear hormone acceptors and a parallel decrease in nuclear RNA polymerase activity. In general, these changes in hormone concentration, acceptor concentration, and polymerase activity take place slowly, declining by only 50% over 48 h of starvation. Refeeding slowly restores the levels to normal. Such changes may affect the synthesis of thyroid-hormone-sensitive enzymes during periods of nutritional adaptation.

In animals fed a normal laboratory diet there is a pronounced diurnal rhythm of glucocorticoid hormone release by the adrenals (see Fig. 13-12a and Ref. 7). The concentration of the hormone decreases during the dark period when the animals (in this case, rats) are active and feeding. By the end of the dark period there is virtually no glucocorticoid hormone in circulation. The concentration of hormone in the blood rises throughout the light period and peaks at the

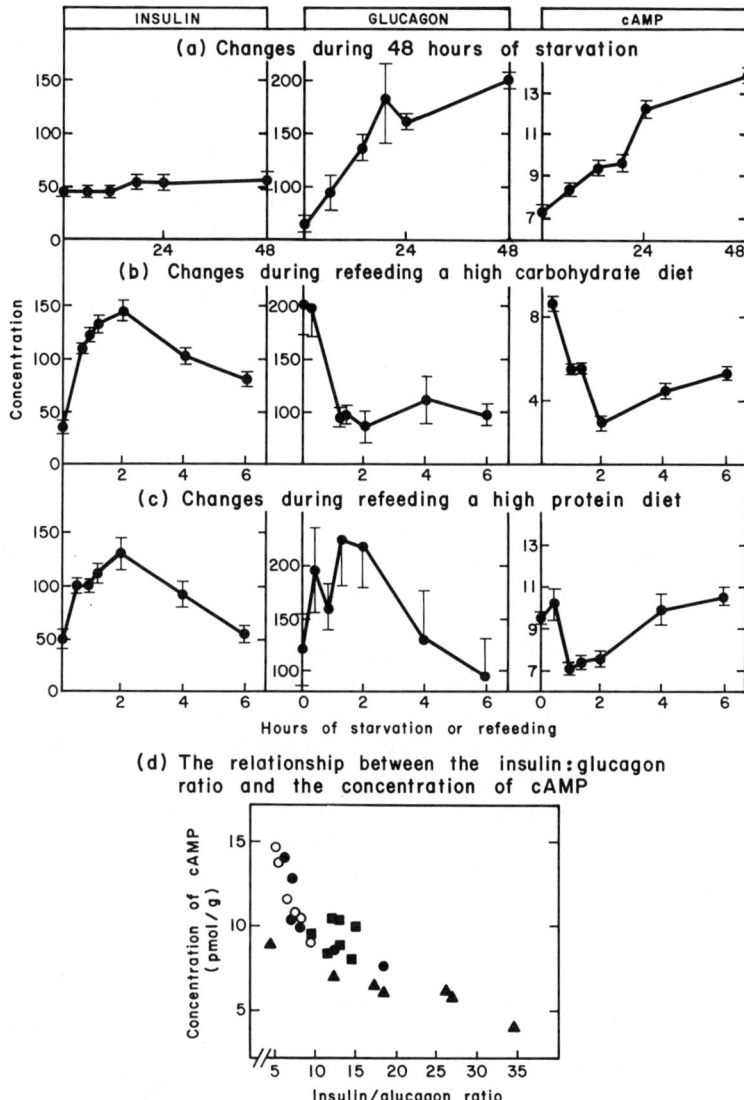

Fig. 12-2 Changes in the concentration of insulin and glucagon in the blood and cAMP in the liver in different nutritional states—(a) during starvation; animals starved for 48 h and then refed either a diet high in carbohydrate (b) or a high-protein (c) diet. In (d) the concentration of cAMP is plotted against the insulin–glucagon ratio for data obtained from (a), (b), and (c). [Reproduced from Seitz et al. (3) with permission from Lippincott/Harper & Row.]

light–dark interface. This rhythm is not affected by minor changes in nutritional status. For example, 48 h of starvation has no effect, and only on the third day of starvation is there a decrease in the amplitude of the rhythm resulting in a higher basal level (Fig. 13-12b).

In summary, the polypeptide hormones, glucagon and insulin, respond most rapidly to nutritional changes, whereas the thyroid hormones respond slowly. The concentration of glucocorticoids is more difficult to predict and in short-term experiments is likely to depend upon the particular protocol used. In general, the steroid is elevated during prolonged periods of starvation-induced stress.

HORMONAL CONTROL OF THE ENZYMES OF GLUCOSE PRODUCTION

General Pathways

During starvation, protein is catabolized in peripheral tissues, and the amino acids released are carried in the blood to the liver, where they are used to synthesize glucose (Fig. 12-3). In the liver, the amino acids are initially catabolized by enzymes that remove the nitrogen group (and direct it toward the urea cycle) and leave a carbon skeleton that can enter the tricarboxylic (TCA) cycle. The TCA cycle functions in a biosynthetic capacity to convert these carbon skeletons into oxaloacetate, which is then converted into phosphoenolpyruvate, an intermediate of the gluconeogenic pathway.

Under these gluconeogenic conditions there are increases in the amino acid transaminases (particularly tyrosine aminotransferase, TAT, and tryptophan oxygenase, TO), the urea cycle enzymes, and some enzymes of gluconeogenesis, most notably phosphoenolpyruvate carboxykinase (PEPCK). The regulation of the synthesis of these enzymes will now be considered in detail.

Enzymes of Amino Acid Catabolism—TAT and TO

Tyrosine aminotransferase is a liver-specific enzyme ($M_r = 100,000$) that catalyzes the first reaction in the catabolism of tyrosine, converting it to p-hydroxyphenylpyruvate, which is then converted to fumarate and acetolactate, permitting the carbon to enter the TCA cycle (Fig. 12-3). Tryptophan oxygenase ($M_r = 42,000$) catalyzes the oxidation of tryptophan to N-formyl-L-kynurenine, which is degraded to formate, alanine, and crotonylCoA (converted to acetyl-CoA, which enters the TCA cycle). Both enzymes respond quickly to gluconeogenic conditions, which is somewhat surprising since both amino acids together contribute only 3–4% of total protein amino acids and their degradation makes a negligible contribution to carbon for gluconeogenesis. However, there is evidence (8, 9) that high concentrations of these amino acids are toxic and that the liver may actually be detoxifying them.

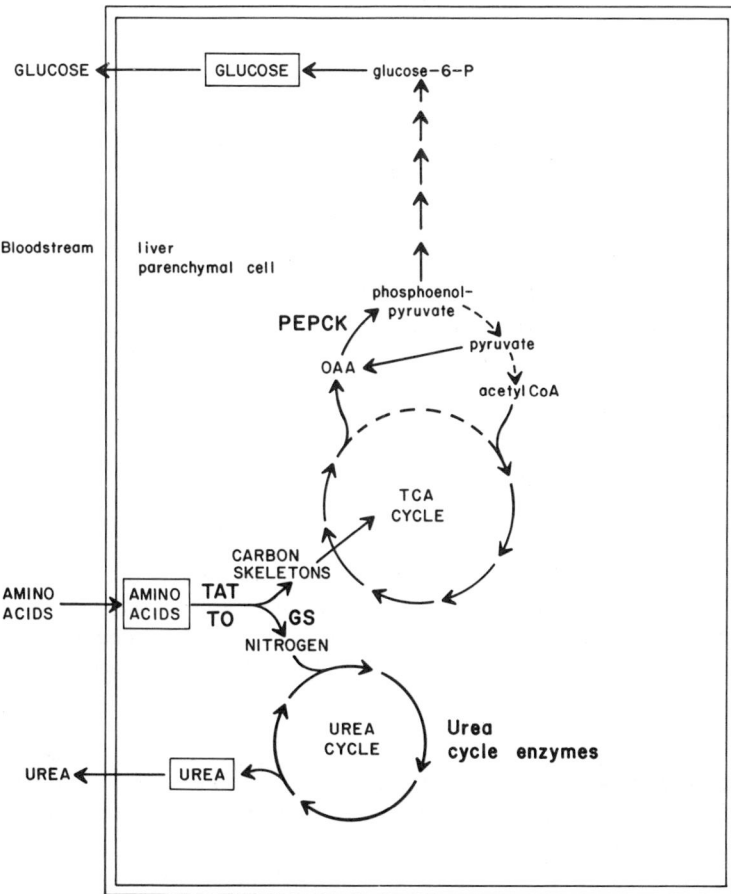

Fig. 12-3 Gluconeogenesis in the liver. Amino acids from peripheral tissues are catabolized, with the carbon skeletons entering anabolic reactions of the tricarboxylic acid cycle and the nitrogen being disposed of by the urea cycle. Oxaloacetate (OAA) formed by the action of the TCA cycle is converted to glucose by the gluconeogenic pathway. Enzyme abbreviations as in Fig. 12-1; PEPCK = phosphoenolpyruvate carboxykinase.

The ease with which it can be assayed, coupled with the rapidity and the magnitude of its response to inducers, has made TAT the most studied enzyme in animal cells. Experiments have been carried out both in liver *in vivo* and in a number of *in vitro* systems, including liver and hepatoma (tumor of the liver) cells in culture, to try to establish the hormonal basis of the regulation of its synthesis. The early work (reviewed in Ref. 10), carried out mostly on the H35 and HTC tumor cell lines, established that three hormones—insulin, glucocorticoids and glucagon—could induce the enzyme, although the induction of the enzyme by glucagon was thought to be due to contamination of commercial preparations of the hormone by insulin. The effects of insulin and glucocorti-

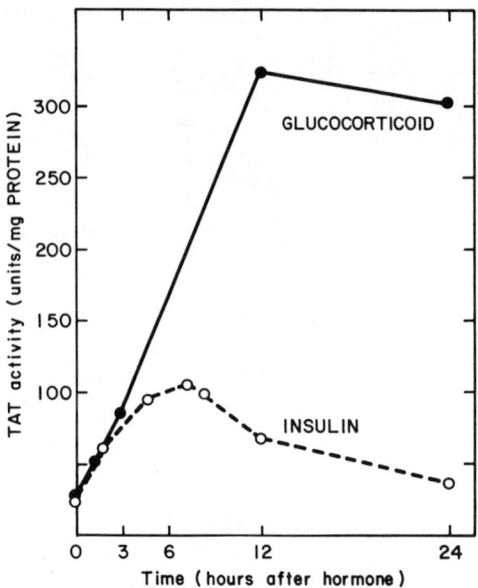

Fig. 12-4 Differences in the abilities of insulin and glucocorticoid hormones to increase the synthesis of tryosine aminotransferase in rat liver. [Reproduced from Figs. 2 and 4 of Reel, Lee, and Kenney (11) by permission of the American Society of Biological Chemists, Inc.]

coids on the enzyme are illustrated in Fig. 12-4. The increase in activity in response to insulin is only about one third that of the response to steroid. The induced level in the presence of glucocorticoid is maintained as long as steroid is in the culture medium. Moreover, when both hormones are added together they act synergistically (i.e., the increase in enzyme synthesis is more than additive), indicating that they act at different steps in the process of gene expression (11).

The increase in activity in response to *glucocorticoids* was shown, using specific antibody for TAT, to be the result of an increase in enzyme protein concentration. Moreover, the increase was prevented by either actinomycin D or cordycepin (inhibitors of transcription and polyadenylation, respectively), which indicates that the hormone acts at the level of transcription. Subsequent work (12, 13) confirmed these observations by showing that the increase in enzyme activity was preceded by an increase in mRNA^{TAT} concentration in the livers of animals injected with hydrocortisone (Fig. 12-5a) and that the rate of enzyme synthesis was directly proportional to the concentration of mRNA^{TAT} (Fig. 12-5b).

The increase in TAT activity in response to *insulin* has also been shown to be preceded by an increase in mRNA^{TAT} concentration [these experiments are carried out in adrenalectomized rats to obviate steroid involvement (14)]. However, this is only a transient increase, and since the hormone appears to induce an increase in *total* cellular mRNA concentration as opposed to only increasing mRNA^{TAT}, the specificity of this effect is questionable.

The discovery that cAMP, the second messenger for glucagon (Chapter 11), can increase the rate of TAT synthesis (15) suggests that glucagon is actually involved in the regulation of the synthesis of the enzyme. The increase in enzyme activity in response to cAMP is due to an increase in enzyme protein synthesis and is preceded by an increase in the concentration of mRNA^{TAT} [Fig. 12-5a (13, 16)]. However, the kinetics of induction are considerably different from those observed with steroid. The concentration of mRNA^{TAT} increases without any lag and peaks at 1 h in response to cyclic nucleotide, but only increases about 10% in response to the steroid hormone during this first hour. Indeed, by the time the concentration of mRNA^{TAT} has increased to 50% of its maximal response to steroid its concentration has returned to basal levels in cyclic nucleotide-treated animals. These results suggest that cAMP acts, quickly, at the transcriptional level (see Fig. 11-13, which shows a correlation between H1 histone phosphorylation and TAT induction), causing an increase in mRNA^{TAT} accumulation and an increase in enzyme synthesis.

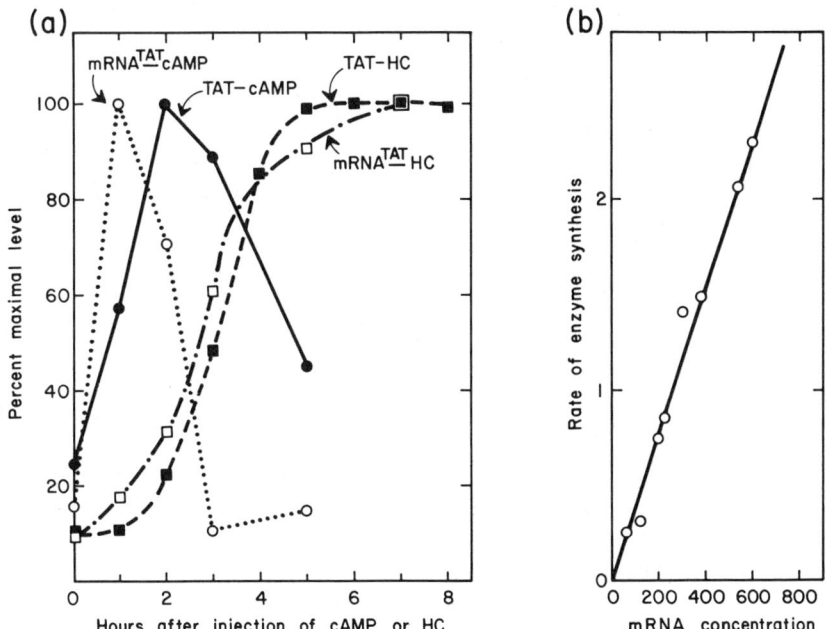

Fig. 12-5 Temporal relationships between increases in TAT synthesis and increases in mRNA^{TAT} in livers of animals injected with either cAMP or the glucocorticoid, hydrocortisone (HC). In (*a*) the concentration of mRNA^{TAT} peaks at 1 h and enzyme activity at 2 h in response to cAMP, but both parameters change much more slowly in response to the steroid. During induction by either effector the rate of enzyme synthesis is directly proportional to the concentration of mRNA (*b*). [Data from Fig. 4 of Nichol, Lee, and Kenney (12) and Fig. 2 of Noguchi, Diesterhaft, and Granner (16) by permission of the American Society for Biological Chemists, Inc.]

However, there are several observations that are not consistent with such an interpretation. For example, the initial increase in enzyme synthesis in response to cAMP is *not* inhibited by actinomycin D, although eventually both actinomycin D and cordycepin prevents enzyme synthesis. This suggests that cAMP acts to stabilize preexisting mRNA molecules, but when they are eventually degraded (and particularly when further synthesis is stopped by actinomycin D or cordycepin) the cyclic nucleotide is no longer effective. Furthermore, studies with primary cultures of hepatocytes (17, 18) have demonstrated that the induction of TAT by glucagon (or cAMP) is absolutely dependent upon the presence of glucocorticoids, which suggests that the steroid acts at the transcriptional level to produce $mRNA^{TAT}$ (the characteristic 2-h lag is observed), and the cyclic nucleotide ensures that the mRNA molecules are efficiently translated. Although other workers (19) have not found such a strict dependence on glucocorticoids for cAMP stimulation [the nutritional status of the animals prior to isolation of the cells (20) and the precise culture conditions (19) have profound effects on the results obtained], the same general conclusions were reached.

Since the concentration of $mRNA^{TAT}$ increases in the presence of cAMP it may act by stabilizing molecules that would otherwise be degraded. However, there are other possible explanations, and a more detailed discussion on the mechanism of action of cAMP will be deferred until its role in the induction of other enzymes has been described.

Tryptophan oxygenase has the distinction of being the first enzyme shown to be induced by a hormone (21). Moreover, it is also the first enzyme whose mRNA was isolated and translated in a cell-free system (22). The enzyme is induced by glucocorticoid hormones, and the kinetics of induction are discussed in Chapter 1 (see Figs. 1-1 and 1-2). This kinetic analysis revealed that glucocorticoid hormones increased the rate of synthesis of the enzyme without changing its rate of degradation. Furthermore, tryptophan, the substrate, was found to stabilize the enzyme and could lead to an increase in enzyme concentration by preventing its degradation (Fig. 1-2). This effect of tryptophan is one of the few examples, in eukaryotic cells, of a metabolite being able to influence enzyme-protein concentration. It is also one of the few examples of the regulation of the concentration of enzyme by modulation of its rate of degradation. Most hormones do not affect the rate of degradation of this or other enzymes.

The induction of TO by hormones has been studied in some detail (21–26), and the results obtained complement those obtained for TAT. Thus the enzyme is induced by glucocorticoids with the typical 1–2-h lag period, and the increase in enzyme activity is preceded by an increase in $mRNA^{TO}$ concentration. Moreover, the kinetics of induction are almost identical to those for TAT and $mRNA^{TAT}$ (Fig. 12-6). The nature of the lag period, typical of enzyme induction by steroid hormones, is uncertain, but since the induction of TO is particularly sensitive to cycloheximide (an inhibitor of protein synthesis) during the lag period, labile protein may be mediators of steroid hormone action (23).

The role of tryptophan in TO induction, predicted by kinetic analysis to be due to stabilization of the enzyme protein, was confirmed by showing that the

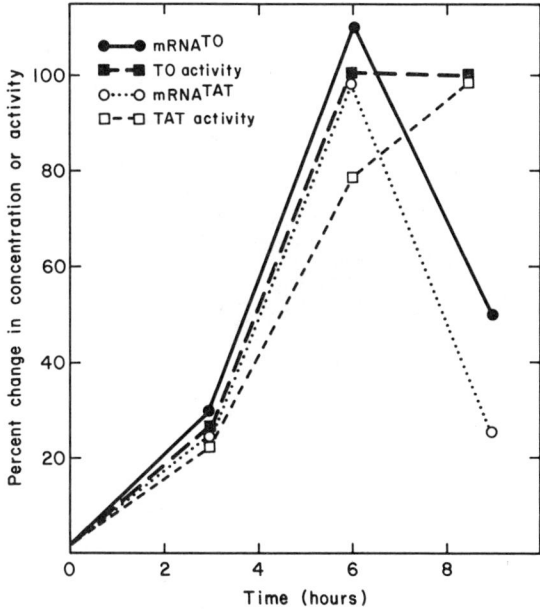

Fig. 12-6 Similarities between the increases in the concentrations of $mRNA^{TO}$, tryptophan oxygenase synthesis, $mRNA^{TAT}$, and tyrosine aminotransferase synthesis in liver following a single injection of hydrocortisone. [Reproduced from Figs. 1 and 2 of Hofer et al. (24) with permission of Dr. C. E. Sekeris and the *European Journal of Biochemistry*.]

metabolite increased enzyme concentration without changing the concentration of $mRNA^{TO}$ (25).

Recent work using primary cultures of freshly isolated hepatocytes (26) has shown that glucagon and cAMP can induce TO and that they act synergistically with glucocorticoids. This indicates a mechanism of interaction for these hormones (and second messenger) in enzyme induction similar to that for TAT. Insulin, on the other hand, was found to prevent steroid- or glucagon-induced enzyme synthesis.

The Urea-Cycle Enzymes

The urea cycle is the major pathway for the removal of excess nitrogen (Fig. 12-3) and is an unusual series of reactions in that half of the cycle is located in the cytoplasm and half of it in the mitochondrial matrix (Fig. 12-7). Under gluconeogenic conditions the activity of the cycle increases to convert the amino groups of the catabolized amino acids into urea, a nonionized form of nitrogen that can be excreted.

Glucocorticoids and glucagon are involved in the induction of these enzymes. In hepatocyte primary cultures (27) enzyme induction only occurs in the presence of *both* glucocorticoids and glucagon (cAMP can substitute for gluca-

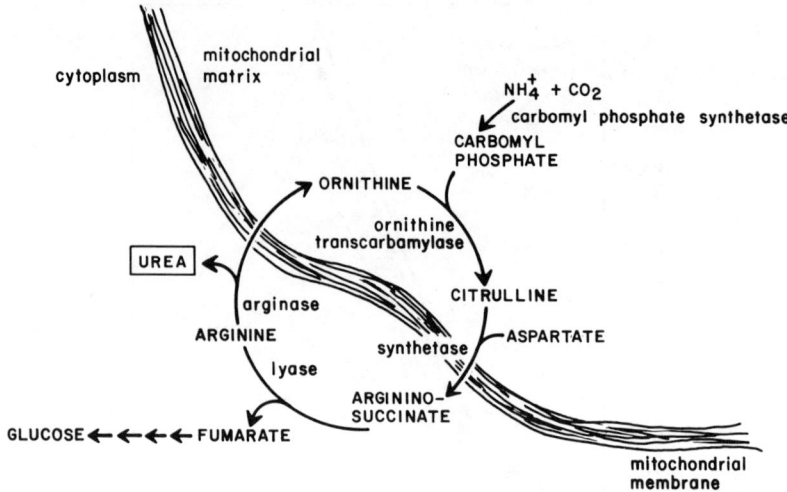

Fig. 12-7 The urea cycle. This cycle catalyzes the conversion of ionic ammonium ions to the nonionized, excretion product urea. The cycle is located partly in the cytoplasm and partly in the mitochondrial matrix.

gon), suggesting that the synthesis of these enzymes may be regulated in a way similar to that of TAT and TO. For most of the enzymes the induction was approximately threefold, but for arginase it was much less, over the time period studied, because of the long half-life (96 h) of this enzyme (see Fig. 1-4).

There are indications that other, nonhormone factors are involved in the regulation of the concentration of these enzymes. For example, in cultured cells (28, 29) the enzyme activities change in response to urea-cycle intermediates. Thus, arginine increases the activity of arginase and decreases the activities of arginino-succinate synthetase and arginino-succinate lyase (Fig. 12-8), whereas a low level of arginine, or the presence of citrulline, in the medium has the opposite effect. These changes were shown to be due to changes in the rates of synthesis of the enzymes, and they represent one of the few examples in animal cells where a small metabolite can increase the synthesis of an enzyme. However, since these experiments were carried out on cells maintained in culture medium containing serum (which has variable amounts of insulin, glucagon, glucocorticoids, and cyclic nucleotides), the participation of hormones in the induction process cannot be excluded.

Glutamine Synthetase

Glutamine synthetase catalyzes the condensation of glutamic acid with ammonium ions to form glutamine. The activity of this enzyme increases in a number of peripheral tissues under gluconeogenic conditions, and glutamine synthesis is one of the major routes by which these tissues export excess nitrogen to the liver. The enzyme is induced by glucocorticoids and cAMP in a number of cells

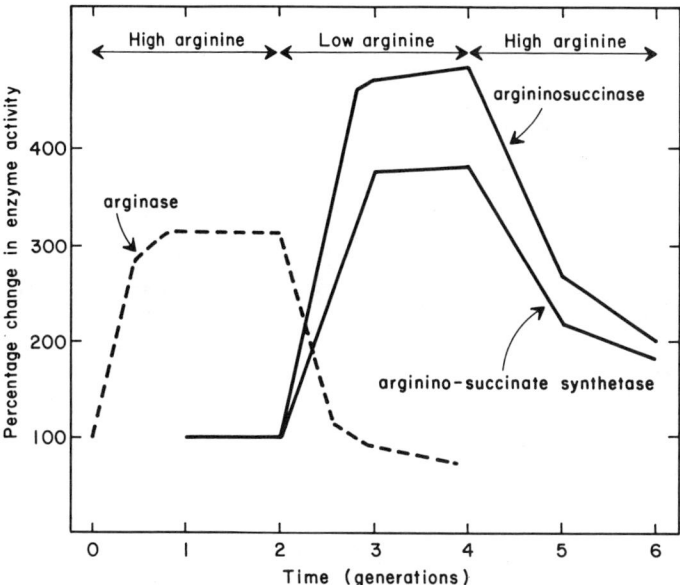

Fig. 12-8 The effect of arginine on the changes in the concentration of urea cycle enzymes in cultured HeLa-S$_3$ cells. Cells grown in a medium containing a high concentration of arginine increase their level of arginase, but decrease their levels of arginino succinase and arginino-succinate synthetase. If the medium contains a low concentration of arginine, arginase levels fall, whereas those of the other urea cycle enzymes increase. Reproduced from P. R. Walker (1977). The Regulation of Enzyme Synthesis *Essays in Biochemistry* **13,** 39–69 with permission.

in culture (30–32), and the increase in response to the steroid is preceded by an increase in mRNA^{GS}. Thus, although not studied in such great detail, this enzyme appears to fall into the category of those enzymes that are regulated by both glucocorticoids and cAMP. Glutamine also regulates the concentration of the synthetase by product repression, but the mechanism has not been studied in detail.

Phosphoenolpyruvate Carboxykinase—A Key Enzyme of Gluconeogenesis

Phosphoenolpyruvate carboxykinase (PEPCK, $M_r = 80,000$) is an important enzyme of the gluconeogenic pathway because it reverses the ATP-generating, pyruvate kinase-catalyzed step of glycolysis (Fig. 12-1). Oxaloacetate, the substrate, is generated from either biosynthetic reactions of the TCA cycle or from pyruvate by the action of pyruvate carboxylase. The activity of PEPCK in liver increases during starvation when there is a need for gluconeogenesis and decreases when glucose is available (33).

Changes in the activity of PEPCK are accompanied by changes in the concentration of mRNA^{PEPCK}. For example, when starved animals are intubated

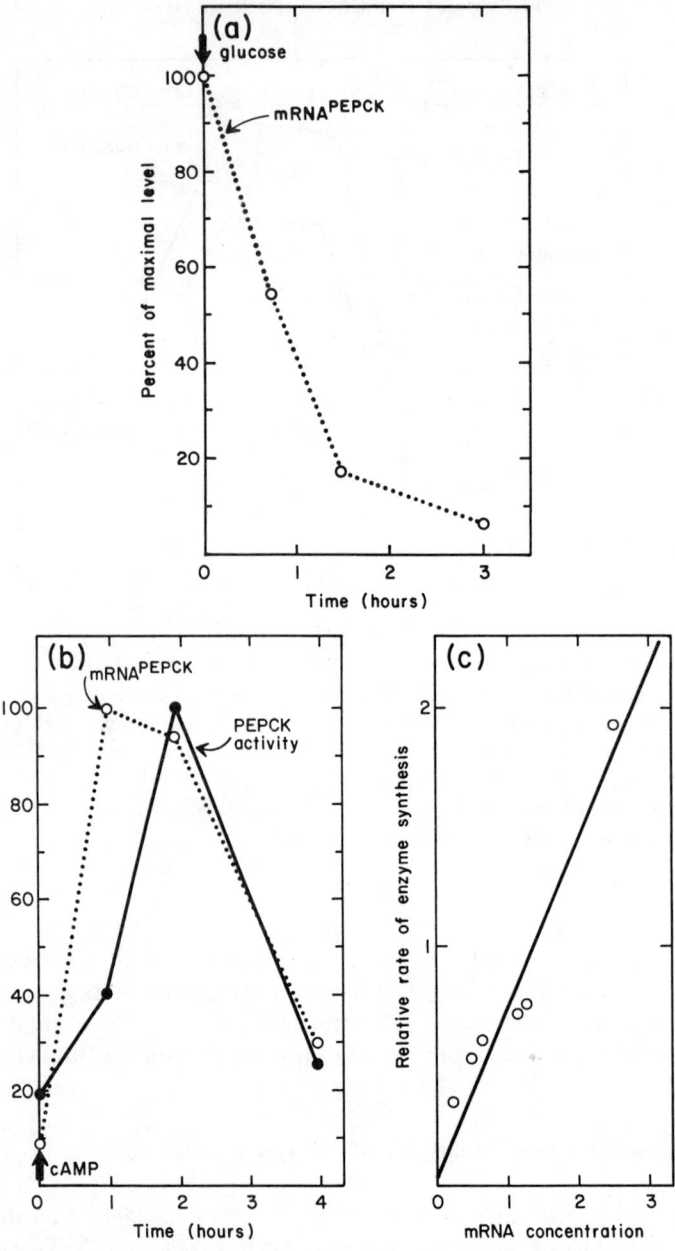

Fig. 12-9 Regulation of the synthesis of phosphoenolpyruvate carboxykinase (PEPCK) and its mRNA. (*a*) The rapid decline in the concentration of mRNA[PEPCK] following glucose administration via stomach tube (intubation) to a starved rat. (*b*) The effect of a single injection of dibutyryl cAMP into a rat treated for 3 h with glucose as in (*a*) on the concentration of mRNA[PEPCK] and PEPCK activity. (*c*) The synthesis of PEPCK in response to cAMP is proportional to the concentration of mRNA[PEPCK]. [Reproduced from Fig. 4 of Ref. 72 and Figs. 2 and 3 of Ref. 34 by permission of the American Society of Biological Chemists, Inc.]

with glucose the activity of the enzyme decreases, and this is preceded by an even more rapid decrease in the concentration of mRNA^{PEPCK} [Fig. 12-9a (34, 35)]. The half-life of mRNA^{PEPCK} under these conditions is 20–40 min. If cAMP (usually as its dibutyryl derivative) is injected into rats when mRNA^{PEPCK} has reached its basal level there is a rapid increase in mRNA^{PEPCK} concentration, followed by an increase in enzyme activity, with the rate of synthesis being directly proportional to mRNA^{PEPCK} content (Fig. 12-9b, 12-9c). Glucagon is as effective as cAMP derivatives, and the kinetics of induction are similar to those for TAT and TO (cf. Fig. 12-6).

The involvement of glucocorticoids in the regulation of the synthesis of enzymes is difficult to study *in vivo* because steroid administration leads to a release of insulin. Since insulin represses PEPCK synthesis any effect of the glucocorticoid on synthesis is nullified. However, experiments carried out on starved, diabetic animals and on H35 cells in culture (36, 37) have established that, whereas the steroid induces the enzyme in liver, it inhibits the synthesis of the enzyme in adipose tissue (38). This is an interesting example of a difference in the regulation of the same enzyme by the same hormone in two different tissues. Insulin inhibits the synthesis of PEPCK in both tissues (38), and cAMP is able to overcome this inhibitory effect of insulin.

Summary of the Regulation of the Synthesis of Gluconeogenic Enzymes

The hormones and other effectors that regulate the synthesis of the enzymes of amino acid catabolism, the urea cycle, and gluconeogenesis are summarized in Table 12-2. A clear pattern emerges. The concentration of all the enzymes is regulated by glucocorticoids, and cAMP (acting as a second messenger for the hormone glucagon) is needed to elicit a proper response. Glucocorticoids act at the transcriptional level to stimulate an increase in the rate of mRNA synthesis. Inhibitor studies indicate that this is due to an increased rate of gene transcription. The mechanism of action of cAMP is less certain. It appears to act at a post-transcriptional site to increase the concentration of mRNA available for translation. Some possible mechanisms are discussed at the end of the chapter.

For all enzymes, the increase in enzyme protein synthesis is directly proportional to the concentration of mRNA. Since most of the mRNA measurements are carried out on *total cellular* RNA there do not appear to be any pools of nontranslatable mRNA in unstimulated cells that could be mobilized by either steroid hormone or cAMP. Furthermore, neither the glucocorticoids nor cAMP have any effect on the rate of enzyme degradation. However, even though the hormones increase the rate of synthesis of the enzymes, the *time course* of the response is determined by the value of K_D (see Figs. 1-3 and 1-4).

Although not studied in great detail, there does appear to be some mechanism for low-molecular-weight metabolites to influence the concentration of enzymes involved in their metabolism. Tryptophan, glutamine, arginine, and citrulline can all change enzyme levels. Moreover, they achieve this via different mechanisms. For example, tryptophan stabilizes TO and increases its concen-

Table 12-2 The Regulation of the Synthesis of the Enzymes of Gluconeogenesis from Amino Acids[a]

ENZYME	$t_{1/2}$	GLUCO-CORTICOID	GLUCAGON	cAMP	ΔmRNA	$t_{1/2}$ mRNA	Δk_S ENZYME	Δk_D ENZYME	OTHER REGULATORS
TAT	1.5h	+	+	+	+	20–40'	+	0	insulin?
TO	2.5h	+	+	+	+	N.D.	+	0	tryptophan
GS	1.75h	+	+	+	+	N.D.	+	0	glutamine
Urea Cycle Enzymes	48–96h	+	+	+	N.D.	N.D.	+	0	arginine\citrulline
PEPCK	5–7h	+	+	+	+	40'	+	0	insulin (repressor)

[a] $t_{1/2}$ = half-life of the enzyme, $t_{1/2}^{mRNA}$ = half-life of mRNA. + = increases enzyme or mRNA synthesis, 0 = no change in rate of degradation of enzyme protein. N.D. = not determined. The effects of other regulators are described in the text.

tration, whereas glutamine represses the synthesis of glutamine synthetase, and arginine is able to increase some enzyme levels and repress others. Much more work is needed to establish the level at which these metabolites interact to effect their changes.

The involvement of insulin in the regulation of the synthesis of these enzymes is still confusing. *In vivo*, insulin does influence the concentration of cAMP in the liver, by altering the insulin:glucagon ratio. During starvation the ratio decreases, and the concentration of cAMP in the liver rises (Fig. 12-2), increasing the concentration of the mRNAs for the gluconeogenic enzymes. In addition the hormone may also influence, more directly, the overall rates of mRNA translation and perhaps amino acid uptake.

HORMONAL CONTROL OF THE ENZYMES OF GLUCOSE CONVERSION TO LIPID

General Pathways

During feeding, the concentration of glucose in the portal blood (the portal vein flows directly from the intestine to the liver) increases. The liver extracts most of this glucose, oxidizing some of it to meet its own energy demands and storing the rest as glycogen or lipid. In this way the animal maintains a constant glucose concentration in the peripheral blood and acquires a short-term energy supply in the form of glycogen and a longer-term energy supply in the form of lipid. These pathways are summarized in Fig. 12-10.

Glucokinase (GK) is the first enzyme of glucose catabolism, and it has such a high K_m (8–10 mM) for glucose that it is only active when there is an excess of glucose in the portal blood (the blood vessel that carries nutrients from the intestine directly to the liver). The product of this reaction, glucose 6-phosphate, is metabolized in three different ways. It may be converted to glycogen, or catabolized by either glycolysis or the pentose phosphate pathway. In general, the enzymes of glycogen metabolism and the enzymes of glycolysis are synthesized constitutively, and the flux of carbon through these pathways is controlled by an elaborate set of allosteric and covalent modification mechanisms. The enzymes of the pentose phosphate pathway, particularly glucose 6-phosphate dehydrogenase (G6PdH) and 6-phosphogluconate dehydrogenase (6PGdH), which produce NADPH for lipogenesis, are inducible and only increase when there is a demand for glucose conversion to lipid.

The acetylCoA formed by glycolysis (Fig. 12-10) either is further oxidized by the TCA cycle to provide energy or is converted to malonylCoA and then into fatty acids by the actions of acetylCoA carboxylase and fatty acid synthetase. The reducing equivalents (NADPH) for fatty acid biosynthesis are supplied by the pentose phosphate pathway enzymes as described above and also by malic enzyme, which catalyzes the oxidation of malate to pyruvate. The regulation of the synthesis of glucokinase, fatty acid synthetase, and the enzyme of NADPH production will now be discussed.

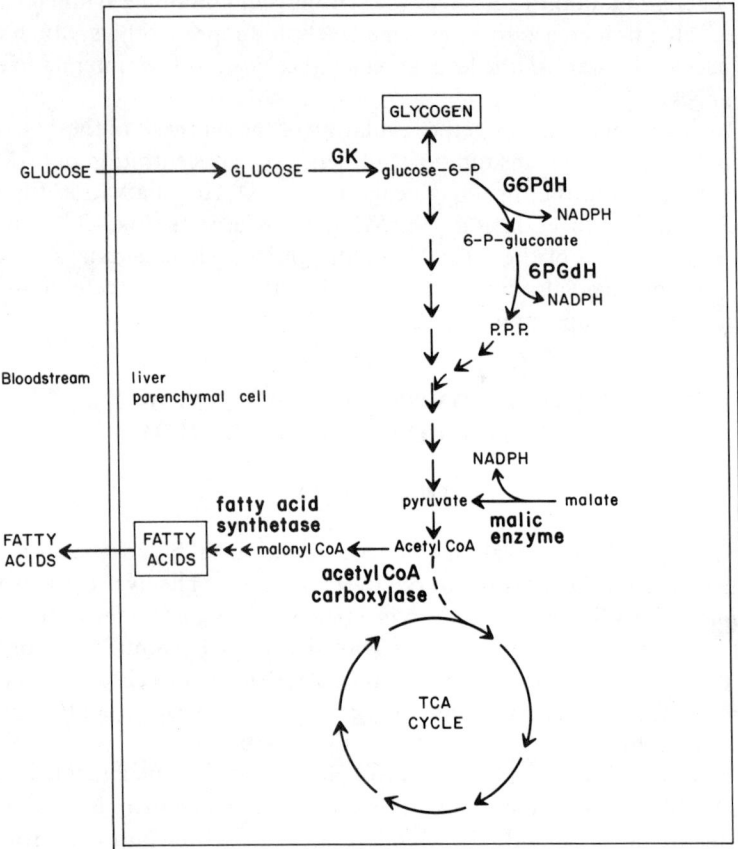

Fig. 12-10 Pathways for the conversion of dietary carbohydrate into fatty acids in liver. Abbreviations are the same as in Fig. 12-1. NADPH is reduced nicotinamide adenine dinucleotide phosphate used as a source of reducing equivalents in lipogenesis.

Glucokinase: The First Enzyme of Glycolysis

Glucokinase (M_r = 50,000–60,000) is located solely in the parenchymal cells of the liver (hepatocytes). The activity of the enzyme changes during starvation–refeeding experiments (Fig. 12-11), and because of its central role in carbohydrate homeostasis the regulation of the synthesis of the enzyme has been studied in great detail (reviewed in Refs. 39 and 40). The early work, carried out *in vivo*, indicated that the enzyme was under quite complex control with glucocorticoids, insulin, thyroid hormones, and possibly glucose itself being involved in the regulation of its synthesis, with glucagon and catecholamines (at pharmacological doses) preventing its induction. More recent work, both *in vivo* and in cultured hepatocytes, has concentrated on trying to establish the role played by these various effectors.

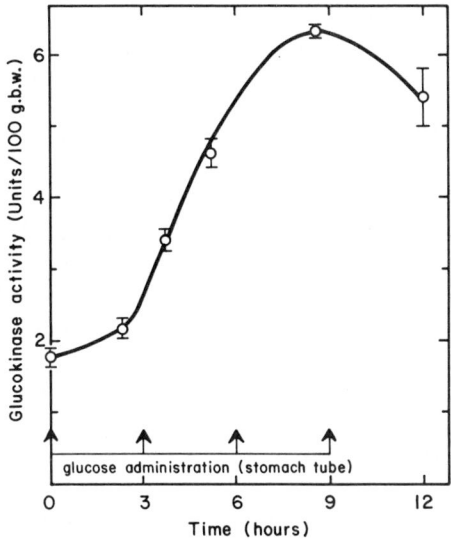

Fig. 12-11 Increase in the synthesis of glucokinase in response to the administration of glucose to a starved rat. The arrows indicate the times of each intubation. [Reproduced from H. Niemeyer et al. (1975). *Mol. Cell Biochem.* **6**, 109–126, with permission.]

Both glucocorticoids (41, 42) and thyroid hormone (43) appear to be necessary for full expression of the enzyme *in vivo*. If either adrenalectomized or hypothyroid animals are starved (to lower glucokinase levels) and then refed a high-carbohydrate diet, there is no increase in glucokinase activity unless the respective hormone is injected. However, if the hormones are injected *without* the carbohydrate there is also no increase in enzyme activity, indicating that both components are essential for a full response. The increased intake of carbohydrate is thought to raise the insulin:glucagon ratio and lower cAMP levels, removing a glucagon-mediated repressive effect.

Experiments carried out on freshly isolated hepatocytes in primary culture (44–47) generally support the interpretation of the data obtained *in vivo*. Thus, insulin and glucocorticoids are essential either to maintain enzyme levels in cultures or to actually increase them. Glucagon, on the other hand, decreases levels or prevents the other hormones from stimulating an increase. Moreover, if hepatocytes are isolated from hypothyroid animals the enzyme can only be induced by glucocorticoids or insulin if thyroid hormones are added to the medium.

An interesting feature of the *in vitro* work is the demonstration (46, 47) that cGMP or carbachol (a compound that stimulates plasma membrane guanylate cyclase) is as effective as glucocorticoids and insulin in eliciting an increase in glucokinase concentrations. At the present time this is the *only* example of the involvement of cGMP in the regulation of the synthesis of an enzyme. There is evidence that cGMP acts as a post-transcriptional site to increase the rate of synthesis of the enzyme. Furthermore, this study ruled out the possibility that cGMP was acting as a second messenger for insulin, although it is possible that cGMP levels are related to the accumulation of products of glucose catabolism.

All of the above studies were carried out with antibody to glucokinase, confirming that changes in activity were due to changes in enzyme protein synthesis. While all the hormones (and cGMP) exert their effects by increasing the rate of enzyme synthesis, there is evidence that both steroids and glucagon can change the rate of degradation of the enzyme (41, 43).

Without any data on changes in the concentration of mRNA^{GK} in these various experimental protocols, it is difficult to make any accurate statements of mechanism. It does appear, however, that the gene for glucokinase lies within the thyroid hormone "domain," and this hormone is necessary to make the gene available for transcription (presumably by a change in chromatin structure, see Chapter 11). Glucocorticoids, and possibly insulin, facilitate its transcription, and cGMP ensures that the product of transcription, mRNA^{GK}, is translated efficiently, whereas glucagon and cAMP prevent its translation.

Fatty Acid Synthetase and AcetylCoA Carboxylase

AcetylCoA carboxylase and fatty acid synthetase act sequentially to catalyze the synthesis of long-chain fatty acids from two-carbon units (acetylCoA—see Fig. 12-10). The two enzymes undergo changes in activity during changes in nutritional status and appear to be under hormonal control. Most of the work has been carried out on the regulation of the synthesis of fatty acid synthetase.

Fatty acid synthetase is a large molecule (M_r = 480,000) that, rather surprisingly considering the complex series of reactions it catalyzes, is only a dimer. Starvation results in a decrease in the activity of the enzyme and refeeding a high-carbohydrate diet results in a *50-fold* increase in activity that has been shown to be due to an increase in the rate of synthesis of the enzyme, rather than to any change in its rate of degradation. Moreover, the increase in fatty acid synthetase synthesis is preceded by an increase in the concentration of both total cellular mRNA^{FAS} and the concentration of mRNA^{FAS} associated with polysomes (Fig. 12-12*a* and Refs. 48 and 49). As with other enzymes the rate of synthesis is directly proportional to the concentration of its mRNA (Fig. 12-12*b*). The enzyme responds particularly slowly because its half-life is 76 h, and it is the degradation constant of the enzyme that determines the time course. This work on the changes in the concentration of mRNA^{FAS} is particularly noteworthy because it is usually extremely difficult to work with large mRNA molecules (M_r = 2.4 × 10^6, and size = 33S).

There is no evidence for the involvement of glucocorticoid hormones in the regulation of the synthesis of either fatty acid synthetase of acetyl CoA carboxylase. There is, however, considerable evidence that thyroid hormones are essential for proper expression of the fatty acid synthetase gene (50, 51). Moreover, there appears to be an interaction between thyroid hormone and some products of carbohydrate metabolism, giving a synergistic response that greatly increases the rate of fatty acid synthetase (and the enzymes of NADPH production) synthesis (51).

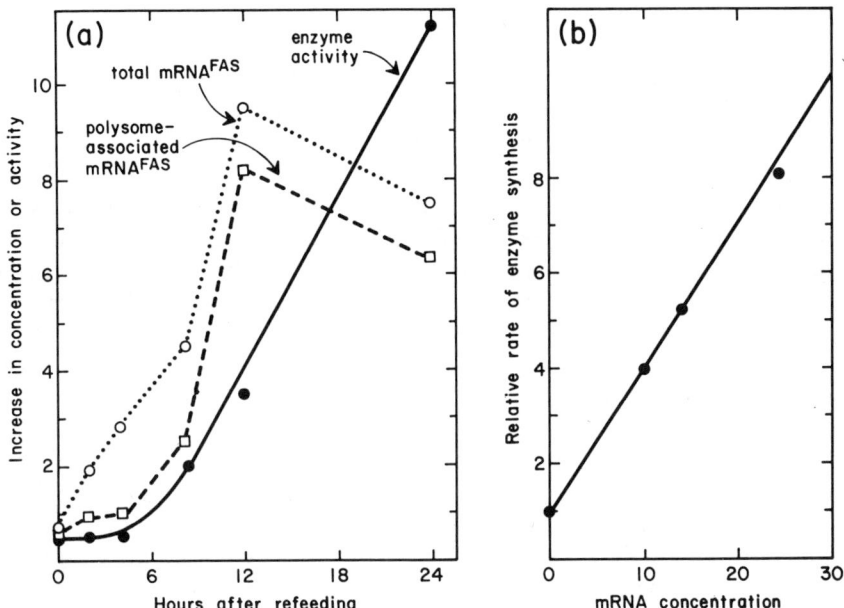

Fig. 12-12 Changes in the concentrations of fatty acid synthetase (FAS), total cellular mRNA^{FAS}, and polysome-associated mRNA^{FAS} in liver during the refeeding of a high-carbohydrate diet (*a*). In (*b*) the rate of enzyme synthesis is shown to be directly proportional to the concentration of mRNA. [Adapted from Ref. 49 by permission of the American Society of Biological Chemists, Inc. and from Ref. 48 with the permission of Dr. Roy Vagelos.]

Insulin, glucagon, and cAMP can also modulate the concentration of mRNA^{FAS} (52). In diabetic rats insulin *increases* the concentration of mRNA^{FAS} more than 10-fold, whereas in normal rats injections of glucagon or cAMP *decrease* the concentration of mRNA^{FAS}. It has not yet been established whether these hormones and the second-messenger cAMP act at the transcriptional or post-transcriptional level. The result obtained with insulin is one of the few demonstrations of an effect of the hormone on the concentration of an enzyme mRNA. It is likely that the insulin:glucagon ratio, which is determined by the carbohydrate content of the diet, plays an important role in the regulation of the synthesis of this enzyme, but the relative contributions of the two hormones to the change in mRNA^{FAS} concentration remain to be established.

The Enzymes of NADPH Production

Three enzymes—malic enzyme, glucose 6-phosphate dehydrogenase, and 6-phosphogluconate dehydrogenase—catalyze reactions that produce the reducing power (NADPH) for lipogenesis (Fig. 12-10). The activities of these enzymes have been found to fluctuate in parallel with the changes in the demand for lipogenesis, and attention has been focused on the regulation of their synthesis.

The synthesis of malic enzyme is controlled by thyroid hormone acting on the synthesis of mRNA^{ME} (53, 54). The *half-life* of mRNA^{ME}, estimated to be 10–12 h (53), is unaffected by the hormone. This long half-life of the mRNA coupled with the long half-life of the enzyme ($t_{1/2}$ = 30 h) explains the "slow" response of this enzyme to changes in hormonal and nutritional status. Insulin and a high-carbohydrate diet also stimulate the rate of malic enzyme synthesis, which at all times is directly proportional to the concentration of its mRNA. As with fatty acid synthetase, the effect of carbohydrate on malic enzyme synthesis is not just mediated by an increased secretion of insulin (51). However, the nature of the stimulation by products of glucose metabolism is not known. It is also not clear whether insulin has a direct effect on the rate of synthesis of the enzyme or whether it merely acts via the insulin:glucagon ratio to change in the concentration of cAMP in the liver. Glucagon and cAMP have been shown to *prevent* the synthesis of malic enzyme.

Glucose 6-phosphate dehydrogenase and 6-phosphogluconate dehydrogenase appear to be regulated along similar lines to malic enzyme and the enzymes of fatty acid biosynthesis (55, 56). Insulin and the carbohydrate content of the diet are able to increase the rate of synthesis of the enzymes by increasing the concentration of their mRNAs. Glucocorticoids are *not* involved in the regulation of the synthesis of these enzymes, and although the influence of thyroid hormones have not been examined recently, the older literature indicates that glucose 6-phosphate dehydrogenase, at least, is controlled by thyroid hormone. Glucagon and cAMP prevent the induction of glucose 6-phosphate dehydrogenase, but an effect on 6-phosphogluconate dehydrogenase is still to be demonstrated.

Summary of the Regulation of the Synthesis of Lipogenic Enzymes

The factors involved in the regulation of the synthesis of the enzymes of carbon storage are summarized in Table 12-3, and once again a clear pattern emerges. With the exception of glucokinase, glucocorticoids are not involved in the regulation of the synthesis of these enzymes. This is consistent with the physiological function of this hormone in the *mobilization* of carbohydrate rather than its storage. The same is true for glucagon, which decreases the concentration of all the enzymes of lipogenesis (with the possible exception of 6-phosphogluconate dehydrogenase). Cyclic AMP is the second messenger for glucagon, and, although its mechanism of action is not definitely established, it seems to act by reducing the concentration of the messenger RNAs.

The thyroid hormones are responsible for the transcriptional activation of the genes for the lipogenic enzymes, and their proper expression requires the presence of insulin or a high concentration of carbohydrate. These mechanisms are not understood, but products of glucose catabolism appear to be involved, and it is not known if insulin increases the rate of formation of these products or whether this hormone has an independent, more direct role in the regulation of the synthesis of the enzymes. It is possible that the redox state of the cyto-

Table 12-3 The Regulation of the Synthesis of the Enzymes of Glucose Conversion to Lipid[a]

ENZYME	$t_{1/2}$	GLUCO-CORTICOID	GLUCAGON	cAMP	INSULIN	THYROID HORMONE	ΔmRNA	$t_{1/2}$ mRNA	Δk_S ENZYME	Δk_D ENZYME	OTHER REGULATORS
GK	17–21h	+	–	–	+?	+	N.D.	N.D.	+	±	cGMP, carbohydrate
FAS	76h	0	–	–	+?	+	+	N.D.	+	0	carbohydrate
Acetyl CoA-CARBOXYLASE		0	–	–	+?	+	N.D.	N.D.	+	0	carbohydrate
MALIC ENZYME	30h	0	–	–	+?	+	+	10–12h	+	0	carbohydrate
G6PdH		0	–	–	+?	+	+	N.D.	+	0	carbohydrate
6PGdH	13–19h	0	0	0	+?	N.D.	+	N.D.	+	0	carbohydrate

[a] Abbreviations are the same as in Table 12-2; — = inhibition of enzyme synthesis, ? = effect not certain. The effects of other regulators are described in the text.

plasm (more reduced in the presence of the products of glucose catabolism) is an important factor. The involvement of cGMP in the regulation of glucokinase is particularly interesting, and the possible involvement of this cyclic nucleotide in the regulation of the synthesis of the other enzymes listed in Table 12-3 should be investigated.

In summary, the enzymes of the opposing pathways of glucose utilization and glucose storage are controlled at the transcriptional level in a very discrete fashion by the two hormones known to interact with the genome. Their domains do not overlap, and neither hormone is inhibitory to the expression of the genes controlled by the other hormone. Cyclic AMP plays a major role in determining the concentration of translatable mRNA for all these enzymes. Insulin, carbohydrate, and some amino acids also appear to play as yet poorly defined roles in the regulation of enzyme synthesis. In general, the enzymes are not regulated by changes in their rates of degradation.

GLUCOCORTICOIDS AND THE SYNTHESIS OF OTHER ENZYMES AND PROTEINS

Glucocorticoids are also involved in the regulation of the synthesis of a number of other enzymes and proteins in the liver, and also some proteins in a number of other cells. Some of these will now be described, including δ-aminolevulinic acid synthetase, alkaline phosphodiesterase, albumin, transferrin, and growth hormone.

Glucocorticoid Induction of Other Liver Enzymes

The first step in heme-porphyrin biosynthesis is catalyzed by the enzyme δ-aminolevulinic acid synthetase, and this enzyme has been shown to be induced by glucocorticoids in liver cells in culture (57), but only in the presence of insulin. Enzyme induction is prevented by actinomycin D, indicating an action of the hormones at the transcriptional level. Insulin, in the absence of glucocorticoids, had a slight stimulatory effect on δ-aminolevulinic acid synthetase synthesis, but general protein synthesis was also increased making it difficult to discern the mechanism of action of the polypeptide hormone.

In addition, glucocorticoids induce certain membrane-associated enzymes. For example, alkaline phosphodiesterase I (58) is increased threefold by the steroid in what appears to be a transcriptional response. The concentration of other membrane-associated enzymes such as alkaline phosphatase and adenosine triphosphatase is unaffected.

Regulation of Albumin and Transferrin Synthesis in Liver

Hydrocortisone stimulates a 2.5-fold increase in albumin synthesis and secretion in mouse Hepa-2 cells in culture (59). The effect of the steroid is amplified by cAMP, which increases the rate of synthesis to 10-fold. This increased rate of albumin synthesis is preceded by an increase in the concentration of mRNA^{ALB}, and the regulation of its synthesis provides another example of the apparent transcriptional–posttranscriptional coupling of glucocorticoids and cAMP in the control of liver-specific proteins. The roles of insulin and thyroid hormone in the synthesis of this important liver protein have not been examined.

Transferrin (also known as conalbumin) is an iron-transport serum protein that is synthesized in the liver of many mammalian species. In birds it is also synthesized in the oviduct and becomes one of the egg proteins. The regulation of its synthesis has been particularly well-studied in chicken liver and oviduct (60, 61). In the unstimulated tissues, the cells of the liver maintain a much higher basal level of transferrin mRNA (700–900 molecules per cell) than cells of the oviduct (30–60 molecules per cell). Glucocorticoids increase the levels of mRNA^{TF} 2-fold in liver and 10-fold in oviduct, indicating a differential action of the hormone on the same gene in two different cell types (the cells also respond differently to estrogen, which induces the synthesis of mRNA^{TF} 50-fold in oviduct, but only 2-fold in liver).

The regulation of transferrin synthesis is particularly interesting because it is controlled by the availability of iron, and the concentration of mRNA^{TF} is directly proportional to the level of dietary iron. The mechanism by which this is achieved is unknown.

Glucocorticoids and the Induction of Growth Hormone in Pituitary Cells

Glucocorticoid receptors are present in the cytoplasm of a pituitary tumor cell line maintained in culture, and dexamethasone produces a 15-fold induction of growth hormone mRNA over a period of about 60–72 h in these cells (62, 63). Moreover, there are also equivalent increases in the concentration of the nuclear primary transcript for mRNA^{GH} (63), confirming the transcriptional activation of the growth hormone gene by the steroid, and excluding an effect on primary transcript processing.

THYROID HORMONES AND THE SYNTHESIS OF OTHER PROTEINS

Although thyroid hormones have been shown to regulate the synthesis of the enzymes of lipogenesis and their mRNAs in liver, there are relatively few other examples of an effect of these hormones on the concentration of specific mRNAs. One example is the induction of growth hormone biosynthesis in the same cell line described above for glucocorticoid induction (62, 63). In these cells, thyroid hormones also activate growth-hormone gene transcription and increase the concentration of the nuclear primary transcript and cytoplasmic mRNA approximately 3-fold. When glucocorticoid and thyroid hormone are added together the response is not additive; in fact, the induction is reduced from 15-fold in the presence of steroid alone to about 10-fold.

cAMP AND THE REGULATION OF THE SYNTHESIS OF ENZYMES IN OTHER TISSUES

Cyclic AMP (functioning as a hormone second-messenger) is involved in the regulation of the synthesis of a number of enzymes in tissues other than liver. Two examples of this second-messenger role of the cyclic nucleotide are the synthesis of lactate dehydrogenase in a brain-tumor cell line and the stimulation of the enzymes of catecholamine biosynthesis in adrenal cells.

Lactate Dehydrogenase in Rat Glial Tumor Cells

The induction of lactate dehydrogenase by catecholamines is mediated by cAMP in rat-brain glioma (C6) cells in culture (64) and also in a clone (C6-2B) of the same cells (65). In the original C6 cell line, the β-agonist isoproterenol induces a 10-fold increase in enzyme activity (and enzyme protein concentration) over a period of 12 h, as shown in Fig. 12-13. Cyclic AMP is equally effective, and both molecules act by increasing the rate of synthesis of the enzyme rather than having any effect on its rate of degradation ($t_{1/2}$ = 6–7 h). In the C6-2B clone the catecholamine noradrenaline increases the concentration of cAMP, resulting in a 40% increase in enzyme activity over a 6-h period. The response is so small because the half-life of the enzyme in these cells is 36–40 h, much longer than in the C6 cells.

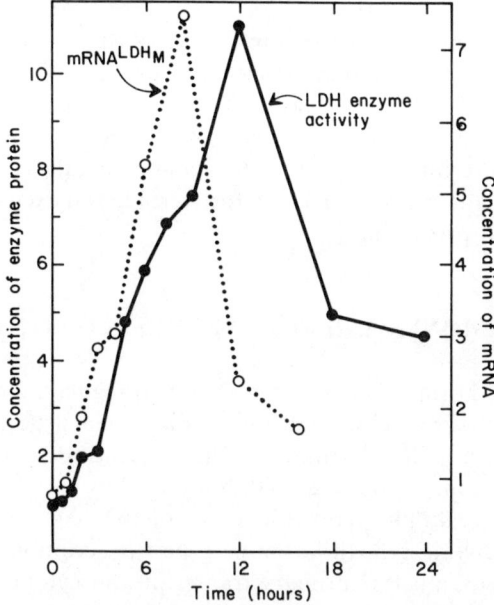

Fig. 12-13 The induction of lactate dehydrogenase synthesis and mRNA^{LDH_M} (the mRNA for M subunits of LDH) by cAMP in C6-glioma cells in culture. [Reproduced from Ref. 64 by permission of the American Society of Biological Chemists, Inc.]

The induction of lactate dehydrogenase is an interesting model system for gene expression, since a functional enzyme is often the product of two genes (see Ref. 1 for details). The enzyme is a tetramer, composed of two different subunits M and H, and this can give rise to five different isoenzymes: H_4, H_3M, H_2M_2, HM_3, and M_4 (designated LDH-1 to LDH-5, respectively). In the uninduced state, both cell types contain all five isoenzymes which indicates that both genes are active. In the C6-2B system noradrenaline induces a fivefold increase in the synthesis of M subunits, but only a twofold increase in the synthesis of H subunits. In the C6 cells only the M subunits were synthesized at a higher rate in the presence of cAMP, and this was found to be caused by an increase in the concentration of $mRNA^{LDH_M}$ as shown in Fig. 12-13.

Enzymes of Catecholamine Biosynthesis in Adrenal Cells

The catecholamines, adrenaline and noradrenaline, are synthesized in the cells of the adrenal gland by the pathway shown in Fig. 12-14. Tyrosine hydroxylase (monooxygenase) is considered to be the rate-limiting enzyme of the pathway, and its synthesis, together with that of dopamine hydroxylase, increases in the presence of the neurotransmitters acetylcholine and carbamyl choline (66, 67).

The increase in the synthesis of tyrosine hydroxylase in response to cAMP has been studied in some detail (67) and found to be highly unusual. Thus a brief, 1-h, pulse of cAMP is sufficient to stimulate the enzyme, but it does not increase until 18–24 h later. The appearance of new enzyme is preceded, by several hours, by an increase in the concentration of the catalytic subunit of cAMP-dependent protein kinase in the nucleus of these cells. This had led to speculation that catalytic subunits translocate into the nucleus and activate the expression of the tyrosine hydroxylase gene. However, with such a protracted time course, secondary phenomena cannot be ruled out. For example, it has recently been demonstrated (68) that tyrosine hydroxylase enzyme protein is activated when phosphorylated by cAMP-dependent protein kinase.

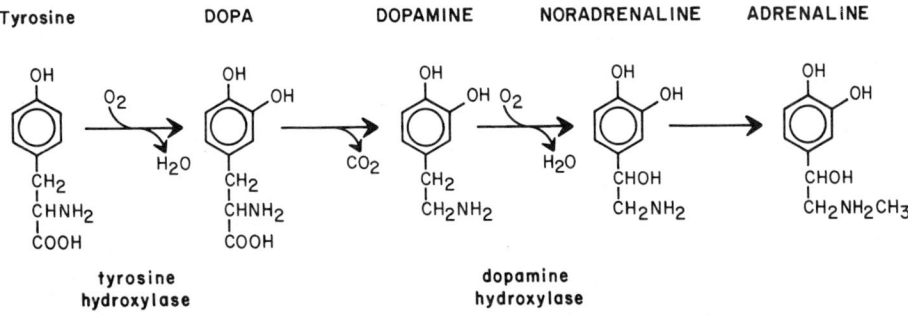

Fig. 12-14 The pathway of adrenaline and noradrenaline biosynthesis from tyrosine in cells of the adrenal medulla.

Table 12-4 Regulation of the Synthesis of Several Enzymes and Proteins by Hormones and Neurotransmitters

ENZYME\PROTEIN	GLUCO-CORTICOID	NEUROTRANSMITTER/ CATECHOLAMINE/ cAMP	INSULIN	THYROID HORMONE	ΔmRNA
LDH		+			+
Tyrosine hydroxylase	+	+			+
Dopamine hydroxylase		+			
Alkaline phosphatase		+			+
Albumin	+	+			+
Transferrin	+				+
δ-amino levulinate synthetase	+				
Alkaline phosphodiesterase	+		+?	+?	
Growth hormone	+			+	+

Other Proteins Controlled by cAMP

Derivatives of cAMP have been shown to induce a *2000-fold* increase in the synthesis of alkaline phosphatase in cultured mouse L cells (69). This phenomenal increase in enzyme synthesis is preceded by an increase in alkaline phosphatase mRNA. In another cell line, the Hepa-2 hepatoma, cAMP interacts synergistically with glucocorticoids in the induction of albumin biosynthesis (59). This induction is also preceded by an increase in the concentration of albumin mRNA.

In summary, the regulation of the synthesis of the enzymes and proteins discussed in the last three sections is presented in Table 12-4. This table illustrates the incompleteness of these studies, since virtually all the proteins shown to be stimulated by glucocorticoids, for example, have not been examined for cAMP effects and vice versa. Indeed, it is only recently (70) that tyrosine hydroxylase synthesis has been found to be increased, at the transcriptional level, by glucocorticoids. In order to formulate coherent theories of hormone actions in the whole body, all cells must be examined for the effects of many hormones, and even negative results are invaluable.

THE MECHANISM OF ACTION OF cAMP IN THE REGULATION OF ENZYME SYNTHESIS

In all the examples of changes in enzyme synthesis (including those in response to cAMP) in which the concentration of the mRNA for the enzyme has also been measured, the rate of enzyme synthesis is directly proportional to the concentration of its mRNA. These observations rule out regulation at the translational level (this would result in an increase in the rate of synthesis of an enzyme without any change in the concentration of its mRNA). Thus, enzyme synthesis is regulated by mechanisms that change the availability of mRNA for translation, and this can be achieved at either the transcriptional level or at some post-transcriptional step that changes the efficiency of processing.

The steroid and thyroid hormones almost certainly act at the transcriptional level to increase the rate of transcription of genes. The steroid or thyroid hormone may contain all the information necessary to achieve this, or it may be a two-stage process in which the steroid or thyroid hormone makes the gene available for transcription (through a conformational change in chromatin structure), and additional factors (perhaps cAMP) are required to actually increase the rate of gene transcription.

Although the site of action of cAMP is not at the translational level, evidence has been presented that it does not act at the transcriptional level either. Thus in isolated liver cells, cAMP cannot induce TAT synthesis unless glucocorticoid hormone is also present (17, 18), implying that the steroid increases transcription of the gene, and the cyclic nucleotide acts at some post-transcriptional step. Moreover in some tumor cells actinomycin D (an inhibitor of transcription) does not prevent the increase in TAT synthesis in response to cAMP (17), which suggests that the cyclic nucleotide converts a pool of nontranslated $mRNA^{TAT}$ into a translatable form. However, such pools of mRNA have not been measured and since most experiments in which mRNA concentrations are measured use *total* cellular mRNA, there is little or no evidence for the existence of such pools. Moreover, since *in vitro* translation experiments are carried out on mRNA and *not* mRNP, any mRNA packaged into mRNP in an inactive form would be liberated (although mRNA inactivated by *covalent modification* would not be detectable).

Cyclic AMP could act at a post-transcriptional step to stabilize specific mRNAs against degradation. This would explain the accumulation of mRNAs in the presence of cyclic nucleotide. Support for this idea came from studies that showed that cycloheximide, but not puromycin, was able to mimic the increase in $mRNA^{PEPCK}$ concentration normally seen with cAMP (34). These results were interpreted as a stabilization of $mRNA^{PEPCK}$ when it is associated with polysomes during translation because cycloheximide (an inhibitor of the *translocation* step of translation) "freezes" ribosomes on the mRNA, whereas puromycin (an inhibitor of *initiation* of translation) does not protect the mRNA. However, it has recently been shown that the half-life of $mRNA^{PEPCK}$ remains unaltered (at about 20') in the presence or absence of cAMP (35), indicating that the cyclic nucleotide must increase $mRNA^{PEPCK}$ levels by increasing its rate of synthesis and not by stabilizing it [the interpretation of the puromycin/cycloheximide data is sound (34), but their effects cannot be extrapolated to the action of cAMP].

If cAMP increases the concentration of $mRNA^{PEPCK}$ and other mRNAs by increasing their rates of synthesis, then it either acts at the transcriptional level after all, or it acts at an early posttranscriptional event to stabilize some *mRNA precursor* and promote its conversion into mature RNA. These two possibilities can only be distinguished by the application of the latest techniques for studying gene expression described in Chapters 2-4.

Since cAMP is believed to act solely through protein kinases the mechanism of gene activation, or transcript stabilization, must be mediated by a phosphorylation reaction. In this regard, it should be recalled (Fig. 11-13) that the increase in TAT synthesis following treatment with cAMP is accompanied by an increase in the phosphorylation of histone H1. It is also important to remember that cAMP "represses" some enzymes by decreasing the concentrations of their mRNAs, and this mechanism of action must also be explained.

REFERENCES

1 C. C. Rider and C. B. Taylor (1980). *Isoenzymes*. Chapman and Hall, London.
2 C. L. Markert (1975). *Isoenzymes*, Vols. I-III. Academic Press, New York.
3 H. J. Seitz, M. J. Müller, P. Nordmeyer, W. Krone, and W. Tarnowski (1976). Concentration of cAMP in rat liver as a function of the insulin/glucagon ratio in blood under standardised physiological conditions. *Endocrinol.* **99**, 1313-1318.
4 M. Tiedgen and H. J. Seitz (1980). Dietary control of circadian variations in serum insulin, glucagon and hepatic cyclic AMP. *J. Nutr.* **110**, 876-882.
5 L. J. Degroot, A. H. Coleoni, P. A. Rue, H. Seo, E. Martino, and S. Refetoff (1977). Reduced nuclear triiodothyronine receptors in starvation-induced hypothyroidism. *Biochem. Biophys. Res. Commun.* **79**, 173-178.
6 C. Wimpfheimer, E. Saville, M. J. Voirol, E. Danforth, Jr., and A. G. Burger (1979). Starvation-induced decreased sensitivity of resting metabolic rate to triiodothyronine. *Science* **205**, 1272-1273.
7 H. Kato, M. Saito, and M. Suda (1980). Effect of starvation on the circadian adrenocortical rhythm in rats. *Endocrinol.* **106**, 918-921.
8 A. Cihak (1979). L-Tryptophan action on hepatic RNA synthesis and enzyme induction. *Mol. Cell. Biochem.* **24**, 131-142.
9 A. A.-B. Badawy (1977). The functions and regulation of tryptophan pyrrolase. *Life Science* **21**, 755-768.
10 T. D. Gelehrter (1971). Regulatory mechanisms of enzyme synthesis: enzyme induction. In M. Rechcigl, Jr. (Ed.). *Enzyme Synthesis and Degradation in Mammalian Cells.* Karger, Basel, pp. 165-199.
11 J. R. Reel, K.-L. Lee, and F. T. Kenny (1970). Regulation of tyrosine α-ketogluterate transaminase in rat liver. *J. Biol. Chem.* **245**, 5800-5805.
12 J. N. Nickol, K.-L. Lee, and F. T. Kenney (1978). Changes in hepatic levels of tyrosine aminotransferase messenger RNA during induction by hydrocortisone. *J. Biol. Chem.* **253**, 4009-4015.
13 M. Diesterhaft, T. Noguchi, and D. Granner (1980). Regulation of rat-liver tyrosine aminotransferase mRNA by hydrocortisone and by $N^6, O^{2'}$-dibutyryladenosine 3',5'-phosphate. *Eur. J. Biochem.* **108**, 357-365.
14 R. E. Hill, L.-L. Lee, and F. T. Kenney (1981). Effects of insulin on messenger RNA activities in rat liver. *J. Biol. Chem.* **256**, 1510-1513.

15 K. Wagner, M. D. Roper, B. H. Leichtling, J. Wimalasena, and W. D. Wicks (1975). Effects of 6- and 8-substituted analogs of adenosine 3':5'-monophosphate on phosphoenolpyruvate carboxykinase and tyrosine amino-transferase in hepatoma cell cultures. *J. Biol. Chem.* **250**, 231–239.

16 T. Noguchi, M. Diesterhaft, and D. Granner (1978). Dibutyryl cyclic AMP increases the amount of functional messenger RNA coding for tyrosine aminotransferase in rat liver. *J. Biol. Chem.* **253**, 1332–1335.

17 M. J. Ernest, C.-L. Chen, and P. Feigelson (1977). Induction of tyrosine aminotransferase synthesis in isolated liver cell suspensions. *J. Biol. Chem.* **252**, 6783–6791.

18 M. J. Ernest and P. Feigelson (1979). Multihormonal control of tyrosine aminotransferase in isolated liver cells. In J. D. Baxter and G. G. Rousseau (Eds.). *Glucocorticoid Hormone Action*. Springer-Verlag, Berlin, pp. 219–241.

19 J. A. Gurr and V. R. Potter (1980). Independent induction of tyrosine aminotransferase activity by dexamethasone and glucagon in isolated rat liver parenchymal cells in suspension and in monolayer cultures in serum free medium. *Exp. Cell Res.* **126**, 237–248.

20 F. A. O. Marston, A. J. Dickson, and C. I. Pogson (1981). Factors affecting induction of tyrosine aminotransferase in isolated rat liver cells. *Mol. Cell. Biochem.* **34**, 59–64.

21 W. E. Knox and V. H. Auerbach (1955). The hormonal control of tryptophan peroxidase in the rat. *J. Biol. Chem.* **214**, 307–314.

22 G. Schutz, M. Beato, and P. Feigelson (1973). Messenger RNA for hepatic tryptophan oxygenase: Its partial purification, its translation in a heterologous cell-free system and its control by glucocorticoid hormones. *Proc. Natl. Acad. Sci. USA* **70**, 1218–1221.

23 L. Delap and P. Feigelson (1978). Effect of cycloheximide on the induction of tryptophan oxygenase mRNA by hydrocortisone *in vivo*. *Biochem. Biophys. Res. Commun.* **82**, 142–149.

24 E. Hofer, H. Land, C. E. Sekeris, and H. P. Morris (1978). Messenger RNA activities for two liver enzymes, tyrosine aminotransferase and tryptophan oxygenase, in Morris Hepatomas 5123C and 9618A and in HTC cells. *Eur. J. Biochem.* **91**, 223–229.

25 P. Feigelson and L. A. Killewich (1979). Hormonal and developmental modulation of tryptophan oxygenase mRNA. In J. D. Baxter and G. G. Rousseau (Eds.). *Glucocorticoid Hormone Action*. Springer-Verlag, Berlin, pp. 243–251.

26 T. Nakamura, H. Shimno, and A. Ichihara (1980). Insulin and glucagon as a new regulator system for tryptophan oxygenase activity demonstrated in primary cultured rat hepatocytes. *J. Biol. Chem.* **255**, 7533–7535.

27 R. Gebhardt and M. Decke (1979). Permissive effect of dexamethasone on glucagon induction of urea-cycle enzymes in perifused primary monolayer cultures of rat hepatocytes. *Eur. J. Biochem.* **97**, 29–35.

28 R. T. Schimke (1964). Enzymes of arginine metabolism in mammalian cell culture, 1. Repression of arginino-succinate synthetase and arginino-succinase. *J. Biol. Chem.* **239**, 139–145.

29 J. D. Irr and L. B. Jacoby (1977). Genetic control of arginino succinate synthetase in human lymphoblasts. In G. Wilcox, J. Abelson, and C. F. Fox (Eds.). *Molecular Approaches to Eucaryotic Genetic Systems*, ICN-UCLA Symp. No. 8. Academic Press, New York, pp. 421–430.

30 R. G. Kulka and H. Cohen (1973). Regulation of glutamine synthetase activity of hepatoma tissue culture cells by glutamine and dexamethasone. *J. Biol. Chem.* **248**, 6738–6743.

31 P. K. Sarkar and B. Griffith (1976). Messenger RNA for glutamine synthetase: its partial purification, translation in a cell-free system and its regulation by hydrocortisone. *Biochem. Biophys. Res. Commun.* **68**, 675–681.

32 R. B. Crook, M. Louie, T. F. Deuel, and G. M. Tomkins (1978). Regulation of glutamine synthetase by dexamethasone in hepatoma tissue culture cells. *J. Biol. Chem.* **253**, 6125–6131.

33 R. W. Hanson, M. A. Cimbala, J. Garcia-Ruiz, K. Nelson, and D. Kioussis (1979). Regulation of phosphoenolpyruvate carboxykinase (GTP) synthesis. In D. E. Atkinson and C. F. Fox

(Eds.). *Modulation of Protein Function*, ICN-UCLA Symp. No. 13. Academic Press, New York, pp. 357–368.

34. K. Nelson, M. A. Cimbala, and R. W. Hanson (1980). Regulation of phosphoenolpyruvate carboxykinase (GTP) mRNA turnover in rat liver. *J. Biol. Chem.* **255**, 8509–8515.

35. E. G. Beale, C. S. Katzen, and D. K. Granner (1981). Regulation of rat liver phosphoenolpyruvate carboxykinase (GTP) messenger ribonucleic acid activity by $N^6,O^{2'}$-dibutyryladenosine 3',5'-phosphate. *Biochemistry* **20**, 4878–4883.

36. J. M. Gunn, R. W. Hanson, O. Meyuhas, L. Reshef, and F. J. Ballard (1975). Glucocorticoids and the regulation of phosphoenolpyruvate carboxykinase (guanosine triphosphate) in the rat. *Biochem. J.* **150**, 195–203.

37. J. M. Gunn, S. M. Tilghman, R. W. Hanson, L. Reshef, and F. J. Ballard (1975). Effects of cyclic adenosine monophosphate, dexamethasone and insulin on phosphoenolpyruvate carboxykinase synthesis in reuber H-35 hepatoma cells. *Biochemistry* **14**, 2350–2357.

38. O. Meyuhas, L. Reshey, F. J. Ballard, and R. W. Hanson (1976). The effect of insulin and glucocorticoids on the synthesis and degradation of phosphoenolpyruvate carboxykinase (GTP) in rat adipose tissue cultures *in vitro*. *Biochem. J.* **158**, 9–16.

39. H. Niemeyer, T. Ureta, and L. Clark-Turri (1975). Adaptive character of liver glucokinase. *Mol. Cell. Biochem.* **6**, 109–126.

40. S. Weinhouse (1976). Regulation of glucokinase in liver. *Curr. Topics Cell. Reg.* **11**, 1–50.

41. W. Sibrowski and H. J. Seitz (1980). Hepatic glucokinase turnover in intact and adrenalectomised rats *in vivo*. *Eur. J. Biochem.* **113**, 121–129.

42. H. J. Seitz, W. Lüth, and W. Tarnowski (1979). Regulation of rat liver glucokinase activity *in vivo*: Predominant role of hepatic cyclic AMP and glucocorticoids. *Arch. Biochem. Biophys.* **195**, 385–391.

43. W. Sibrowski, M. J. Müller, and H. J. Seitz (1981). Effect of different thyroid states on rat liver glucokinase synthesis and degradation *in vivo*. *J. Biol. Chem.* **256**, 9490–9494.

44. C. Schudt (1979). Hormonal regulation of glucokinase in primary cultures of rat hepatocytes. *Eur. J. Biochem.* **98**, 77–82.

45. N. R. Katz, M. A. Nauck, and P. T. Wilson (1979). Induction of glucokinase by insulin under the permissive action of dexamathasone in primary rat hepatocyte cultures. *Biochem. Biophys. Res. Commun.* **88**, 23–29.

46. J. T. Spence and H. C. Pitot (1979). Hormonal regulation of glucokinase in primary cultures of adult rat hepatocytes. *J. Biol. Chem.* **254**, 12,331–12,336.

47. J. T. Spence, M. J. Merill, and H. C. Pitot (1981). Role of insulin, glucose and cyclic GMP in the regulation of glucokinase in cultured hepatocytes. *J. Biol. Chem.* **256**, 1598–1603.

48. P. K. Flick, J. Chen, A. E. Alberts, and P. R. Vagelos (1978). Translation of rat liver fatty acid synthetase mRNA in a cell-free system derived from wheat germ. *Proc. Natl. Acad. Sci. USA* **75**, 730–734.

49. C. P. Nepokroeff and J. W. Porter (1978). Translation and characterisation of the fatty acid synthetase messenger RNA. *J. Biol. Chem.* **253**, 2279–2283.

50. D. K. Das (1980). Regulation of hepatic fatty-acid synthesizing enzymes of diabetic animals by thyroid hormone. *Arch. Biochem. Biophys.* **203**, 25–36.

51. J. H. Oppenheimer, C. N. Mariash, H. C. Towle, H. L. Schwartz, and F. E. Kaiser (1981). Interaction of T_3 and carbohydrate in the induction of lipogenic enzymes. *Life Sci.* **28**, 1693–1699.

52. T. A. Pry and J. W. Porter (1981). Control of fatty acid synthetase mRNA levels in rat liver by insulin, glucagon, and dibutyl cyclic AMP. *Biochem. Biophys. Res. Commun.* **100**, 1002–1009.

53. H. C. Towle, C. N. Mariash, H. L. Schwartz, and J. H. Oppenheimer (1981). Quantitation of rat liver messenger ribonucleic acid for malic enzyme during induction by thyroid hormone. *Biochemistry* **20**, 3486–3492.

54. U. A. Siddiqui, T. Goldflam, and A. G. Goodridge (1981). Nutritional and hormonal regulation of the translatable levels of malic enzyme and albumin mRNAs in avian liver cells *in vivo* and in culture. *J. Biol. Chem.* **256**, 4544–4550.
55. J. S. Hutchison and D. Holten (1978). Quantitation of messenger RNA levels for rat liver 6-phosphogluconate dehydrogenase. *J. Biol. Chem.* **253**, 52–57.
56. J. D. Sun and D. Holten (1978). Levels of rat liver glucose-6-phosphate dehydrogenase messenger RNA. *J. Biol. Chem.* **253**, 6832–6836.
57. S. Sassa, L. Bradlow, and A. Kappas (1979). Steroid induction of δ-amino-levulinic acid synthase and porphyrins in liver. *J. Biol. Chem.* **254**, 10,011–10,020.
58. G. G. Rousseau, A. Amar-Costesec, M. Verhaegen, and D. K. Granner (1980). Glucocorticoid hormones increase the activity of plasma membrane alkaline phosphodiesterase I in rat hepatoma cells. *Proc. Natl. Acad. Sci. USA* **77**, 1005–1009.
59. P. C. Brown and J. Papaconstantinou (1979). Coordinated modulation of albumin synthesis and mRNA levels in cultured hepatoma cells by hydrocortisone and cyclic AMP analogs. *J. Biol. Chem.* **255**, 9379–9384.
60. G. S. McKnight, D. C. Lee, D. Hemmaplordh, C. D. Finch, and R. D. Palmiter (1980). Transferrin gene expression: Effects of nutritional iron deficiency. *J. Biol. Chem.* **255**, 144–147.
61. G. S. McKnight, D. C. Lee, and R. D. Palmiter (1980). Transferrin gene expression: Regulation of mRNA transcription in chick liver by steroid hormones and iron deficiency. *J. Biol. Chem.* **255**, 148–153.
62. J. A. Martial, P. H. Seeburg, D. T. Matulick, H. M. Goodman, and J. D. Baxter (1979). Regulation of growth hormone messenger RNA. In J. D. Baxter and G. G. Rousseau (Eds.). *Glucocorticoid Hormone Action.* Springer-Verlag, Berlin, pp. 279–289.
63. R. P. Dobner, E. S. Kawasaki, L.-Y. Yu, and F. C. Bancroft (1981). Thyroid or glucocorticoid hormone induces pre-growth-hormone mRNA and its probably nuclear precursor in rat pituitary cells. *Proc. Natl. Acad. Sci. USA* **78**, 2230–2234.
64. D. F. Derda, M. F. Miles, J. S. Schweppe, and R. A. Jungman (1980). Cyclic AMP regulation of lactate dehydrogenase. *J. Biol. Chem.* **255**, 1112–1121.
65. S. Kumar, J. F. McGinnis, and J. de Villis (1980). Catecholamine regulation of lactate dehydrogenase in rat brain cell culture. *J. Biol. Chem.* **255**, 2315–2321.
66. R. D. Ciaranello, G. F. Wooten, and J. Axelrod (1975). Regulation of dopamine β-hydroxylase in rat adrenal gland. *J. Biol. Chem.* **250**, 3204–3211.
67. E. Costa, A. Kurosawa, and A. Guidotti (1976). Activation and nuclear translocation of protein kinase during transsynaptic induction of tyrosine 3-monooxygenase. *Proc. Natl. Acad. Sci. USA* **73**, 1058–1062.
68. P. R. Vulliet, T. A. Langan, and N. Weiner (1980). Tyrosine hydroxylase: A substrate of cyclic AMP-dependent protein kinase. *Proc. Natl. Acad. Sci. USA* **77**, 92–96.
69. G. L. Firestone and E. C. Health (1981). The cyclic AMP-mediated induction of alkaline phosphatase in mouse L-cells. *J. Biol. Chem.* **256**, 1396–1403.
70. E. E. Baetge, B. B. Kaplan, D. J. Reis, and T. H. Joh (1981). Translation of tyrosine hydroxylase from poly (A)-mRNA in pheochromocytoma cells is enhanced by dexamethasone. *Proc. Natl. Acad. Sci. USA* **78**, 1269–1273.
71. G. T. Snoek, H. O. Voorma, and R. Van Wijk (1981). Further evidence for translational regulation of tyrosine aminotransferase synthesis by dibutyryl cyclic AMP in Reuber H35 hepatoma cells. *Biochim. Biophys. Acta* **655**, 107–112.
72. D. Kioussis, L. Reshef, H. Cohen, S. M. Tilghman, P. B. Iynedjian, F. J. Ballard, and R. W. Hanson (1978). Alterations in translatable messenger RNA coding for phosphoenolpyruvate carboxykinase (GTP) in rat liver cytosol during deinduction. *J. Biol. Chem.* **253**, 4327–4332.

CHAPTER THIRTEEN

Regulation of the Diurnal Rhythms of Enzyme Synthesis

In Chapter 12 we established that a number of enzymes are adaptable and can change their rates of synthesis in response to changes in the external environment. Usually in the laboratory, these environmental changes are the result of rather gross, unphysiological manipulations of the dietary or endocrine status of the animal, designed to maximize the response of the enzyme being studied. Animals are also faced with daily, monthly, seasonal, and annual changes in the environment to which they must adapt, although these are not usually so extreme. Of these rhythmic changes, the 24-h cycle of light and darkness that determines the periodicity of such major physiological responses as sleeping,

Table 13-1 Physiological and Biochemical Responses That Are Known to Undergo Diurnal Rhythms of Activity

PHYSIOLOGICAL RESPONSES	HORMONES
Running activity	Glucocorticoid
Sleep	ACTH
Feeding activity	Melatonin
Drinking activity	GH
	TSH
	Prolactin
	LH
BIOCHEMICAL PARAMETERS	**ENZYMES**
Chromatin template activity	ODC
RNA synthesis	TAT
DNA synthesis	GK
Glycogen synthesis	TO
Lipid deposition	PEPCK
Drug metabolism	N-acetyltransferase
Mitochondrial activity	glycogen synthesis
Mitosis	glycogen phosphorylase
Amino acid uptake	HMG CoA reductase

locomotor activity, and feeding is the most studied. For example, nocturnal animals, such as the rat, actively search for food during the period of darkness and sleep during the light period. Other animals are active in the daytime and sleep during the dark period. These rhythms are referred to as *diurnal rhythms* and sometimes as *circadian rhythms*. However, since the term *circadian* implies a specific mechanism (that the rhythm will still persist within the organism when the environmental stimulus is removed), it should be used with some caution unless the actual mechanism of the rhythm is understood.

These rhythms have attracted a great deal of attention recently from biochemists, neurologists, pharmacologists, and physiologists, who are attempting to define the rhythms in organismic and molecular terms. It is now known that many biochemical and physiological parameters undergo pronounced rhythmic changes, a partial listing of which is given in Table 13-1. The first part of this chapter is devoted to a description of some of these rhythms of enzymes and hormones, followed by a discussion on the molecular basis of the rhythms and some speculation on the nature of the "biological clock." The implications of these rhythms in the design of experimental protocols are also discussed.

DIURNAL RHYTHMS OF ENZYME SYNTHESIS

Hepatic Tyrosine Aminotransferase

Tyrosine aminotransferase was the first enzyme to be shown to exhibit a diurnal rhythm of activity (1). In laboratory rats subject to alternating cycles of 12 h of light followed by 12 h of dark, with food freely available, the activity of the enzyme increased during the dark period, when the animals were feeding, and returned to basal levels during the light period when they were resting.

In order to investigate the biochemical basis of the rhythm, Potter introduced a number of controlled lighting and feeding schedules so that more consistent and reproducible changes could be observed. Typically, the cycle of light and darkness is held constant at 12 h of light followed by 12 h of dark in each 24-h day, whereas the feeding period is modified to suit individual experimental protocols. A typical rhythm of TAT activity for animals adapted to the "8 + 16" feeding schedule (8 h of food availability in each 24-h day, with the start of the feeding period being coincident with the start of the dark period) is shown in Fig. 13-1. This schedule simulates the natural nocturnal feeding response of the animals, and the activity of the enzyme increases immediately after the start of feeding and peaks toward the end of the feeding period, before rapidly returning to basal levels.

By manipulating the timing and extent of the feeding period it has been possible to assess the relative contributions of food, and of light and darkness to the enzyme rhythm (2-4). A number of these results are summarized in Fig. 13-2, demonstrating that food plays a major role in determining the amplitude and periodicity of the rhythm. Figure 13-2*a* shows that the protein content of

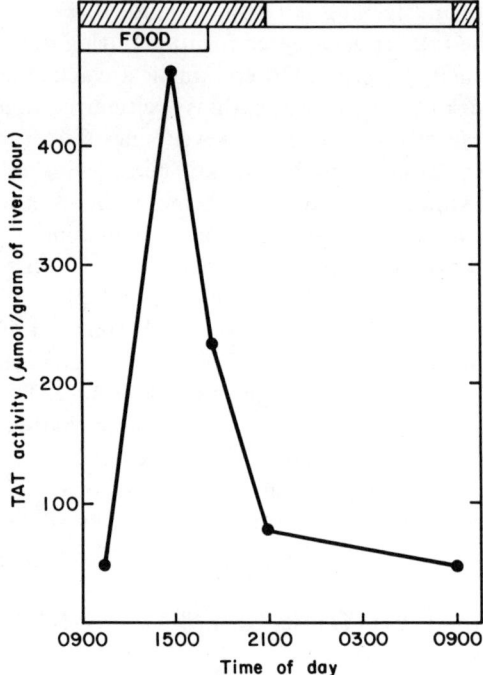

Fig. 13-1 The diurnal rhythm of tyrosine aminotransferase activity in the livers of animals adapted to the "8 + 16" feeding schedule. In this and all subsequent diagrams the L : D cycle is 12 h of light and 12 h of dark (shaded). [Reproduced from Fig. 1 of Ref. 4, with permission.]

the diet determines the *amplitude* of the rhythm, with the *periodicity* remaining unaltered. If the duration of the feeding period is decreased from 8 h to 2 h it has little or no effect on the rhythm as long as the feeding period still commences at the onset of the dark period (Fig. 13-2b). However, if the feeding period is moved to a later time, as shown in Fig. 13-2b ("2 + 22 PM" schedule), then it has a marked effect on both the amplitude and periodicity of the rhythm. The initial increase in activity is food-related, but the onset of the light period attenuates this initial response and returns the enzyme level to near-basal activities. This is followed by a secondary rise during the light period.

In view of the seemingly important role of glucocorticoids in the induction of this enzyme (see Chapter 12) the effect of adrenalectomy on the TAT rhythm was investigated. As shown in Fig. 13-2c *adrenalectomy has little or no effect* on the diurnal rhythm. Furthermore, it has been shown that the rhythm still persists in the absence of the pancreas, thyroid gland, and pituitary (5), which suggests, somewhat surprisingly, that neither insulin, glucagon, nor the thyroid or pituitary hormones are involved in maintaining the rhythm. Since the change in TAT synthesis is paralleled by an increase in the concentration of $mRNA^{TAT}$ (6), the rhythm does appear to require some effector capable of eliciting a change in gene expression at the transcriptional level.

When animals are adapted to an 8-h feeding period followed by 40 h without food (the "8 + 40" feeding schedule, Fig. 13-2d) there is an interesting change in the rhythm. On the day when food is available the increase in TAT activity is very similar to that of animals adapted to the "8 + 16" feeding schedule. However, instead of the usual rapid decline to basal levels, the "8 + 40" animals maintain a substantial level of enzyme throughout the early light period, followed by a second peak late in the light period before the day without food. The activity of the enzyme then rapidly declines to basal levels. The amplitude of the first peak is greatly decreased if the protein content of the diet is lowered,

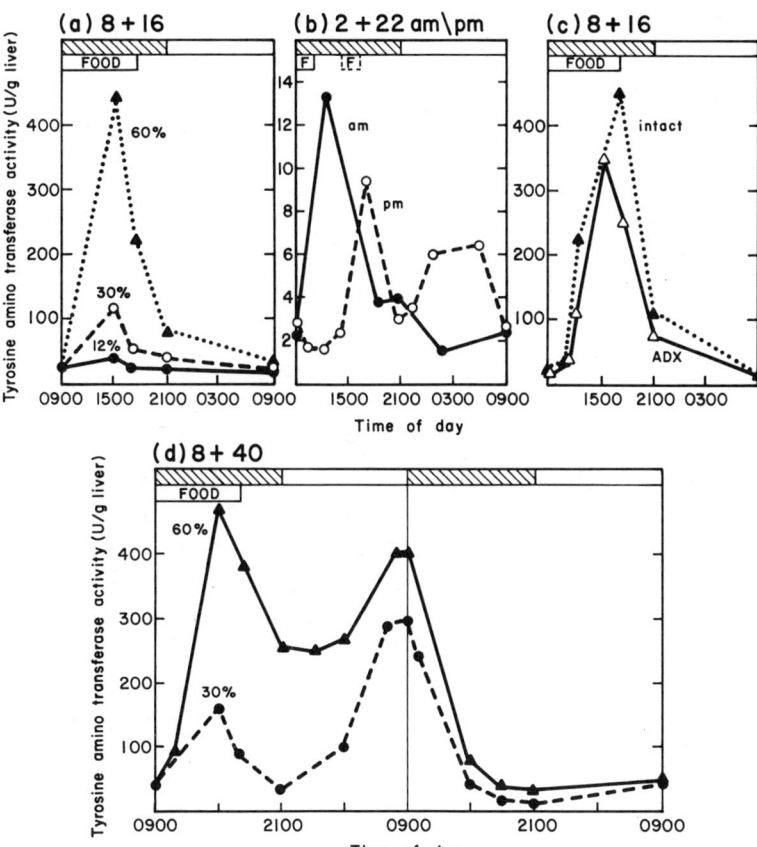

Fig. 13-2 Factors affecting the diurnal rhythm of TAT synthesis in the liver. (*a*) The effect of percentage of protein in the diet for animals on the "8 + 16" feeding schedule. (*b*) Animals were adapted to the "2 + 22" feeding schedule, in which the 2-h feeding period occurred at the beginning of the dark period (2 + 22 AM) or midway through it (2 + 22 PM). (*c*) The lack of an effect of adrenalectomy (ADX) on the TAT rhythm. (*d*) The effect of the protein content of the diet on the rhythm of TAT in animals adapted to the "8 + 40" feeding schedule. [Data reproduced from Watanabe et al. *J. Nutr.* 1968, **95**, 207–227, by permission of the American Institute of Nutrition.]

whereas the second peak is essentially unaffected. Moreover, adrenalectomy has no effect on the peak of the first day but abolishes the second peak, indicating differences in the regulation of the synthesis of the enzyme at these times. This secondary rise probably reflects a release of glucocorticoids during this prolonged fasting period, stimulating a gluconeogenic response to supply glucose to vital organs. Under these conditions this second peak is probably regulated by the factors described in Chapter 12 (see Table 12-2), whereas the first peak certainly is not.

Glucokinase, Glycogen Synthetase, Phosphorylase, and Glycogen Rhythms

Hepatocytes deposit glycogen throughout the dark period using carbon derived from food (Fig. 13-3), with the amount deposited being directly related to the carbohydrate content of the diet (2, 4, 5). This glycogen, which can amount to 10% of the wet weight of the tissue, is used as an energy reserve during the 16 h without food.

The increase in glycogen accumulation is paralleled by an increase in the activity of glycogen synthetase a and a decrease in the activity of glycogen phosphorylase (Fig. 13-3 and Refs. 7–9). These changes are preceded by an increase in the activity of glucokinase, the first enzyme of glucose metabolism,

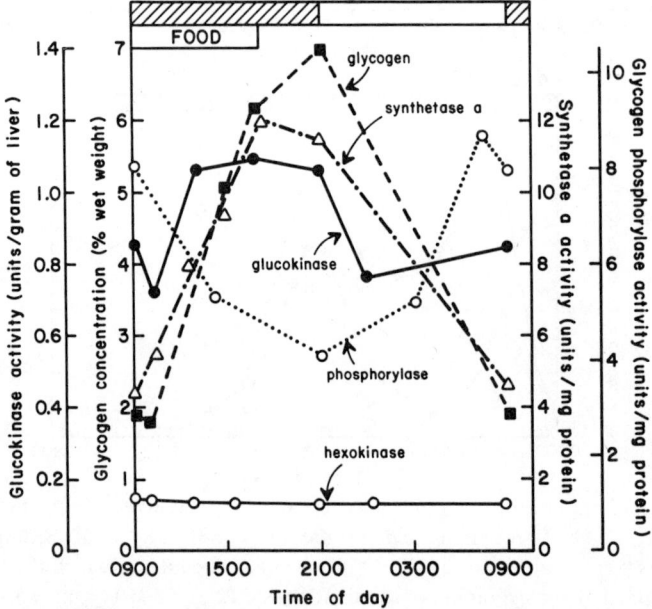

Fig. 13-3 Diurnal changes in the synthesis of enzymes of carbohydrate metabolism and of liver glycogen content in animals adapted to the "8 + 16" controlled feeding schedule. [Data adapted from Refs. 4 and 7 with permission of the author and from Fig. 1 of Shimazu (9) with the permission of Elsevier/North-Holland Biomedical Press.]

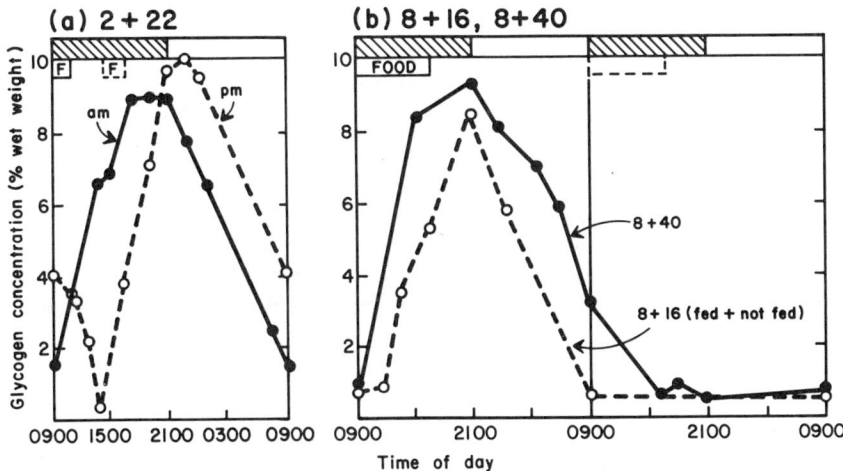

Fig. 13-4 Factors affecting the diurnal rhythm of glycogen accumulation. (*a*) The rhythm in animals adapted to either "2 + 22 AM" or the "2 + 22 PM" feeding schedule. (*b*) Comparison of the rhythm of glycogen in animals adapted to the "8 + 40" schedule with animals adapted to the "8 + 16" schedule that were not fed on the second day. [Reproduced from Fig. 4 of Ref. 3 with permission.]

which remains elevated throughout the period of glycogen deposition. The uninducible, "low-K_m" group of hexokinase isoenzyme activities does not show any diurnal rhythms (Fig. 13-3).

The rhythm of glycogen accumulation is clearly related to feeding activity (Fig. 13-4). The rhythm for animals fed on the "2 + 22 AM" feeding schedule (with the 2-h feeding period commencing at the onset of darkness) is very similar to that for rats on the "8 + 16" feeding schedule (cf. Fig. 13-3), whereas for rats on the "2 + 22 PM" schedule accumulation does not begin until they commence feeding. On both these schedules glycogen reserves are depleted within 24 h. When rats adapted to the "8 + 16" feeding schedule are not fed there is no accumulation of glycogen on that day (Fig. 13-4*b*), further illustrating the strict dependence of the glycogen rhythm upon food intake. Moreover, for rats adapted to the "8 + 40" feeding schedule there is no accumulation of glycogen during the dark period of the second day when the animals are fasting (Fig. 13-4*b*).

The behavior of the rhythm of glycogen metabolism in rats adapted to the "8 + 40" schedule provides an interesting insight into the nature of these rhythms. The rhythm has evolved as an efficient means of storing dietary carbohydrate over a 24-h period, but the animals are unable to *adapt* the mechanism to the more demanding "8 + 40" feeding schedule, and all the glycogen is consumed during the first day even though there is a further 24 h without food. Thus, there is a severe constraint upon the capacity of the organism to store energy in this form (4).

Rhythms of Other Hepatic Enzymes

A number of other liver enzymes exhibit pronounced diurnal rhythms in rats either fed *ad libitum* or adapted to the "8 + 16" controlled feeding schedule (Fig. 13-5 and Refs. 4, 10, and 11). In this figure the results are shown as a percentage of their maximum value. Ornithine decarboxylase, tryptophan oxygenase, and hydroxymethylglutarylCoA reductase show rhythms similar to that for TAT except that they reach their peaks somewhat earlier. The rhythm of PEPCK, on the other hand, is quite different. The activity of this enzyme increases toward the end of the light period and remains high for most of the period of food intake, only declining as glycogen levels reach their maximum value. A similar rhythm is observed for the urea-cycle enzyme argininosuccinate synthetase (11).

All these rhythms are abolished if the animals are denied access to food, and they shift when the feeding period is shifted, indicating that they are entrained to the feeding cycle, rather than the light:dark cycle.

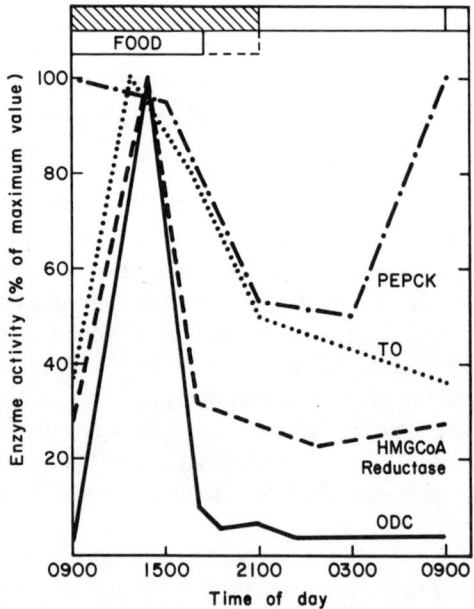

Fig. 13-5 Diurnal rhythms of hepatic enzymes in animals either adapted to the "8 + 16" feeding schedule (HMGCoA reductase, ODC) or fed *ad libitum* (PEPCK, TO). HMGCoA = hydroxymethylglutaryl CoA, ODC = ornithine decarboxylase, PEPCK = phosphoenolpyruvate carboxykinase, TO = tryptophan oxygenase. [Data reproduced from Fig. 1 of Ref. 4, Fig. 8 of Ref. 10, with permission of Karger, Basel, and Fig. 6 of Ref. 11 with permission of Elsevier/North-Holland, Amsterdam.]

The Rhythm of N-Acetyltransferase in the Pineal Gland

The pineal gland is a small endocrine tissue located on the third ventricle of the brain, and its principle secretory product is the indoleamine melatonin. The pathway of melatonin biosynthesis from tryptophan is illustrated in Fig. 13-6. The synthesis and release of melatonin by the pineal gland exhibit a pronounced diurnal rhythm, being high during the dark period and negligible during the light period. This rhythm of melatonin synthesis and release is paralleled by a rhythm in the activity of serotonin N-acetyltransferase, a key enzyme in the biosynthetic pathway (Fig. 13-7; see Ref. 12 for a recent review of pineal gland metabolism). Since the rhythm is abolished by both cycloheximide and actinomycin D, a change in gene expression is required to maintain the rhythm.

The regulation of the rhythm of N-acetyltransferase synthesis has been studied in detail in rats and in chickens, and important differences have been found both in the regulation of the rhythm and in apparent function of the gland between the species.

In the rat, the pineal gland is innervated by nerves originating in various parts of the brain, including the superior cervical ganglion [which carries electrical signals from an area of the hypothalamus called the suprachiasmatic nuclei (SCN), Fig. 10-2] and the retinohypothalamic tract (which carries information concerning the light:dark cycle from the eye). In turn, the pineal gland sends electrical signals back to the hypothalamus. Surgical destruction of the fibers of the superior cervical ganglion abolishes the rhythm of enzyme synthesis in the gland. Moreover, treatment with reserpine (to deplete the nerve endings of norepinephrine) or treatment with β-adrenergic antagonists (to prevent endogenous neurotransmitters from binding to the pinealocytes) also abolishes the rhythm. These data suggest that the superior cervical ganglia carry information that governs the rhythmic behavior of the tissue. This information is carried by neurotransmitters interacting with β-adrenergic receptors, leading to an increase in cAMP [Figs. 13-8 and 11-14 (13)] and induction of enzyme synthesis.

The pineal gland receives signals from the retinohypothalamic tract, either directly or via the SCN, and it is believed that these signals entrain the rhythm of N-acetyltransferase synthesis to the light:dark cycle (Fig. 13-8). The acute sensitivity of this rhythm to the light:dark cycle was demonstrated in experiments that showed that 90% of the induced level of N-acetyltransferase activity was lost within 10 min of the lights being switched on during the dark period (injection of β-adrenergic antagonists have the same effect). Starvation or changes in the composition of the diet have no effect on the enzyme rhythm, indicating that, unlike liver enzymes, the rhythm of the pineal enzyme is entrained to the light:dark cycle and not feeding activity.

The rhythm of N-acetyltransferase synthesis in chickens and other avian species is similar to that of rats (see Ref. 12 for references); however, destruction of the superior cervical ganglion does *not* abolish the rhythm. Since blinding abolishes the rhythm it still appears to be entrained by light:dark cycles, but the

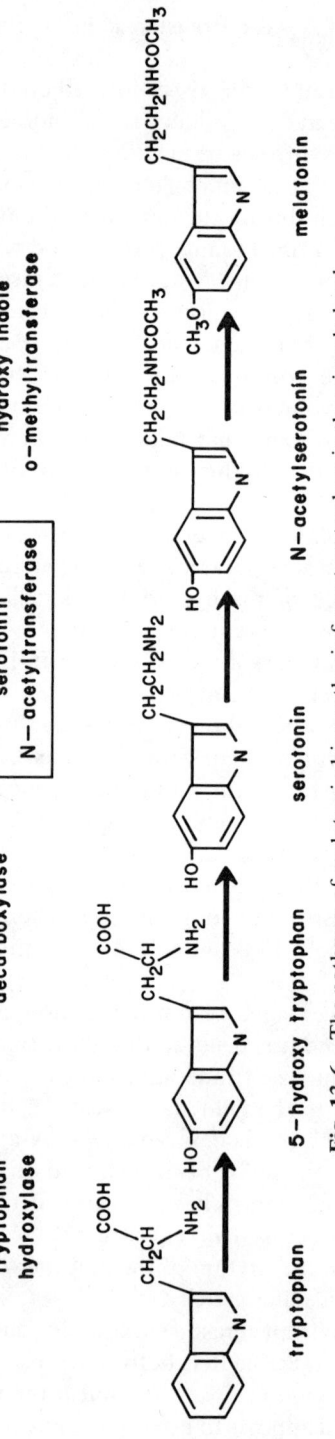

Fig. 13-6 The pathway of melatonin biosynthesis from tryptophan in the pineal gland.

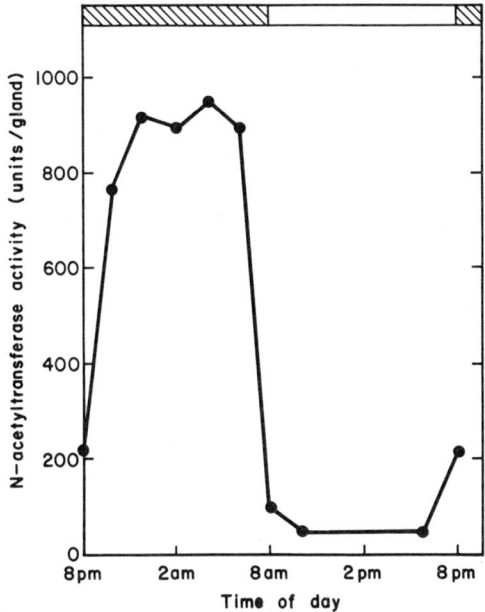

Fig. 13-7 The diurnal rhythm of serotonin n-acetyltransferase activity in the pineal glands of unfed rats. [Reproduced, with permission, from T. Deguchi (1979). *Mol. Cell. Biochem.* **27**, 57–66.]

information is not transmitted via the SCN and superior cervical ganglion. Moreover, the neurotransmitters that induce *N*-acetyltransferase in rat pineal glands have no effect on chicken glands, which suggests that the mechanism for the generation of the rhythm is different. Indeed, it has recently been shown that the increase in enzyme synthesis in the chicken pineal gland is preceded by an increase in cGMP concentration, not cAMP (14).

Chicken pineal glands can be excised and maintained in organ culture for a period of several days. Studies on these cultures led to the fascinating discovery that the glands were responsive to the light:dark cycle of the laboratory, and rhythms of melatonin biosynthesis and release identical to those *in vivo* could be observed for as long as 5 days (Fig. 13-9 and Ref. 15). This *in vitro* rhythm is sensitive to both actinomycin D and cycloheximide, indicating that changes in transcriptional activity result in the synthesis of new enzyme protein. Thus the chicken pineal gland must contain its own photoreceptors and appears capable of generating its own rhythm that becomes entrained to the light:dark cycle. Attempts to maintain pinealocytes in *cell culture* have been less successful, but there are indications that the ability to generate a rhythm resides in *individual* pinealocytes rather than being solely a property of the whole gland (12). These cells, therefore, have all the properties of a circadian oscillator, whereas the pinealocytes of rats and other mammals are dependent upon the SCN to drive the rhythm of melatonin synthesis.

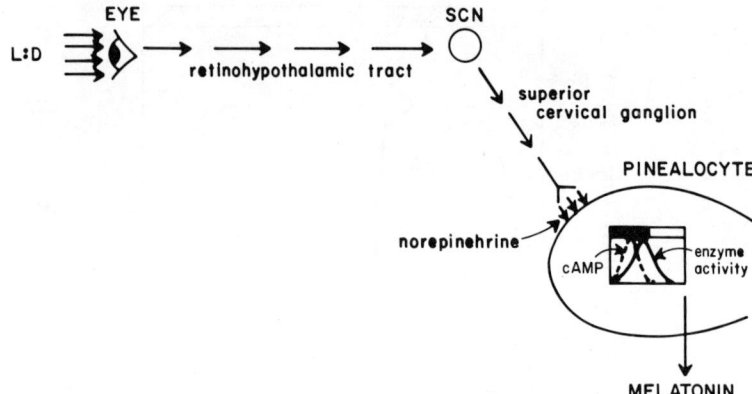

Fig. 13-8 Neural pathways involved in the synchronization of the pineal gland's rhythm of melatonin production to the light:dark cycle by the SCN (suprachiasmatic nucleus) in the rat.

Diurnal Rhythms in Other Tissues

Studies on rhythms in other tissues are very incomplete, but there are indications that they do exist. For example, PEPCK has a rhythm of synthesis in the kidney as well as in the liver (11), with synthesis taking place in the light period and the early part of the dark period. Enzyme activities in the brain also appear

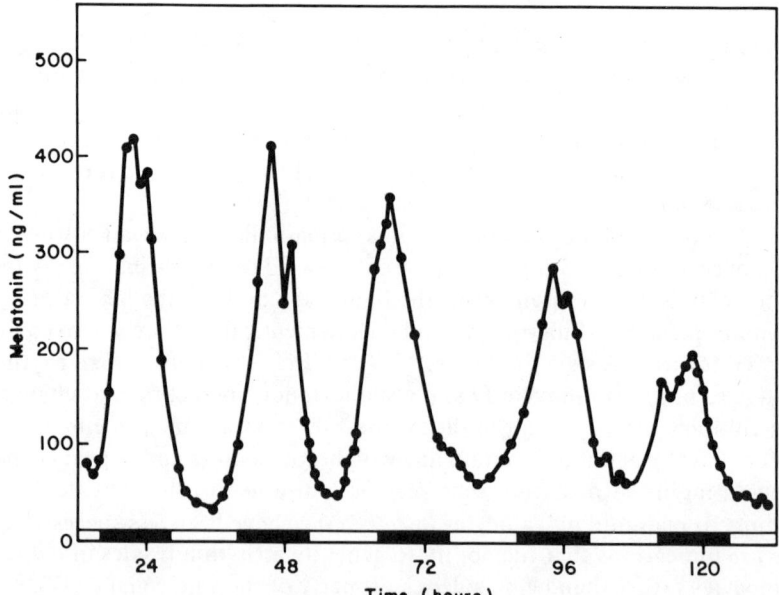

Fig. 13-9 Diurnal rhythms of melatonin production by chicken pineal glands maintained *in vitro*. The alternating 12-h periods of light and dark are those maintained in the incubator in which the glands were cultured. [Reproduced, with permission, from J. S. Takahashi et al. (1980). *Proc. Natl. Acad. Sci. USA* **77**, 2319–2322.]

to exhibit rhythms (16). There are also rhythms in the synthesis of several hydrolytic enzymes in the cells of the intestinal mucosa (17). These rhythms appear related to food intake, with peaks during the feeding period. Interestingly, synthesis commences shortly *before* the start of the feeding period, which indicates that the animals anticipate each feeding period. This produces an interesting response when the animals are starved, since the anticipatory increase has already occurred. Indeed, it takes about 3 days without food before the rhythms run down. Thus, although the *amplitude* of the rhythms is food-related, some other mechanisms must be involved in the *generation* of the rhythms to produce the anticipatory response.

DIURNAL RHYTHMS OF OTHER BIOCHEMICAL FUNCTIONS

Chromatin Template Activity, RNA Synthesis, DNA Synthesis, and Mitosis

There is a diurnal rhythm of chromatin template activity in the nuclei of liver parenchymal cells (18). Template activity, measured as the fraction of total chromatin available for transcription by *E. coli* RNA polymerase, increases shortly after the onset of feeding, as indicated in Fig. 13-10. There is a corresponding increase in *in vivo* total RNA synthesis at this time (19, 20), as well as increases in the synthesis of the mRNAs for the enzymes such as TAT, ODC, and TO

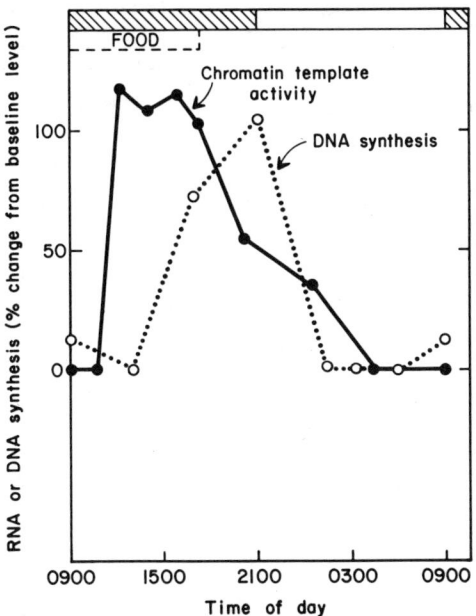

Fig. 13-10 Diurnal rhythms of hepatic chromatin template activity and of the replication of DNA in hepatocytes of young rats. The rats were either adapted to the "8 + 16" feeding schedule (replication) or fed *ad libitum* (template activity). [Reproduced from Fig. 1 of Earp (18) and Fig. 1 of Hopkins et al. (21) with permission.]

that undergo pronounced diurnal rhythms. Template activity does not increase when food is withheld, but it is only slightly reduced by adrenalectomy.

In young rats, the liver is still growing, and new cells are produced on a rhythmic basis. The S phase (DNA replication) of the cell cycle commences approximately halfway through the feeding period and peaks at the dark-light interface (Fig. 13-10 and Ref. 21). A rhythm of cells in mitosis also occurs in these and other tissues.

Glycogen and Lipid Deposition

The diurnal rhythm of glycogen accumulation in animals adapted to controlled feeding schedules has already been discussed (see Fig. 13-3). Similar rhythms are observed in rats fed *ad libitum*. In addition to glycogen, the liver also synthesizes fatty acids, and its synthesis exhibits a marked diurnal rhythm [Fig. 13-11*a* (22)]. Moreover, a similar rhythm is observed in the fat pads, which also synthesize and store lipids. Thus the major pathways of carbon storage are entrained to the rhythm of food intake so that the animal can most efficiently extract the carbon from the diet and store it.

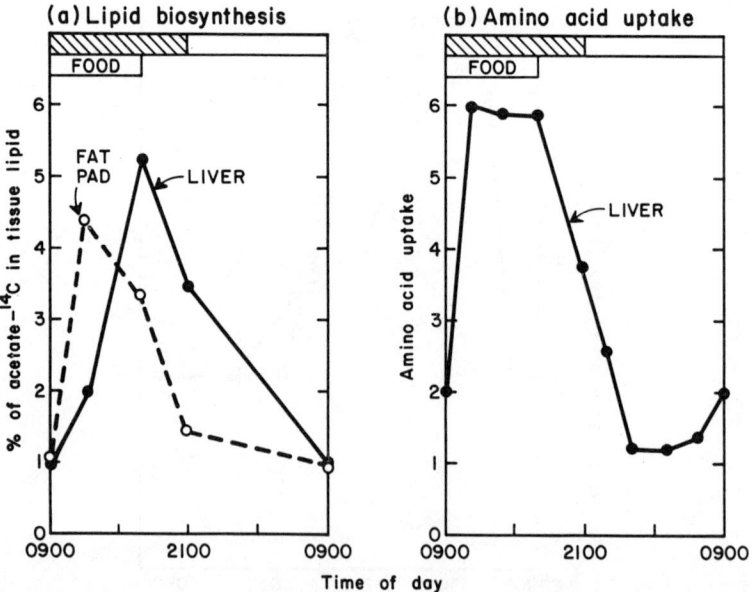

Fig. 13-11 Diurnal rhythms of (*a*) lipid biosynthesis in liver and fat pad and (*b*) liver amino acid uptake in rats adapted to the "8 + 16" feeding schedule. [Reproduced from Figs. 1 and 3 of Scott and Potter (1970). *Fed. Proc.* **29**, 1553–1559, with permission of *Federation Proceedings*.]

Amino Acid Uptake and Protein Synthesis

Amino acid uptake (measured using a radioactively labeled, nonmetabolizable amino acid such as cycloleucine) into the liver is elevated throughout the feeding period of rats adapted to controlled feeding schedules (Fig. 13-11b and Refs. 23 and 24). This influx of amino acids stimulates the enzymes of amino acid catabolism, such as TAT and TO, as well as the urea-cycle enzymes. Rather surprisingly, there is no detectable rhythm of protein biosynthesis in liver (25). However, since the liver synthesizes approximately 50% of its weight of protein each day (most of it is released as serum proteins), the protein-synthesizing machinery is likely to be working near maximum capacity 24 h a day, and the extra synthesis required to produce the inducible enzymes is negligible.

DIURNAL RHYTHMS OF HORMONE SECRETION

Since the induction of most enzymes is controlled by hormones, considerable effort is being applied to the measurement of hormone levels in the blood to establish which ones may be involved in the maintenance of enzyme rhythms. Unfortunately, there having been virtually no work carried out on the changes in receptor concentration during each 24-h day, it is still not possible to assess the contribution of this important level of regulation to the responses of cells to rhythmic changes in the environment.

Corticosterone and ACTH

There is a pronounced diurnal rhythm of corticosterone (the principle glucocorticoid in the rat) in the blood of rats either fed *ad libitum* or adapted to controlled-feeding schedules (2, 26–31). The rhythm in animals fed *ad libitum* is shown in Fig. 13-12a. Steroid levels increase throughout the light period and peak at the light:dark interface, before falling to negligible levels by the end of the dark period. The rhythm of the hormone appears to be related to food intake since, in meal-fed rats, moving the feeding period to different times of day produces a corresponding shift in the steroid rhythm (27–29). However, if rats are starved the rhythm persists (Fig. 13-12b) for at least 2 days (29). Thus, although the rhythm is *entrained* by food intake, it is not absolutely dependent upon it. Furthermore, the rhythm is not dependent upon, or entrained by, the light:dark cycle, because the rhythm persists in animals exposed to constant light and in blinded animals (29).

The rhythm also persists in starved animals that are maintained in constant light, a situation where there is no input from the external environment. Such a rhythm is deemed to be *free-running* since it can be generated and maintained by biochemical mechanisms within the organism. Usually such endogenous rhythms free-run with a periodicity of approximately, but seldom exactly, 24 h.

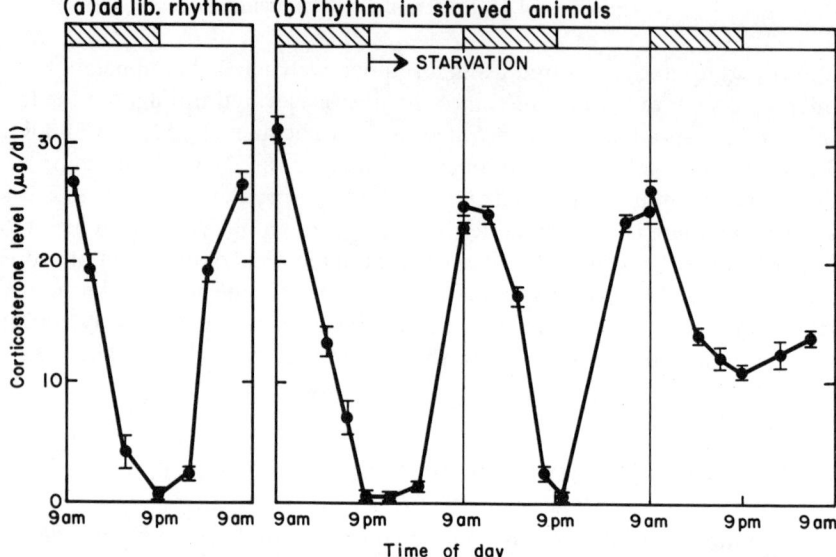

Fig. 13-12 Diurnal rhythms of serum corticosterone levels in rats either fed *ad libitum* (*a*) or starved as indicated (*b*). [Reproduced from Figs. 1 and 3 of Kato et al. (29), with permission. Copyright © 1980 The Endocrine Society.]

The rhythm can be *entrained* to a periodic change in the environment (in this case food intake) so that its periodicity becomes *exactly* 24 h.

The release of glucocorticoids from the adrenal glands is under the control of corticotrophin-releasing hormone (CRH) of the hypothalamus, which drives a diurnal rhythm of adrenocorticotrophin hormone (ACTH) release from the anterior pituitary (Table 10-1). Therefore, attention has been focused on the brain to try to determine the factors involved in the regulation of the endogenously driven glucocorticoid rhythm. An examination of various centers of the brain that may be involved (26) concluded that the suprachiasmatic nuclei (SCN) of the hypothalamus were essential. The SCN also drives the rhythm of melatonin release (Fig. 13-8), indicating a fundamental role for this group of cells in the origination and maintenance of endogenous rhythms.

Melatonin

The biochemistry of melatonin biosynthesis in the pineal gland and the diurnal rhythms of associated enzymes have already been discussed. The release of hormone parallels the rhythm of N-acetyltransferase activity (Fig. 13-7), occurring only during the dark period. If the dark period is extended the synthesis and release of melatonin remain elevated, whereas in constant light there is no melatonin release.

In any 24-h day the amount and duration of melatonin release are functions of the length of the dark period, and this may be the key to the function of the pineal gland. Thus, the gland is sensitive to incremental changes in the light:dark cycle on a seasonal basis when the amount of daylight varies.

Pituitary Hormones

A number of the hormones released by the anterior pituitary have been recently shown to undergo small, statistically significant diurnal variations, including prolactin, luteinizing hormone, thyroid-stimulating hormone, and adrenocorticotrophic hormone (27, 28, 32–36).

Prolactin and luteinizing-hormone release from the pituitary is higher in the dark period than in the light period (Figs. 13-13a, 13-13b). In contrast, the release of adrenocorticotrophic hormone and thyroid-stimulating hormone is higher during the light period than in the dark period (the ACTH rhythm is similar to that of corticosterone shown in Fig. 13-12a). It is likely that the release of these hormones is controlled by rhythmic changes in the secretory activity of the cells of the hypothalamus, indicating that higher centers of the brain are involved in the generation of these diurnal rhythms.

Growth hormone, the other large polypeptide hormone released from the pituitary, has been found to undergo 20-fold changes in the concentration in the blood at various times of day. However, attempts to establish a definite diurnal rhythm for the secretion of this hormone have been unsuccessful (28, 33–35). This is because growth-hormone release is episodic or pulsatile. That is, release of the hormone from the pituitary occurs in a series of discrete episodes or

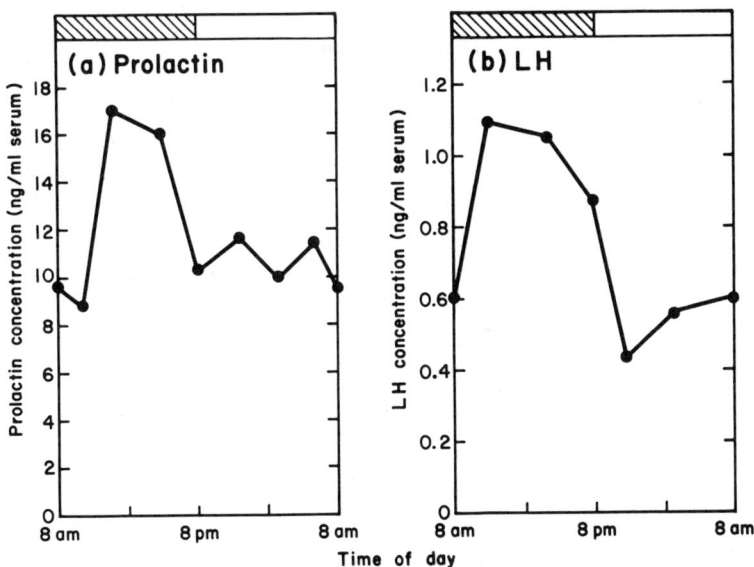

Fig. 13-13 Diurnal variations in (a) prolactin and (b) luteinizing hormone in the serum of rats fed *ad libitum*. [Reproduced, with permission of Lippincott/Harper & Row, from Figs. 1 and 2 of Dunn et al. (32).]

pulses, with some of the episodes leading to 20-fold changes in concentration of the hormone for a few minutes. In the face of changes of this magnitude it has not been possible to discern any significant diurnal rhythm. However, the pulses of secretion do show a regular periodicity. For example, there is usually a large, double burst of secretion in the light period and another large burst during the dark period (34).

Rhythms of Insulin, Glucagon, and Hepatic cAMP

Insulin and glucagon release by the pancreas is not under direct control of either the pituitary or the hypothalamus (Chapter 10). The hormones are released in response to changes in feeding activity and are modulated by the carbohydrate and protein content of the diet, reflected in the levels of glucose and amino acids in the portal blood. Since rats feed mostly in the dark period, rhythms of hormone secretion are likely to occur. Moreover, the composition of the diet plays a major role in the observed hormone rhythms (37). For example, glucagon is only elevated during the feeding period in animals fed a high-protein (82%) diet (insulin remains at basal levels), and insulin is only elevated in animals fed a high-carbohydrate (60–75%) diet (glucagon remains at basal levels).

Since the concentration of cAMP in the liver is inversely proportional to the insulin:glucagon ratio (Fig. 12-2), the concentration of the cyclic nucleotide increases during the feeding period of animals on a high-protein diet and decreases during the feeding period of animals on a high-carbohydrate diet (37).

Thyroid Hormones

The diurnal variation in the release of thyroid-stimulating hormone from the anterior pituitary drives a diurnal rhythm of thyroid hormone release from the thyroid gland (36). Thyroid hormone levels in the blood are low during the dark period and increase throughout the light period to give maximum levels at the light–dark interface. Thus the rhythm is similar to that of the glucocorticoids, except the amplitude of the rhythm is much smaller.

RELATIONSHIPS BETWEEN DIURNAL RHYTHMS OF HORMONES AND ENZYME SYNTHESIS

In Chapter 12 the hormones involved in the regulation of the synthesis of several liver enzymes are discussed (see Tables 12-2 and 12-3). These interpretations were based on data obtained by chronic manipulations of the dietary or endocrine status of the animals that may be unphysiological. In this section we reconsider the roles of some of these hormones, based upon their rhythms in relation to the rhythms of the enzymes they may control.

Glucocorticoids, Template Activity, and Enzyme Rhythms

There is a close temporal relationship between the peak in the concentration of glucocorticoids and the increase in template activity of liver chromatin (Figs. 13-12a and 13-10). Furthermore, the increase in template activity precedes the synthesis of a number of liver enzymes (Figs. 13-1 and 13-5), supporting the view that the steroid could be responsible for producing the enzyme rhythms (see Table 12-2). However, there are a number of observations that are inconsistent with this view. For example, starvation abolishes the rhythms of enzymes (Fig. 13-4), but has no immediate effect on the glucocorticoid rhythm (Fig. 13-12b). Moreover, adrenalectomy reduces, but does not abolish, the rhythm of template activity and has little or no effect on the rhythm of TAT synthesis (Fig. 13-2c).

The rhythm of TAT synthesis is paralleled by an increase in $mRNA^{TAT}$ concentration, which suggests that the liver cells respond to a signal acting at the transcriptional, or early post-transcriptional, level. Despite the evidence presented in Chapter 12 for a role of glucocorticoids in the transcriptional activation of the TAT gene the steroid does not appear to be involved in the regulation of the diurnal rhythms of this enzyme and its mRNA.

The increase in template activity (a measure of changes in chromatin structure) is partially independent of glucocorticoids, suggesting that the steroid acts synergistically with some other factor(s) in the diurnal control of chromatin structure and activity in the liver. Presumably, in adrenalectomized animals these other factor(s) can change chromatin structure and permit the synthesis of $mRNA^{TAT}$. For example, thyroid hormones, the only other molecules shown to induce changes in liver chromatin structure, have a rhythm that parallels the glucocorticoids, and they may be involved in the diurnal rhythm of template activity.

Insulin and Carbon Storage

The diurnal rhythm of food intake from a diet of high-carbohydrate stimulates an increase in the concentration of insulin in the general circulation and particularly in the portal vein. It is likely that this increase in insulin is responsible for the stimulation of glucokinase and glycogen synthetase activities and the increased flow of carbon into glycogen (Fig. 13-3). It is also likely that the hormone (either alone or via a decrease in the concentration of cAMP) is responsible for activation of the enzymes of the pathways of lipid biosynthesis in liver and other tissues (Fig. 13-11 and Table 12-3).

Insulin acts as a signal to *entrain* the rhythms of certain enzymes to the periodicity and composition of food intake. However, there is the question whether insulin is also responsible for the *generation* of these rhythms or whether other factors, perhaps endogenous to the liver cells, actually generate the rhythm, and insulin merely alters its amplitude and phase. At present this question cannot be answered, but some clues about the generation of hepatic enzyme rhythms have come from a study on the enzymes synthesized in response to glucagon.

Glucagon, cAMP, and Rhythms of Enzyme Synthesis

Glucagon, acting via its second-messenger cAMP, is involved in the entrainment of the rhythm of TAT biosynthesis to food intake (4). This was demonstrated by the somewhat complex experiment described in Fig. 13-14, which was carried out over a period of 48 h on rats adapted to either the "8 + 16" feeding schedule or the "8 + 40" feeding schedule. For rats adapted to the "8 + 16" feeding schedule there is no increase in TAT synthesis if food is withheld on the second day (in contrast to the biphasic rhythm in rats *accustomed* to having no food every second day, i.e., the "8 + 40" group). However, if glucagon and theophylline (an inhibitor of phosphodiesterase) are injected at the time when the 8-h feeding period would normally begin there is an increase in the synthesis of TAT. Moreover, the amplitude of the rhythm is proportional to the dose of hormone plus theophylline, and it may even exceed that observed in fed rats.

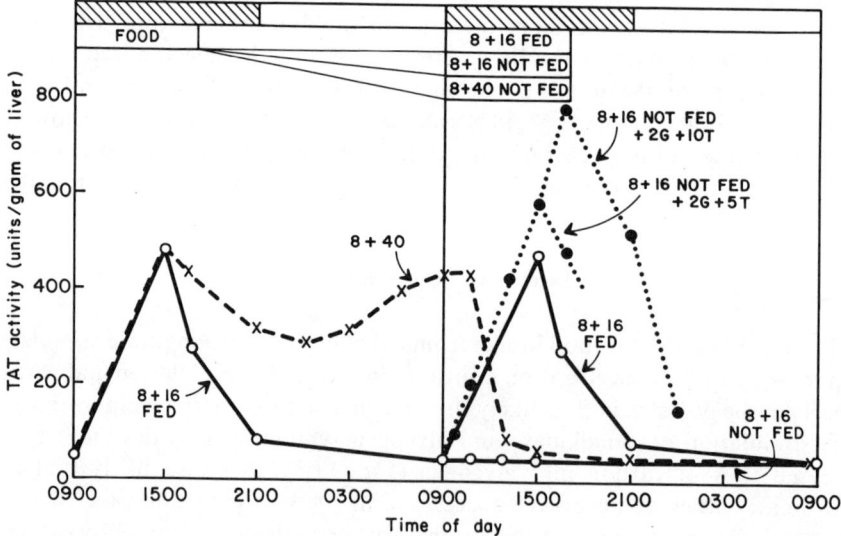

Fig. 13-14 The effect of feeding schedule and hormonal status on the diurnal rhythm of hepatic TAT biosynthesis. The rhythm for animals adapted to the "8 + 16" feeding schedule is shown over a 2-day period (o—o) in which the animals were either fed or not fed on the second day as indicated. Groups of animals that were not fed on the second day were given injections of glucagon plus theophylline (G + T with the numbers referring to the milligrams of each compound injected per 100 g of body weight) at the start of the dark period (●···●). Changes in TAT activity in rats adapted to the "8 + 40" feeding schedule are shown for comparison. [Reproduced, with permission, from Fig. 3 of Walker and Potter (4).]

The results of this experiment, together with the knowledge that rhythms of glucagon and cAMP are related to the protein content of the diet (discussed earlier), suggest that glucagon is responsible for entraining the TAT rhythm to food intake (and composition). Since the rhythmic increase in TAT synthesis is preceded by an increase in mRNA^{TAT}, it appears that cAMP is acting either at the transcriptional level or at an early post-transcriptional event.

However, pancreatectomy does not abolish the rhythm [although it shifts it 6 h earlier (38)], indicating that glucagon does not *generate* the rhythm of TAT synthesis in the liver. This shift in the rhythm is probably caused by free-running in the absence of entrainment signal.

In a further series of experiments, Ehret and Potter (39) shifted rats that were adapted to the "8 + 40" feeding schedule with 12 h of light and 12 h of darkness, to constant darkness with no food. Under these conditions TAT activity remained at a low basal level. However, when challenged with an injection of theophylline these animals maintained their ability to synthesize TAT. Moreover, they exhibited a pronounced diurnal rhythm of *"inducibility."* In other words, even though TAT was not being expressed in the starved animals in constant darkness, the mechanisms responsible for enzyme synthesis maintain their rhythm. Since there is an absence of environmental signals or *Zeitgeber* this rhythm of inducibility must be endogenous to the animal (although not necessarily just to the liver cells).

Melatonin, Indoleamines, and Enzyme Rhythms

The pineal gland releases melatonin with a rhythm that is very much in phase with the rhythm of TAT and other enzymes in the liver. Moreover, the rhythm of melatonin is independent of food intake, and, although entrained to the light:dark cycle, it is not abolished in constant darkness. It is possible, therefore, that melatonin, or a similar indoleamine, may be responsible for the generation of the rhythms of enzyme synthesis in the liver. It was shown some time ago (2) that tryptophan, hydroxytryptophan, and serotonin (see Fig. 13-6) all are capable of inducing the synthesis of TAT in the liver (melatonin was not examined). Clearly, these compounds should be examined in more detail, not just for their effect on the *induction* of TAT, but also on the *inducibility* of the enzyme.

Evidence for the Involvement of Higher Centers of the Brain in Enzyme Rhythms

There is some evidence that, in addition to the SCN, two other areas of the brain are involved in either generating or entraining enzyme rhythms (9, 11). These are the ventromedial-hypothalamic nuclei (VMH) and the lateral-hypothalamic nuclei (LH), two areas of the hypothalamus that regulate food intake (see Fig. 10-2). For example, lesion of the VMH nuclei extends the feeding period of rats and consequently abolishes the liver glycogen rhythm (lesion of the LH nuclei is lethal and cannot be studied). These two areas of the brain play an important role in determining the diurnal rhythm of food intake, which, in turn, entrains a number of enzyme rhythms.

The VMH and the LH nuclei play a particularly important role in the entrainment of feeding activity *in meal-fed* rats (including rats adapted to controlled feeding schedules) where there is an element of regularity in the environmental food stimulus. Indeed, lesion of the SCN in meal-fed animals does not abolish glycogen or enzyme rhythms, since the VMH and LH nuclei have taken over the control of feeding activity. In animals fed *ad libitum*, the SCN plays a major role in entraining feeding activity to the light:dark cycle, and SCN lesion abolishes nocturnal feeding causing the animals to eat uniformly throughout the 24-h day. These observations have important implications for work on animals that are meal-fed, since the very act of giving food on a periodic basis changes the way in which the brain controls other rhythms in the body.

THE BIOCHEMICAL NATURE OF THE BIOLOGICAL CLOCK

It is now widely accepted that animals have an endogenous clock that oscillates with a periodicity close to 24 h. This clock is responsible for driving rhythms in a number of tissues such as the pineal gland and discrete areas of the hypothalamus responsible for the secretion of neurotransmitters and hypothalamic hormones (such as CRH, which drives the glucocorticoid rhythm). It is highly likely that this endogenous clock, or circadian oscillator, resides in the higher centers of the brain (see Refs. 40–42 for reviews) and that the SCN is an important component of it.

In addition to a clock that entrains the behavior of an animal to the events of the 24-h day, there is also the clock that sensitizes the animal to the longer-term changes occurring on a seasonal basis. The pineal gland may play a central role in this phenomenon. In this section we will examine the roles of the SCN and the pineal gland in biological rhythms.

The Suprachiasmatic Nucleus

The SCN consists of two compact nuclei, each containing 10,000 neurons that are essential for the existence of the following rhythms:

(a) Locomotor activity
(b) Feeding activity
(c) Drinking behavior
(d) Melatonin biosynthesis
(e) Corticosterone biosynthesis
(f) Some aspects of the estrous cycle

It is not clear whether the SCN is actually the source of circadian oscillations or whether it is in intimate contact with another, separate, circadian oscillator.

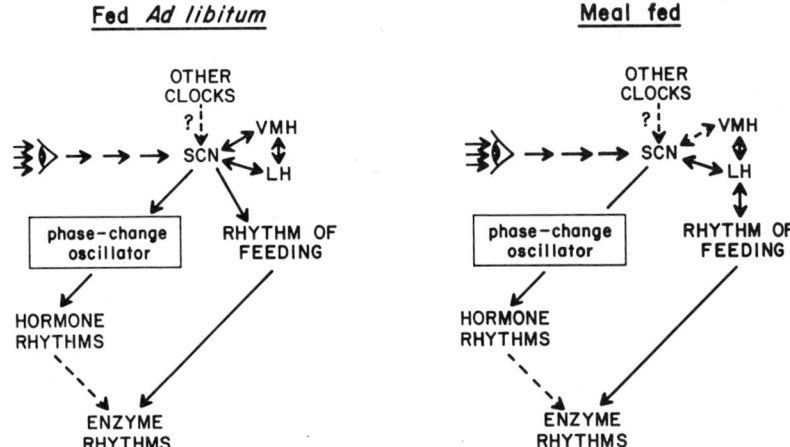

Fig. 13-15 Factors involved in the regulation of diurnal rhythms of liver enzymes. In animals fed *ad libitum* the SCN, receiving input from the eyes, the ventromedial nuclei, the lateral nuclei of the hypothalamus (VHM and LH), and possibly other clocks, entrains the feeding rhythm and hormone rhythms, and both factors interact with the liver to generate enzyme rhythms. In meal-fed animals the environmentally induced rhythms of feeding produce changes in the activity of the VMH and LH nuclei that are monitored by the SCN, which, in turn, drives rhythms of hormone secretion. Both the hormones and feeding activity influence liver enzyme rhythms. The phase-change oscillators produce rhythms of hormone secretion out of phase with the light:dark cycle.

The SCN shows diurnal changes in metabolism reflected in increases in glucose consumption during the light period (43). The fact that this rhythm of glucose uptake is abolished by constant light (but not constant darkness) suggests that the SCN does not generate its own endogenous oscillation. It is possible that the SCN, which receives direct neural input from the eyes (Fig. 13-15), also receives input from a free-running circadian oscillator and integrates these two signals into an entrained rhythm that is then transmitted to other tissues.

In rats fed *ad libitum* the rhythm originating from the SCN entrains feeding activity to the photoperiod and also entrains a number of hormone rhythms. Biochemical phase-change oscillators must exist to create rhythms, such as the glucocorticoid rhythm, that are 180° out of phase with feeding activity (Fig. 13-15). In meal-fed rats the rhythm of feeding activity is environmentally directed, and the VMH and LH nuclei maintain this rhythm. However, the SCN is still responsible for the rhythms of hormones, and many other functions in meal-fed rats and the signals from SCN and VMH/LH nuclei can sometimes conflict as described below.

Diurnal rhythms of liver enzymes are produced by hormonal signals reaching the tissue on a rhythmic basis multiplexed with signals reaching the tissue from the intestine during feeding rhythms. In animals fed *ad libitum* or in animals fed a meal at the start of the dark period a highly synchronized response is produced. If the feeding period is moved, then the liver enzymes have broader peaks or even two distinct peaks (Fig. 13-2) because the signals become separated. Indeed, if animals are meal-fed during the light period some enzyme rhythms are abolished (44) because the signals are completely out of phase, and each signal, individually, is insufficient to trigger a rhythm.

The Pineal Gland

Since animals can survive quite well following pinealectomy, the role played by this gland was overlooked until the discovery of melatonin and the rhythmic nature of its biosynthesis and release. The functions of the pineal gland have recently been reviewed (12, 45).

In birds, the pineal gland is a circadian oscillator capable of generating a rhythm that is usually entrained to the light:dark cycle either directly through light penetrating the skull or via the optic nerves. Although the rhythm appears to be generated within individual cells its biochemical mechanism is unknown. In birds, the pineal controls locomoter activity and body temperature and probably many more functions. This control of locomotor activity was convincingly demonstrated in birds entrained to two different light:dark cycles. When their pineal glands were "exchanged" each group of birds immediately adopted rhythms of locomotor activity characteristic of the other group.

In animals it is fairly well-established that the pineal gland is not *the* circadian oscillator or biological clock, since it is now known to be controlled by the SCN. It is likely that the gland senses the *length* of the dark period, and it appears to be involved in the control of certain seasonal activities, particularly breeding activity (pinealectomy results in atrophy of the gonads). Moreover, the release of FSH and luteinizing hormone appears to be controlled by melatonin.

DIURNAL RHYTHMS AND EXPERIMENTAL PROTOCOLS

The existence of diurnal rhythms has tremendous implications for the design of experiment protocols, particularly those that interfere with the environmental signals that entrain these rhythms (food, light and dark). Moreover, studies involving hormone injections should be planned with an appreciation of the large oscillations in endogenous levels that are now known to take place. Clearly,

---

Fig. 13-16 The influence of diurnal rhythms on the adaptive behavior of liver enzymes in the rat. Animals were adapted to the "2 + 22 AM" feeding schedule on a diet containing 12.5% protein. On day 1 this food was replaced with one containing 60% protein, and the changes in the activities of tyrosine aminotransferase (TAT), ornithine decarboxylase (ODC), ornithine aminotransferase (OAT), and serine dehydratase (SDH) followed for a total of 21 days. [Reprinted with permission from *Life Sciences* 17, S. Yanagi, H. A. Campbell, and V. R. Potter, Diurnal variations in activity of four pyridoxal enzymes in rat liver during metabolic transition from high carbohydrate to high protein diet. Copyright 1975, Pergamon Press, Ltd.]

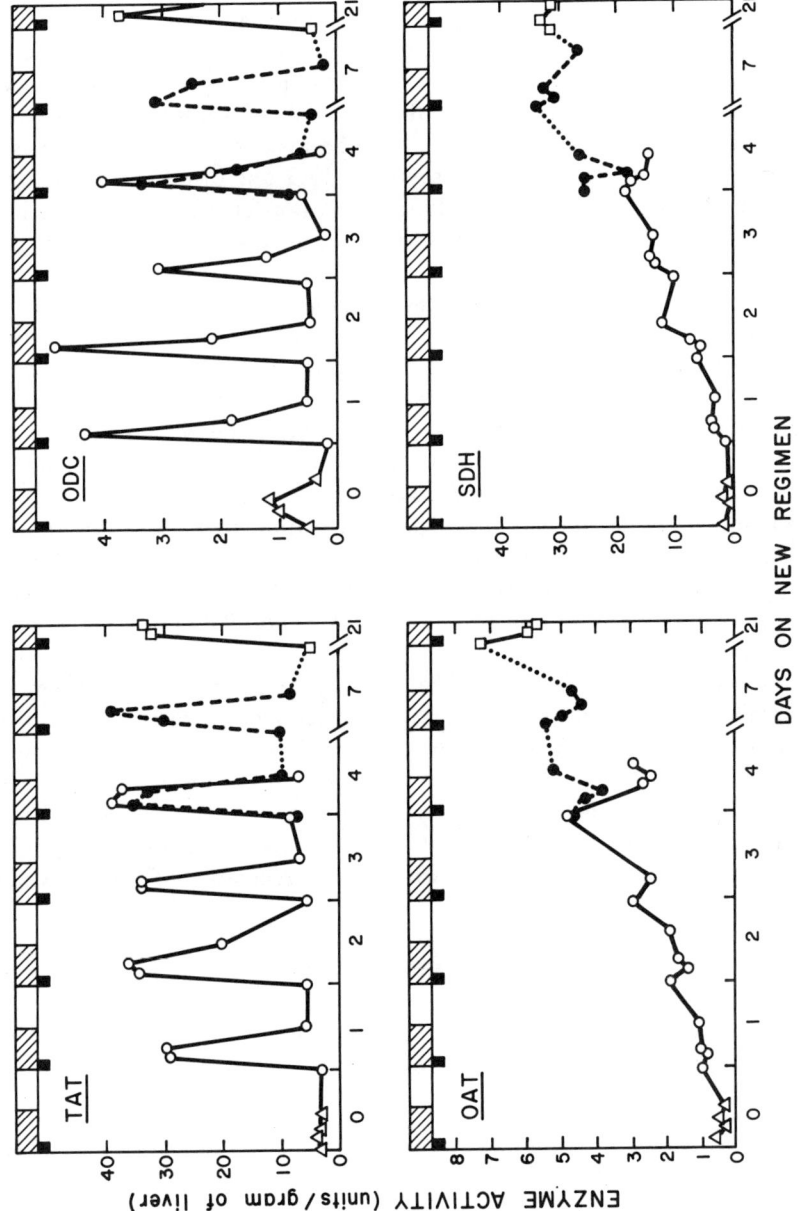

experiments carried out on cultured cells (except very short-term primary cultures) that have lost their rhythmic cues may have little relevance to the behavior of enzymes *in vivo*.

On the other hand, a knowledge of diurnal rhythms can be used to investigate the mechanisms involved in the adaptation of inducible enzymes. For example, Yanagi et al. (46) examined the relationship between inducibility and rhythmicity in animals entrained to the "2 + 22 AM" feeding schedule by shifting them from a high-carbohydrate to a high-protein diet. Four enzymes were studied—two that exhibit diurnal rhythms (TAT and ornithine decarboxylase, ODC) and two that do not (ornithine aminotransferase, OAT, and serine dehydratase, SDH). The results are shown in Fig. 13-16. In animals on a high-carbohydrate diet (day 0), all four amino acid-catabolizing enzymes are at very low levels, and only ODC has a demonstrable diurnal rhythm. The *first* (2-h) feeding period on the high-protein diet produces a marked increase in the amplitude of the ODC rhythm, and there is also a pronounced rhythm of TAT activity (adrenalectomy does not prevent this). In contrast, there is only a small increase in the activities of OAT and SDH. Thus there is a remarkable difference in the adaptive capacity of enzymes that undergo rhythmic changes and those that do not. Both TAT and ODC respond almost immediately, whereas OAT and SDH do not complete their adaptation to the new diet until *1-3 weeks* after it was initiated. The new steady-state eventually reached by these enzymes is some 20-30-fold higher than on the high-carbohydrate diet, a change similar in magnitude to that reached by TAT and ODC at the peaks of their rhythms.

This experiment demonstrates the clear adaptive advantage of enzymes that maintain a "rhythm of inducibility" even when fed a high-carbohydrate diet. Moreover, these experimental protocols provide a natural system for studying both the regulation of the synthesis of individual enzymes and the nature of diurnal rhythms at the molecular level. It is clear that as yet unknown factors interact with the genome of liver cells to create this rhythm of inducibility, and they play a major role in gene expression and the regulation of enzyme synthesis.

REFERENCES

1. V. R. Potter, R. A. Gebert, H. C. Pitot, C. Peraino, C. Lamar, Jr., S. Lesher, and H. P. Morris (1966). Systematic oscillations in metabolic activity in rat liver and in hepatomas. *Cancer Res.* **26**, 1547-1553.
2. M. Watanabe, V. R. Potter, and H. C. Pitot (1968). Systematic oscillations in tyrosine aminotransferase and other metabolic functions in liver of normal and adrenalectomized rats on controlled feeding schedules. *J. Nutr.* **95**, 207-227.
3. H. A. Hopkins, R. J. Bonney, P. R. Walker, J. D. Yager, Jr., and V. R. Potter (1973). Food and light as separate entrainment signals for rat liver enzymes. *Adv. Enz. Reg.* **11**, 169-191.
4. P. R. Walker and V. R. Potter (1974). Diurnal rhythms of hepatic enzymes from rats adapted to controlled feeding schedules. In L. E. Scheving, F. Halberg and J. Pauly (Eds.). *Chronobiology*. Igaku Shoin, Tokyo, pp. 17-22.

REFERENCES

5 R. D. Reynolds, V. R. Potter, and H. C. Pitot (1971). Response of several hepatic adaptive enzymes to a shift from low to high protein diet in intact and adrenalectomized rats. *J. Nutr.* **101**, 797–802.

6 M. Diesterhaft, D. K. Granner, and V. R. Potter (1980). Oscillation of mRNA coding for tyrosine aminotransferase in meal fed rats. *Fed. Proc.* **39**, p. 1804.

7 P. R. Walker (1977). The influence of diurnal rhythms of carbohydrate metabolism in adult rat liver on metabolic characteristics of isolated liver parenchymal cells. *Biochim. Biophys. Acta* **496**, 255–263.

8 P. McVerry and K.-H. Kim (1972). Diurnal rhythm of rat liver glycogen synthetase. *Biochem. Biophys. Res. Commun.* **46**, 1242–1246.

9 T. Shimazu (1979). Hypothalamic regulation of the circadian rhythm of liver glycogen metabolism. In M. Suda, O. Hayaishi, and H. Nakagawa (Eds.). *Biological Rhythms and Their Central Mechanism.* Elsevier/North-Holland, Amsterdam, pp. 283–294.

10 R. W. Fuller (1971). Rhythmic changes in enzyme activity and their control. In M. Rechcigl, Jr. (Ed.). *Enzyme Degradation and Synthesis.* Karger, Basel, pp. 311–338.

11 H. Nakagawa, K. Nagai, K. Ida, and T. Nishio (1979). Control mechanisms of circadian rhythms of feeding behaviour and metabolism influenced by food intake. In M. Suda, O. Hayaishi, and H. Nakagawa (Eds.). *Biological Rhythms and Their Central Mechanisms.* Elsevier/North-Holland, Amsterdam, pp. 283–294.

12 T. Deguchi (1979). Circadian rhythms of indoleamine and serotonin N-acetyltransferase activity in the pineal gland. *Mol. Cell. Biochem.* **27**, 57–66.

13 T. Deguchi (1973). Role of the *Beta* adrenergic receptor in the elevation of adenosine cyclic 3',5'-monophosphate and induction of serotonin N-acetyltransferase in rat pineal glands. *Mol. Pharmacol.* **9**, 184–190.

14 S. D. Wainwright (1980). Diurnal cycles in serotonin acetyltransferase activity and cyclic GMP content of cultured chick pineal glands. *Nature* **285**, 478–480.

15 J. S. Takahashi, H. Hamm, and M. Menaker (1980). Circadian rhythms of melatonin release from individual superfused chicken pineal glands *in vitro*. *Proc. Natl. Acad. Sci. USA* **77**, 2319–2322.

16 C. North, R. J. Feuers, L. E. Scheving, J. E. Pauly, T. H. Tsai, and D. A. Casciano (1981). Circadian organization of thirteen liver and six brain enzymes of the mouse. *Am. J. Anat.* **162**, 183–199.

17 M. Zaviacic and M. Brozman (1979). Effects of feeding a single daily meal and of changes in lighting schedule on the circadian rhythms of hydrolases in rat gastric mucosa. *Cell. Mol. Biol.* **25**, 223–231.

18 H. S. Earp III (1974). Glucocorticoid regulation of transcription: The role of physiological concentrations of adrenal glucocorticoids in the diurnal variation of rat liver chromatin template availability. *Biochim. Biophys. Acta* **340**, 95–107.

19 E. D. Whittle and V. R. Potter (1968). Systematic oscillations in the metabolism of orotic acid in the rat adapted to a controlled feeding schedule. *J. Nutr.* **95**, 238–246.

20 J. Seifert (1980). Circadian variations in pyrimidine nucleotide synthesis in rat liver. *Arch. Biochem. Biophys.* **201**, 194–198.

21 H. A. Hopkins, H. A. Campbell, B. Barbirolli, and V. R. Potter (1973). Thymidine kinase and deoxyribonucleic acid metabolism in growing and regenerating livers from rats on controlled feeding schedules. *Biochem. J.* **136**, 955–966.

22 D. F. Scott and V. R. Potter (1970). Metabolic oscillations in lipid metabolism in rats on controlled feeding schedules. *Fed. Proc.* **29**, 1553–1559.

23 V. R. Potter, E. F. Baril, M. Watanabe, and E. D. Whittle (1968). Systematic oscillations in metabolic functions in liver from rats adapted to controlled feeding schedules. *Fed. Proc.* **27**, 1238–1245.

24. V. Ehrhardt (1979). Daily rhythm of AIB uptake in rat liver: kinetic characteristics of transport. *J. Inter. Cycle Res.* **10**, 125–131.
25. P. J. Garlick, D. J. Millward, and W. P. T. James (1973). The diurnal response of muscle and liver protein synthesis *in vivo* in meal-fed rats. *Biochem. J.* **126**, 935–945.
26. R. Y. Moore and V. B. Eichler (1972). Loss of circadian adrenal corticosterone rhythm following suprachiasmatic lesions in the rat. *Brain Res.* **42**, 201–206.
27. C. Gomes-Sanchez, O. B. Holland, J. R. Higgins, D. C. Kem, and N. M. Kaplan (1976). Circadian rhythms of serum renin activity and serum corticosterone, prolactin, and aldosterone concentrations in the male rat on normal and low-sodium diets. *Endocrinol.* **99**, 567–572.
28. G. P. Moberg, L. L. Belinger, and V. E. Mendel (1975). Effect of meal feeding on daily rhythms of plasma corticosterone and growth hormone in the rat. *Neuroendocrinol.* **19**, 160–169.
29. H. Kato, M. Sato, and M. Suda (1980). Effect of starvation on the circadian adrenocortical rhythm in rats. *Endocrinol.* **106**, 918–921.
30. Y. Morimoto and Y. Yamamura (1979). Regulation of circadian adrenocortical periodicities and of eating–fasting cycles in rats under various lighting conditions. In M. Suda, O. Hayaishi, and H. Nakagawa (Eds.). *Biological Rhythms and Their Central Mechanism.* Elsevier/North-Holland, Amsterdam, pp. 177–188.
31. K. Takahashi, K. Hanada, and Y. Takahashi (1979). Factors setting the phase of the circadian adrenocortical rhythm in rats. In M. Suda, O. Hayaishi, and H. Nakagawa (Eds.). *Biological Rhythms and Their Central Mechanism.* Elsevier/North-Holland, Amsterdam, pp. 189–198.
32. J. D. Dunn, A. Arimura, and L. E. Scheving (1972). Effect of stress on circadian periodicity in serum LH and prolactin concentration. *Endocrinol.* **90**, 29–33.
33. D. Cacchi, I. Gil-Ad, A. E. Panerai, V. Locatelli, and E. E. Miller (1976). Circadian variations in plasma growth hormone and prolactin in the infant rat: comparison with the adult pattern. *Life Sci.* **19**, 825–836.
34. J. B. Martin, G. Tarnenbaum, J. O. Willoughby, L. P. Renaud, and B. Brazeau (1976). Functions of the central nervous system in regulating pituitary GH secretion. In M. Motta, P. G. Crosignani, and L. Martine (Eds.). *Hypothalamic Hormones,* Vol. 6, Serons Symposium. Academic Press, London, pp. 217–235.
35. R. E. Poland, R. T. Rubin, and M. E. Weichsel, Jr. (1980). Circadian patterns of rat anterior pituitary and target gland hormones in serum: determination of the appropriate sample size by statistical power analysis. *Psychoneuroendocrinol.* **5**, 209–224.
36. D. Jordan, B. Rousset, F. Perrin, M. Fournier, and J. Orgiazzi (1980). Evidence for circadian variations in serum thyrotropin, 3,5,3'-triiodothyronine, and thyroxine in the rat. *Endocrinol.* **107**, 1245–1248.
37. M. Tiedgen and H. J. Seitz (1980). Dietary control of circadian variations in serum insulin, glucagon and hepatic cyclic AMP. *J. Nutr.* **110**, 876–882.
38. R. W. Fuller, G. T. Jones, H. D. Snoddy, and I. H. Slater (1969). Daily rhythms of tyrosine aminotransferase and glucose after removal of the pancreas of rats. *Life Sci.* **8**, 685–691.
39. C. F. Ehret and V. R. Potter (1974). Circadian chronotypic induction of tyrosine aminotransferase and depletion of glycogen by theophylline in the rat. *Int. J. Chronobiology* **2**, 321–326.
40. M. Menaker, J. S. Takahashi, and A. Eskin (1978). The physiology of circadian pacemakers. *Ann. Rev. Physiol.* **40**, 501–526.
41. B. Rusch and I. Zucker (1979). Neural regulation of circadian rhythms. *Physiol. Rev.* **59**, 449–526.
42. M. C. Moore-Ede, F. M. Fulzman, and C. A. Fuller (1982). *The Clocks That Time Us.* Harvard University Press, Boston.
43. W. J. Schwartz, C. B. Smith, and L. C. Davidson (1979). *In vivo* glucose utilisation of the suprachiasmatic nucleus. In M. Suda, O. Hayaishi, and H. Nakagawa (Eds.). *Biological Rhythms and Their Central Mechanism.* Elsevier/North-Holland, Amsterdam, pp. 355–367.

44 K. B. Ekelman and C. Peraino (1981). Effects of feeding and lighting stimuli on the synthesis of ornithine aminotransferase and serine dehydratase in rat liver. *Arch. Biochem. Biophys.* **209**, 677–681.
45 R. J. Reiter (1973). Comparative physiology: pineal gland. *Ann. Rev. Physiol.* **35**, 305–329.
46 S. Yanagi, H. A. Campbell, and V. R. Potter (1975). Diurnal variations in activity of four pyridoxal enzymes in rat liver during metabolic transition from high carbohydrate to high protein diet. *Life Sci.* **17**, 1411–1422.

SECTION FIVE

Regulation of Enzyme Synthesis During Differentiation

The higher mammals and other multicellular organisms depend, for their survival, upon the coordinated interaction of a number of differentiated tissues, each of which carries out a specialized function. Many of these functions are now quite well-understood in molecular terms, and attention is currently being focused *upon the formation of these specialized tissues during the course of differentiation*. The goal is to be able to explain how individual genes are readied for their role in the mature adult.

Although the differentiation of many of the higher mammals is well-understood in morphological terms, we still know relatively little about the biochemical changes that take place within the various groups of cells that will develop into individual tissues. In this section some of the work that has been carried out on the differentiation of the liver (Chapter 14) and of the reproductive tissues (Chapter 15) will be described. The work on the liver has been mostly descriptive, although recently progress has been made toward an understanding of the hormonal control of the appearance of some of the enzymes characteristic of the differentiated tissue. In contrast, the work on the reproductive tissues (which can be more conveniently studied *ex utero*), notably chicken oviduct and rat mammary gland, has progressed rapidly and is currently being studied in terms of the regulation of the expression of individual genes. The results obtained from all of these studies are providing insight into this most challenging area of molecular biology.

CHAPTER FOURTEEN

Regulation of Enzyme Synthesis During Development

The last two chapters dealt in some detail with the way in which the synthesis of enzymes is regulated in adult tissues. For the most part, these studies dealt with the regulation of the synthesis of enzymes that are already present in the cells as the result of a basal level of transcription of an already active gene (although some of these genes *may be inactive* at certain times during the 24-h day). Of considerable importance for a complete knowledge of the nature of gene expression in animal cells is an understanding of the way in which individual genes first become active during differentiation. It is important to establish whether activation of the genes is under the control of specific developmental factors, or whether the same hormones that control enzyme synthesis in the adult are also responsible for switching these genes on during development.

This chapter concentrates mostly on the development of the liver and examines the regulation of changes in the synthesis of enzymes, with particular emphasis on those enzymes discussed in detail in Chapters 12 and 13. Since hormones do play some role in the regulation of the appearance of these enzymes, changes in hormone levels are also described, together with the development of the components of information-transfer systems such as hormone receptors and second messengers. Finally, the development of some of the more complex aspects of enzyme synthesis such as the development of diurnal rhythms of enzyme inducibility is described. Unfortunately, only a limited amount of material can be discussed here; a number of books and reviews should be consulted for additional information (1–6).

ENZYME SYNTHESIS DURING DEVELOPMENT IS DISCONTINUOUS

Physiological Considerations

Changes in enzyme synthesis in the livers of *adult* animals usually take place in response to changes in the external environment, especially changes (either rhythmic or longer-term) in the nutritional status of the animal. In the developing animal there are also marked changes in nutritional status. For example, at birth the neonate switches from the predominantly carbohydrate-rich maternal blood supply to the high-fat content of the mother's milk. Furthermore, at weaning, a diet rich in carbohydrates usually replaces the milk supply. There are also changes in nitrogen metabolism as the neonate switches to nitrogen removal through the kidneys rather than through exchange across the placenta. In addition, there is the stage, midway through the suckling period, when the animal first becomes aware of the cycles of light and darkness. We can anticipate that all of these environmental changes play important roles in enzyme synthesis during development, but the question is—do these environmental stimuli actually determine the course of differentiation by provoking the expression of new genes, or does the animal anticipate and prepare itself for these changes?

Patterns of Enzyme Synthesis in Liver

The earliest work on changes in enzyme activities in liver during development established that enzymes were synthesized at discrete times rather than being gradually accumulated throughout the period of development. Greengard (2) placed liver enzymes into groups, or "clusters," according to the time at which they were first synthesized. Three main clusters were identified: late fetal, neonatal, and late suckling, with 15–20 enzymes being organized into each group, although some enzymes may actually belong to more than one cluster. Typical examples of enzymes in each of the three clusters are given in Table 14-1, and the developmental pattern for one enzyme from each cluster is illustrated in Fig. 14-1.

Some enzymes, pyruvate kinase, for example, were initially (7) shown to undergo quite unusual changes during development when total enzyme activity was measured (Fig. 14-2a). However, when changes in the individual isoenzymes of pyruvate kinase were measured three distinct developmental patterns were obtained (Fig. 14-2b and Ref. 8). One of the isoenzymes, PK II, is present in fetal lever, but absent from adult liver, whereas the isoenzyme characteristic of the adult tissue, PK I, is absent or at very low levels before birth.

A close examination of the change in isoenzyme PK II illustrates the need to interpret changes in enzyme activities during development in terms of the morphological changes in the tissue occurring at the same time. In Fig. 14-3 the changes in *total* PK II activity are shown, along with changes in another isoenzyme, aldolase A [the glycolytic enzyme aldolase exists in two isoenzymic forms, A and B, and only aldolase B is found in adult liver (8)]. The isoenzymes PK II and aldolase A have similar developmental profiles, and both disappear from the liver about 4 days after birth. When the activity of these isoenzymes is compared to the changes in the percentage of hemopoietic cells in the liver there is a good correlation (Fig. 14-3), indicating that these isoenzymes are confined to this population of cells. Hemopoietic cells are small, blood-forming elements

Table 14-1 The Appearance of Enzymes during Development of the Liver[a]

Enzyme	Late fetal	neonatal	late suckling
	(activity expressed as % of adult value)		
Glycogen synthetase	100		
Glycogen phosphorylase	100		
G-l-P uridyl transferase	100		
Phosphoglucomutase	100		
Phosphofructokinase	100		
Isocitrate dehydrogenase	60-100	100	
Carnitine palmitoyl transferase	60-100	100	
Malate dehydrogenase	70-80	100	
Aspartate amino transferase	70	100	
ATP citrate lyase	50	10	100
G6P dehydrogenase	50	30	100
FDPase	40	100	
Aldolase B	30-40	100	
PEPCK	30	100	
Pyruvate carboxylase	30	100	
Glucose 6-phosphatase	30	100	
Glutaminase	20-40	70-80	100
Urea cycle enzymes	30	50-100	100
Threonine dehydratase	20	100	
Fumarase	10	80	100
Serine dehydratase	<10	100	
HMGCoA synthetase	10	100	
Histidase	10	10	100
AcetylCoA carboxylase	5	10	100
Glucokinase			100
Malic enzyme			100
Tryptophan oxygenase			100

[a] Activities are depressed as a percentage of their adult level to indicate the amount of enzyme produced at each stage of liver development.

present in fetal liver. Before birth these cells account for more than 50% of the liver cell number, but they decline rapidly after birth when the liver ceases to function as a blood-forming tissue. Studies of this nature are quite important because they show that the genes for PK II and aldolase A are not expressed in the liver parenchymal cells. They also permit a more accurate description of the development of each cell type since we are dealing with the products of *individual* genes.

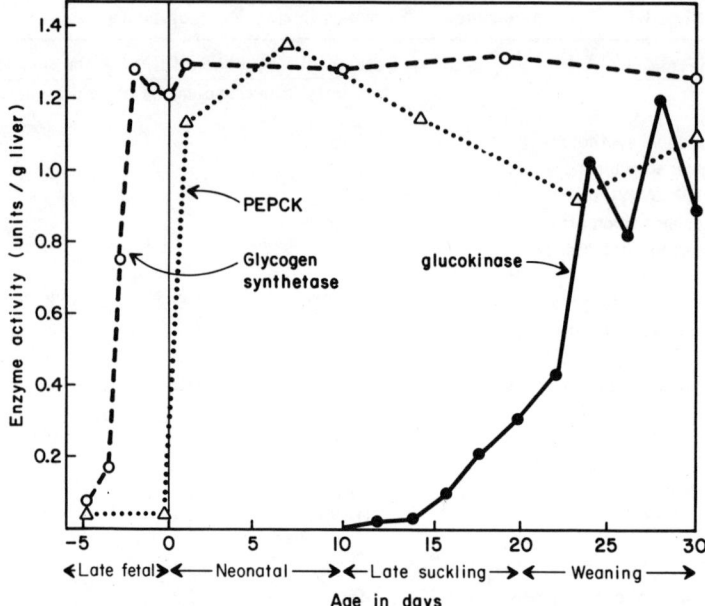

Fig. 14-1 Changes in the activities of glycogen synthetase, phosphoenolpyruvate carboxykinase (PEPCK), and glucokinase during liver development. The age (in days) is given, with 0 being the day of birth and −5 being 5 days before birth.

ENZYMES OF THE LATE-FETAL CLUSTER IN LIVER

In this section the regulation of the first appearance of two enzymes of this cluster, TAT and PEPCK, is considered in some detail, followed by a brief account of the regulation of the appearance of some of the other enzymes of this group.

Phosphoenolpyruvate Carboxykinase

Activity of this enzyme is first detected immediately after birth, which suggests that the process of birth and its associated metabolic changes are responsible for the stimulation of the PEPCK gene (9, 10) (Fig. 14-4a). If the young are delivered 1–2 days before term, the enzyme also appears immediately after delivery, showing that the liver has already *developed the capacity* to induce the enzyme 2 or 3 days before birth. Studies on the regulation of the expression of the PEPCK gene have focused primarily on the metabolic and humoral changes that accompany birth, and to some extent on the nature of the factors that give the liver the capacity to respond to birth with changes in enzyme synthesis.

Birth is accompanied by a marked drop in the blood glucose concentration (Fig. 14-4b), leading to a transient decrease in the insulin:glucagon ratio. Since

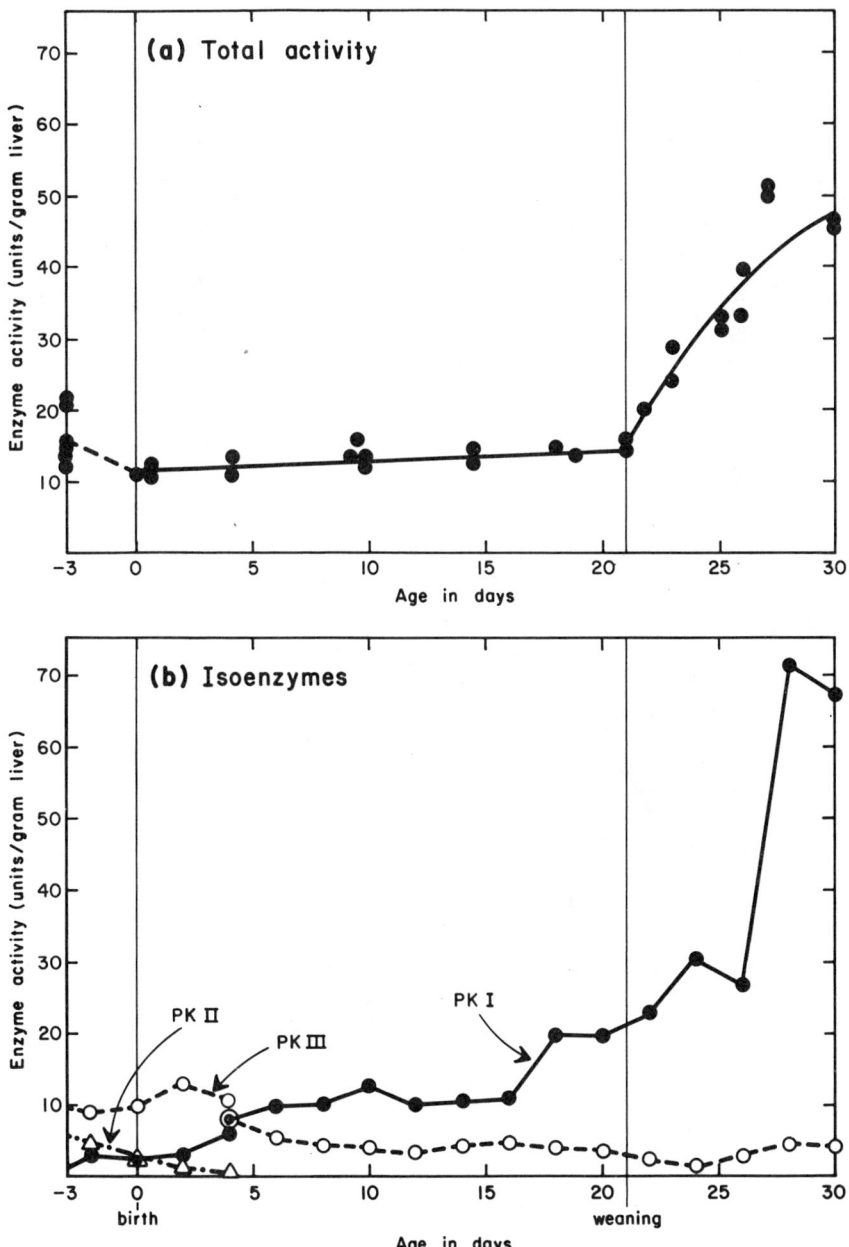

Fig. 14-2 Changes in *total* pyruvate kinase activity (*a*) and in the activities of individual isoenzymes of pyruvate kinase (*b*) during liver development. [The data in (*a*) are reproduced from Fig. 5 of Taylor, Bailey, and Bartley (7) and that in (*b*) from Fig. 1 of Walker (8) with permission.]

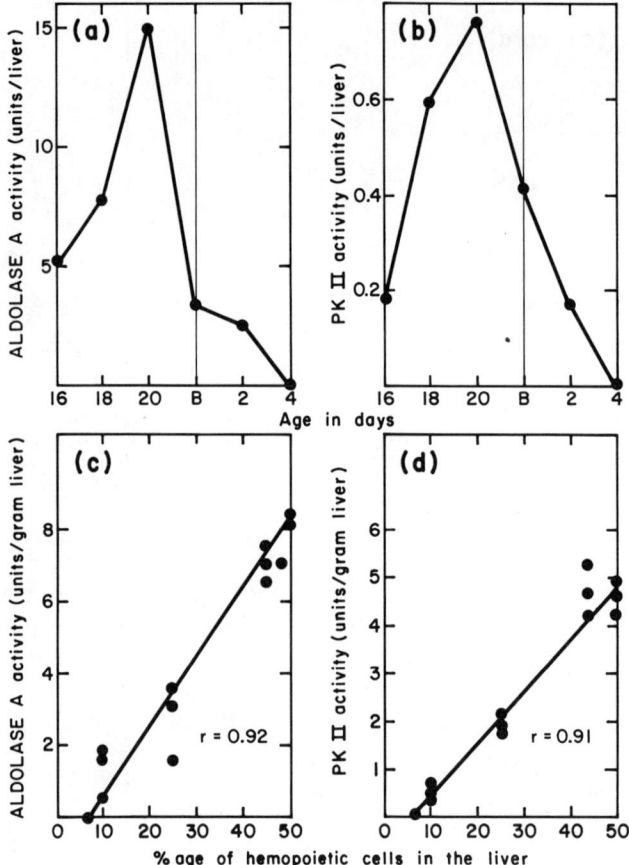

Fig. 14-3 Changes in the liver content of aldolase A and of pyruvate kinase II isoenzymes from 6 days before birth until 4 days after birth. These changes appear to be directly related to the changes in the amount of hemopoietic cells in the tissue at this time. r = correlation coefficient.

phosphoenolpyruvate carboxykinase is a key gluconeogenic enzyme, it is probably synthesized in order to increase flux through the gluconeogenic pathway to provide glucose to relieve the hypoglycemia. If either glucagon or its second-messenger cAMP is injected into fetuses *in utero* 1–2 days before birth there is an immediate, precocious increase in the activity of the enzyme. A simultaneous injection of insulin prevents this precocious increase, supporting the view that the change in the insulin:glucagon ratio at birth is the key event in the expression of the PEPCK gene (10, 11). Measurements on the changes in the concentration of mRNA^{PEPCK} and the inhibitory effect of actinomycin D (12) confirm that the increase in PEPCK activity is a result of gene activation.

A detectable level of mRNA^{PEPCK} is present in fetal liver, indicating that the gene is already being transcribed at a basal level, but it is not known whether

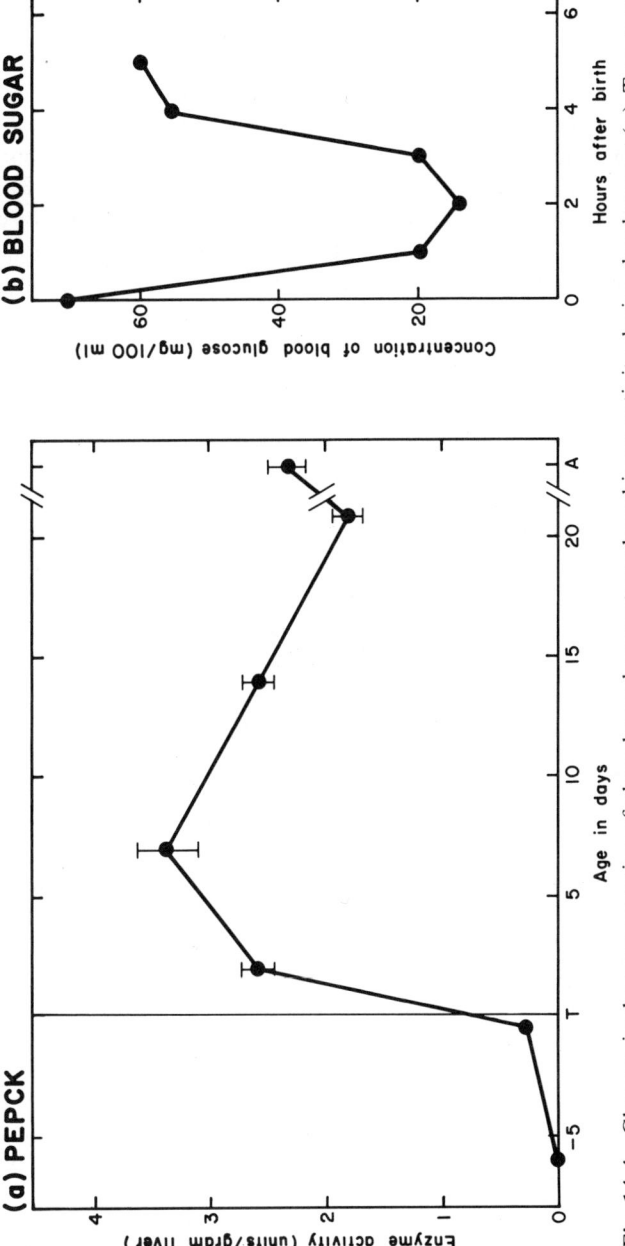

Fig. 14-4 Changes in the concentration of phosphoenolpyruvate carboxykinase activity during development (*a*). T = term (birth) and A = adult levels of activity. The pronounced hypoglycemia that accompanies birth is shown in (*b*). [Reproduced from Fig. 1 of Ballard and Hanson (9) and Table 3 of Yeung and Oliver (10) with the permission of the Biochemical Society.]

this mRNA is translated. Enzyme protein cannot be detected, but since mRNA assays are more sensitive than enzyme assays, the presence of some enzyme molecules cannot be eliminated. The mRNA determinations were carried out on *total cellular* mRNA. Since only a few copies of the sequence were found before birth, the study excluded the existence of a previously synthesized, inactive store of mRNA that is activated and translated at birth.

Since the gene does appear to be expressed to some extent before birth the effects of glucocorticoid hormone (a possible inducer of the enzyme in adult animals, Chapter 12) injections into fetuses have been examined (13), but no stimulatory effect could be demonstrated. However, since the level of endogenous glucocorticoids are quite high at this time, the liver receptors may be saturated before injection, thus the participation of glucocorticoids in the activation of the gene late in fetal development cannot be ruled out.

Work carried out on fetal hepatocytes in primary culture (14) support the results obtained *in utero*. Glucagon and cAMP can induce the enzyme in these cells, and the effect is inhibited by insulin. Glucocorticoids had no effect, but since the culture medium contained fetal calf serum, which already contains near optimal amounts of glucocorticoids, the results are equivocal.

When cells were taken from fetuses of different age some interesting insight into the development of information-transfer capability was uncovered. Thus both glucagon and cAMP can induce PEPCK in cells obtained from 14- and 16-day-old fetuses (6–8 days before birth), but only cAMP can induce the enzyme in 12-day-old fetuses (10 days before birth), which suggests that either the glucagon receptor or some component of the adenylate cyclase system is first synthesized between day 12 and day 14. Furthermore, the cyclic AMP-dependent protein kinase and other components of the induction system must already be present by day 12.

Tyrosine Aminotransferase

Tyrosine aminotransferase activity is first detectable in the liver 1–2 days before birth, but it remains at very low levels until birth, when it increases rapidly in a similar manner to PEPCK (15). Glucagon and cAMP can "preinduce" the enzyme in fetuses *in utero* and in liver explants or primary cultures (16–18), indicating that regulation of the synthesis of the enzyme may be similar to that for PEPCK. However, unlike PEPCK, glucocorticoids can induce the enzyme in hepatocytes obtained from 19-day-old fetuses (16, 19). An effect *in utero* cannot be demonstrated, and this is probably because of either high levels of endogenous glucocorticoids or high levels of insulin. Insulin has recently (20) been shown to inhibit the induction of TAT by glucocorticoids in fetal hepatocytes in culture.

In addition to the stimulatory effects of hormones, tyrosine, the substrate of the enzyme, which increases during the transient tyrosinemia that accompanies birth, also increases the concentration of the enzyme *in utero* and *in vitro* (15).

Other Enzymes in the Late-Fetal Cluster

Many of the enzymes involved in nitrogen excretion, such as those of the urea cycle, appear in the liver during the last few days before birth (Table 14-1). The regulation of the synthesis of the urea-cycle enzymes has been studied in some detail (21), and they have been found to be controlled by glucocorticoids. If the levels of endogenous fetal glucocorticoids are lowered by hypophysectomy, the levels of the urea-cycle enzymes decline. A normal developmental profile can be restored by the injection of glucocorticoids (this approach has not yet been applied to the study of PEPCK and TAT synthesis). Threonine dehydratase, another enzyme of amino acid metabolism, also appears shortly after birth (22), and since it can be preinduced by cAMP it falls into the category of enzymes of the late-fetal cluster regulated by glucagon.

In summary, the insulin:glucagon ratio appears to play a major role in the increase in the synthesis of the enzymes at the time of birth. Furthermore, there is evidence that glucocorticoids play some role during the last few days before birth to prepare the cells for the shifts in carbohydrate and nitrogen metabolism that occur at birth.

ENZYMES ALREADY PRESENT IN MAXIMAL AMOUNTS AT BIRTH

The ability to synthesize, store, and mobilize glycogen is one of the first differentiated functions to be formed during liver development. Indeed, the store of glycogen that is accumulated just before birth is of vital importance to the survival of the animal during its first few hours after birth.

Glycogen is undetectable in the livers of fetuses at day 16 of gestation. However, between day 17 and the time of birth, glycogen levels increase rapidly, reaching 10% of the wet weight of the tissue. All of this is used within the first few hours of birth, but it is not sufficient to prevent a substantial hypoglycemia developing (Fig. 14-4b). Since the liver contains less than 50% hepatocytes at birth the tissue cannot store enough glycogen to meet the demand for glucose.

Glycogen synthetase is already fully induced at 3 days before birth (Fig. 14-5), and a number of other enzymes of glycogen biosynthesis (see Table 14-1) reach optimal levels around the same time (23). Glycogen phosphorylase activity is lower than that of synthetase during the period of glycogen accumulation before birth, but increases rapidly within a few hours after birth to catalyze the rapid mobilization of this glycogen reserve (Fig. 14-5).

The regulation of the synthesis of these enzymes has not been studied in detail, even though they are the first enzymes to express a differentiated function. Moreover, since they are expressed constitutively in adult animals the regulation of their synthesis in the mature animal is almost impossible to study. It has been reported (24) that the increase in the concentration of glucocorti-

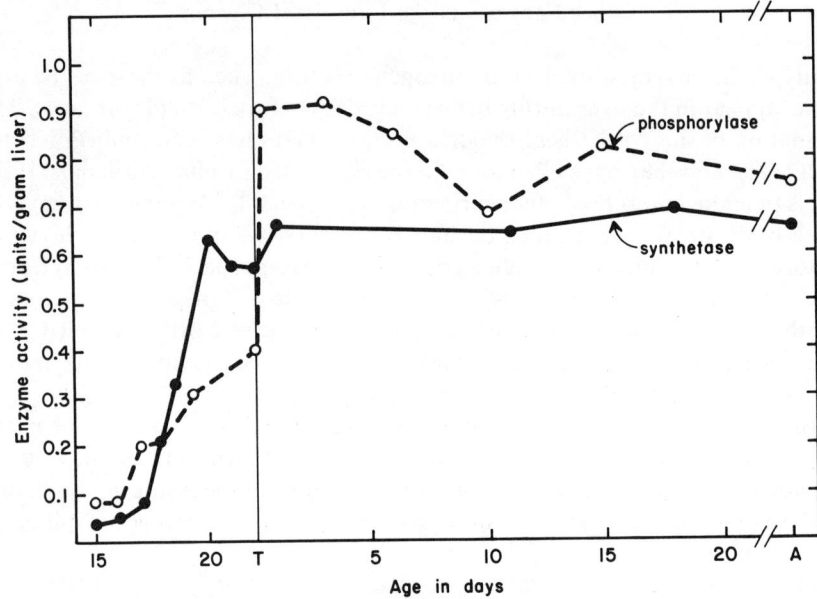

Fig. 14-5 Developmental profiles for glycogen synthetase and glycogen phosphorylase in the liver. T = term, and the numbers before term refer to the age of the fetus in days from conception. A = adult values. [Reproduced from Figs. 2 and 3 of Ballard and Oliver (23) with permission of Elsevier/North-Holland Biomedical Press.]

coids during late-fetal developmental is responsible for both the induction of the enzymes and the deposition of glycogen.

ENZYMES OF THE LATE-SUCKLING PERIOD IN LIVER

Neonatal rats spend approximately 21 days suckling before becoming weaned to a solid diet. During this period, many of the enzymes that first appear during the late-fetal stage of development undergo further periods of synthesis to consolidate the differentiated state of the tissue (Table 14-1). However, there are a number of enzymes not present in the fetal or early neonatal liver that make their first appearance shortly before weaning. Examples of this cluster of gene products are glucokinase, tryptophan oxygenase, and malic enzyme.

Glucokinase

The high-k_m glucokinase isoenzyme first appears in liver between 12 and 14 days after birth (Fig. 14-6) (8). Before this time only the constitutively expressed hexokinase group of isoenzymes is detectable. Glucokinase is synthesized only by the parenchymal cells of the liver (hepatocytes), making it a good model for the study of specific gene expression in liver. However, despite a great deal of effort the factors responsible for the initial appearance of this enzyme are still not completely understood.

Although the enzyme is not useful to the animal until it is weaned onto a high-carbohydrate diet, there is little doubt that it does appear in the parenchymal cells *before* weaning (day 21 in Fig. 14-6). Consequently, emphasis has been placed upon the possible involvement of hormones in the induction of the enzyme, but since the animal is generally quite highly differentiated by this time it is difficult to change the hormonal status of the animal without causing multiple effects. This has resulted in considerable controversy over the role of various hormones. For example, an early report that glucocorticoids could prematurely induce the enzyme, if given along with glucose (25), has not been confirmed by others (26, 27). Moreover, a combination of insulin and glucose was once considered to be the activation signal, even though high levels of both components are present in the fetus, where the enzyme is not synthesized.

There is evidence (26, 27), that thyroid hormones are involved in the regulation of the first appearance of the enzyme. A mixture of triiodothyronine and glucose is able to induce detectable levels of glucokinase as early as the fourth postnatal day. However, this is not an *absolute requirement*, because hypothyroid animals develop glucokinase activity at the proper time, although the levels are much reduced. Therefore, some other stimulus determines the actual

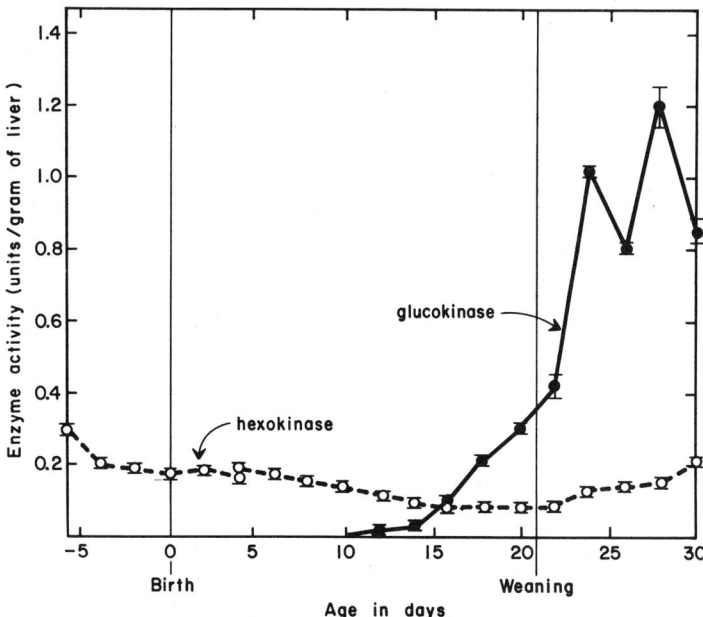

Fig. 14-6 Changes in the activities of hexokinases and glucokinase in the liver during development of the rat. Animals were weaned onto the "8 + 16" feeding schedule at the start of the dark period on day 21 after birth. Before weaning the pups were kept in cages that prevent access to solid food. [Reproduced from Fig. 2 of Walker (8) with permission.]

time course of synthesis, and insulin, glucose, triiodothyronine, and possibly even glucocorticoids may modulate the amount of enzyme synthesized once the gene is activated.

Recently, some work has been carried out on the synthesis of the enzyme in liver cells isolated from 13–14-day-old neonatal animals (28–30). In these cells, a combination of insulin and glucose (30) was an effective inducer of the enzyme. Glucagon or cAMP prevented the induction of the enzyme by either of these combinations. Moreover, cells isolated from hypothyroid animals could not be induced with insulin and glucose unless the cells were exposed to triiodothyronine. Further work with these cultures may clear up some of the mystery surrounding the appearance of this enzyme during development.

In summary, thyroid hormones and a high insulin:glucagon ratio appear to be important for the full expression of glucokinase during development. Glucose is able to further stimulate synthesis when the animals gradually wean themselves onto solid food. A role for glucocorticoids has not been definitively established. Thus the regulation of the synthesis of this enzyme is mediated by the same factors that control the enzyme in the adult (Table 12-3). However, there appears to be an as-yet-unidentified factor that is responsible for the initial activation of the glucokinase gene.

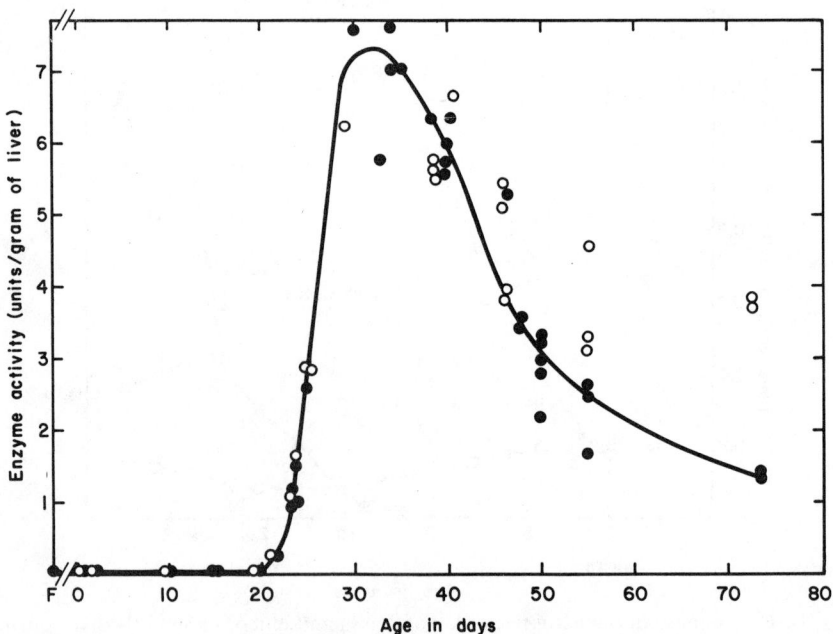

Fig. 14-7 Changes in the activity of malic enzyme in the liver during development. Animals were weaned on day 21. F = fetal enzyme activity (2–3 days before birth). The open circles are values for female rats, whereas the closed circles are values for male rats. [Reproduced from Fig. 3 of Taylor, Bailey, and Bartley (7) with permission.]

Further discussion of this enzyme will take place when we consider the development of diurnal rhythms of enzyme activity in a later section. This is of considerable importance because the function of glucokinase, which has long been overlooked in many of the studies on this enzyme, is to metabolize the periodic surges of portal blood glucose that occur as a result of diurnal feeding behavior in adult animals (Fig. 13-3) (31). Quite clearly the appearance of this enzyme is likely to be a manifestation of the emergence of the capacity of the liver to respond to diurnal rhythms of feeding activity rather than to a simple dietary shift.

Malic Enzyme

Malic enzyme is generally considered to be a lipogenic enzyme because of the role it plays in the generation of reducing equivalents (NADPH) for fatty acid synthesis (Fig. 12-10). It first appears in the liver at the time of weaning, coincident with the high demand for the conversion of dietary carbohydrate into fat reserves (Fig. 14-7) (7). However, it is not clear why, among all the enzymes responsible for the conversion of carbohydrate to fat, only malic enzyme does not appear until this time. It is possible that malic enzyme only contributes NADPH during periods of very high rates of lipogenesis that may occur with the onset of diurnal feeding activity.

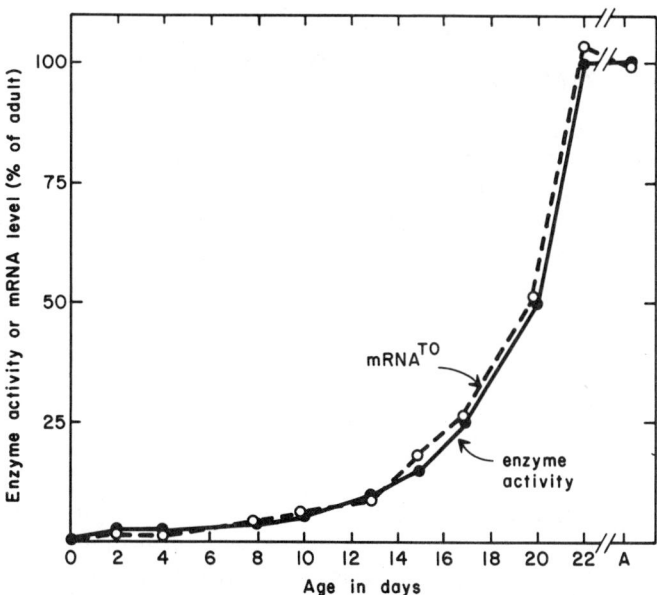

Fig. 14-8 Developmental profiles for tryptophan oxygenase activity and mRNA^{TO} concentration in liver following birth. [Reproduced from Fig. 4 of Ref. 32 with the permission of Dr. P. Feigelson.]

Although not studied in great detail, the first appearance of this enzyme seems to be regulated by thyroid hormones. Thus hypothyroid neonatal rats fail to develop any detectable malic enzyme using the same protocol as that described earlier for glucokinase (27). This experiment shows that the control of the developmental appearance of malic enzyme is more strictly dependent upon thyroid hormone than glucokinase. Thyroid hormone also regulates the synthesis of malic enzyme in adult animals (Table 12-3).

Tryptophan Oxygenase

Tryptophan oxygenase is the third example of a gene product absent from the liver until shortly before weaning. The developmental profiles of this enzyme and its mRNA are illustrated in Fig. 14-8 (32). The concentration of $mRNA^{TO}$ is virtually undetectable in the young neonate and the increase in enzyme activity is paralleled by an increase in $mRNA^{TO}$ concentration, indicating that the increase is a result of transcriptional activation of the tryptophan oxygenase gene.

Glucocorticoids can induce premature synthesis of the enzyme as early as 4 days after birth, particularly when injected along with tryptophan to stabilize the newly synthesized enzyme protein. The regulation of the synthesis of this enzyme during development has not been studied further. For example, it is not known if thyroid hormones have any role in its synthesis.

CHANGES IN THE SECRETION OF HORMONES DURING DEVELOPMENT

There can be little doubt that hormones play a major role in the regulation of the synthesis of inducible liver enzymes both during development and in the adult animal. In the adult, the concentrations of hormones vary, usually on a rhythmic basis, in response to changes in the external environment. In the developing fetus endocrine glands differentiate and begin active secretion of hormones at specific times; the released hormone may then stimulate changes in gene expression and differentiation in other tissues. In this section changes in the concentration of the major metabolic hormones during development are described.

Glucocorticoids

The results of a number of studies designed to measure the changes in glucocorticoid concentration in the blood at various stages of development in the rat are summarized in Fig. 14-9 (33–35). In fetal and early neonatal animals the concentrations of both total glucocorticoid and free glucocorticoid have been determined. The hormone is present 4 days before birth, and its concentration increases rapidly until about 1 day before birth. The concentration of free

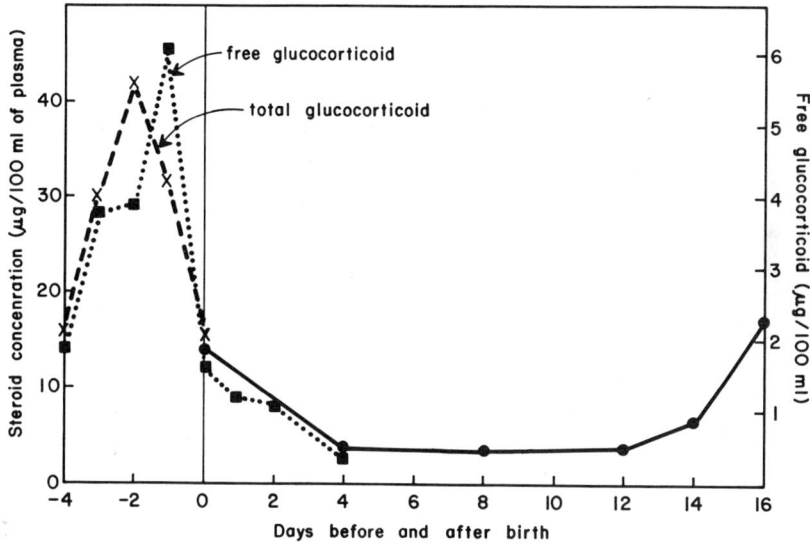

Fig. 14-9 Changes in the concentration of free glucocorticoids and total glucocorticoids in the blood during development of the rat. [The data are redrawn from Table 1 of Ref. 33, Fig. 2 of Ref. 34, and Fig. 2 of Ref. 35 with permission.]

hormone, which is a better indicator of the concentration of the hormone in the liver, peaks slightly later than the total hormone concentration. During the 24-h period immediately before birth the concentration of the hormone falls, and this decline continues after birth. The neonatal period is characterized by relatively low glucocorticoid concentrations, until 12–14 days of age. Between day 12 and day 16 there is a second surge of glucocorticoid release by the adrenals.

A comparison of this pattern of hormone change with those of the enzymes discussed earlier shows that the hormone is synthesized and secreted at elevated levels at times that would be compatible with a role of the hormone in the induction of some of the enzymes. However, it is far from established that this hormone actually activates *previously unexpressed* genes at any of these stages.

Insulin and Glucagon

In the rat, the pancreas is formed at about day 11 of fetal life and during the next 4 days there is considerable growth and differentiation of the tissue, but no hormone secretion (see Ref. 36 for references). Between day 15 and day 20 (1–2 days before birth) there is a *1000-fold* increase in the concentration of serum insulin. Thus the late-fetal period is characterized by a very high concentration of circulating insulin. However, there is a precipitous drop in the level of insulin at birth (37), and low levels persist throughout the neonatal period (Fig. 14-10a). The steady-state levels of the hormone increase again after weaning when the intake of carbohydrates increases.

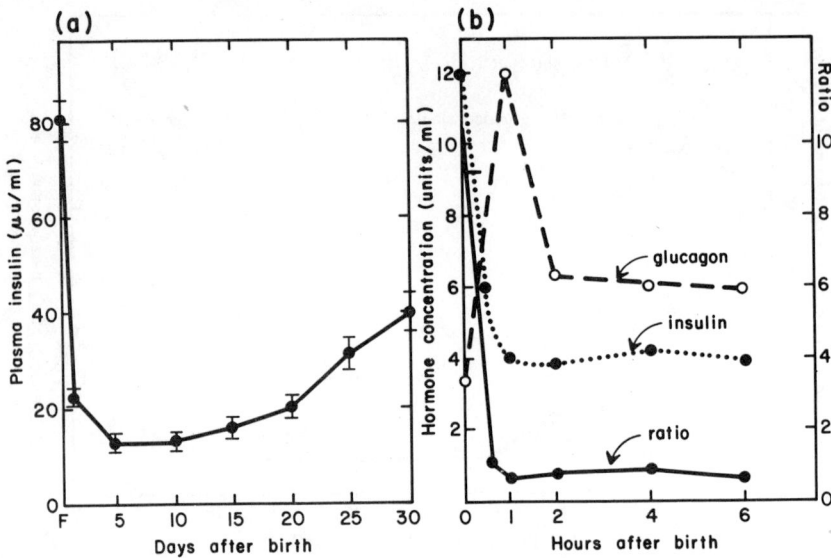

Fig. 14-10 Changes in the concentration of insulin in the plasma from immediately before birth until after weaning (a) and changes in insulin, glucagon, and the insulin: glucagon ratio during the first 6 h after birth (b). [Reproduced from Blázquez, Montoya, and Quijada (37) with permission of the *Journal of Endocrinology*.]

Changes in the concentration of glucagon have only been determined around the time of birth (38, 39) as shown in Fig. 14-10b. The concentration of glucagon increases markedly during the first hour after birth, while that of insulin falls, producing a 20–30-fold decrease in the insulin:glucagon ratio. This shift in the ratio can be related to the pronounced hypoglycemia in the newborn animal (Fig. 14-4b) and the appearance of gluconeogenic enzymes such as PEPCK and TAT (Fig. 14-4a).

Thyroid Hormones

Thyroid hormone has been implicated in the appearance of two enzymes, malic enzyme and glucokinase, during the late suckling period. Measurement of the changes in the concentration of thyroid hormones has revealed increases in body thyroxine and triiodothyronine in the blood immediately before these enzymes appear (Fig. 14-11) (27, 40). Thyroid hormones may be present at low levels in fetal liver, but a role for these hormones in the regulation of enzyme synthesis at this time has not been established.

It is particularly interesting that there are increases in the content of thyroid-releasing hormone (TRH) in the hypothalamus and the concentration of thyroid-stimulating hormone (TSH) in the blood preceding the increases in the concentration of thyroid hormones in the neonate (41). TSH and TRH regulate the activity of the thyroid gland (Fig. 10-1), and the increase in their concentration

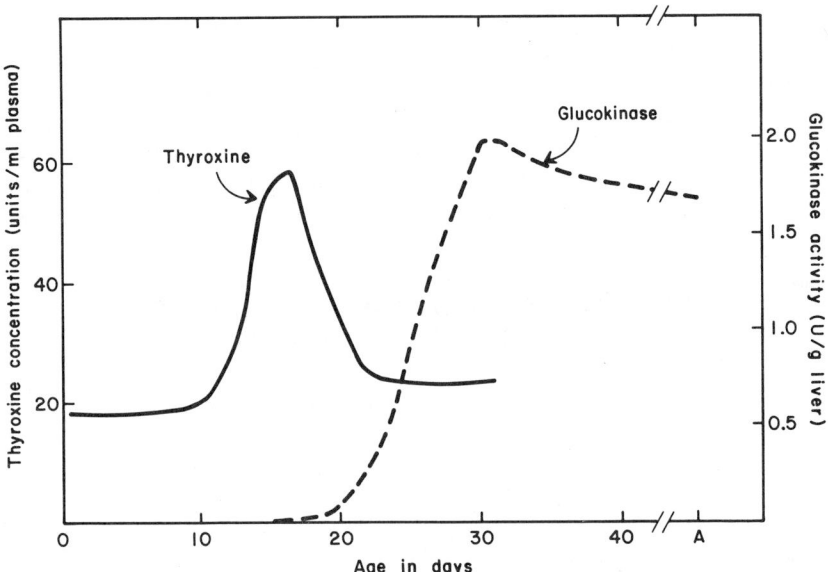

Fig. 14-11 Changes in the concentration of serum thyroxine levels during the neonatal period. The curve for the induction of glucokinase in the same animals is included for a temporal comparison. [Reproduced from Fig. 1 of Partridge et al. (26) with permission of the *European Journal of Biochemistry*.]

coincides with the maturation of the hypothalamic–pituitary axis as a functional regulatory element in the brain (TRH and TSH are not present in the fetus). This is of major importance because the maturation of this key, humoral regulatory system will alter the activity of the thyroid gland (and also the adrenals in the synthesis of glucocorticoids) even through both tissues may have been active at earlier stages. This change may be a stimulus for the further differentiation of liver and a number of other tissues.

Summary of Developmental Changes in Hormone Concentration

The late-fetal stage of development is characterized by high concentrations of insulin and glucocorticoid hormones and low concentrations of glucagon and thyroid hormones. The precipitous fall in the insulin:glucagon ratio at birth is the stimulus for the synthesis of PEPCK, TAT, and probably glycogen phosphorylase. Since low levels of TAT and appreciable levels of phosphorylase (and a low level of mRNA PEPCK) are present before birth, other hormones or factors are probably responsible for the initial activation of these genes. Thyroid hormones do not appear to be involved, but there are indications that glucocorticoids play some role, with insulin having a repressive effect.

The late-suckling period, on the other hand, is characterized by low levels of insulin and glucocorticoids, a peak of thyroid-hormone release and undetermined glucagon levels. While the concentrations of both insulin and glucocorticoids increase *during* weaning, the enzymes of the late-suckling cluster precede these changes. Thyroid hormones induce malic enzyme and facilitate glucokinase synthesis at this time, suggesting that they are important at this stage of development. However, changes in hormone concentrations at this time are likely to be complex, since the hypothalamic–pituitary axis matures at this time, and the animals also first become aware of the cycles of light and darkness.

CHANGES IN THE DEVELOPMENT OF RECEPTORS AND SECOND-MESSENGER SYSTEMS

Although hormones may be present in the blood, they can only interact with tissues that possess appropriate receptors. Moreover, even if receptors are present a fully functional signal transduction, or second-messenger, system must also be present to transmit the information to the nucleus. This section discusses some of the work that has been carried out on the development of these important signal molecules.

Hormone Receptors

The developmental patterns of the receptors for glucocorticoids and insulin in liver are shown in Fig. 14-12. High-affinity glucocorticoid binding can be detected in fetal liver as early as 6 days before birth (42, 43). However, since hemopoietic cells have a high concentration of the glucocorticoid receptor, these cells probably account for most of the binding at these early times, and it is believed that the receptor first appears in the hepatocytes about 4 days before birth (44). Hepatocytes isolated from fetuses 3–4 days before birth can be stimulated to synthesize TAT when incubated with glucocorticoid. Thus the absence of the enzyme *in utero*, until shortly before birth, is not due to a lack of either glucocorticoids, the receptor, or the ability of the hormone–receptor complex to translocate to the nucleus and activate the TAT gene.

The apparent decrease in glucocorticoid binding (which is the technique for measuring receptor concentration) at birth (Fig. 14-12a) may be due to the saturation of the receptors with endogenous hormone, rather than a real loss of receptors. After birth there is a marked increase in the concentration of glucocorticoid receptors despite low levels of glucocorticoids in the blood (cf. Fig. 14-9).

The concentration of insulin receptors also undergoes marked changes shortly before birth (Fig. 14-12b) (45). Four days before birth only a very low level of high-affinity insulin binding can be detected. However, during the last 3 days before birth the number of receptors increase, in parallel with the increase in the concentration of insulin in the blood. At birth the liver cells possess adult levels of receptors, and despite the low levels of circulating insulin (Fig. 14-10) this high concentration of receptors is maintained throughout the neonatal period.

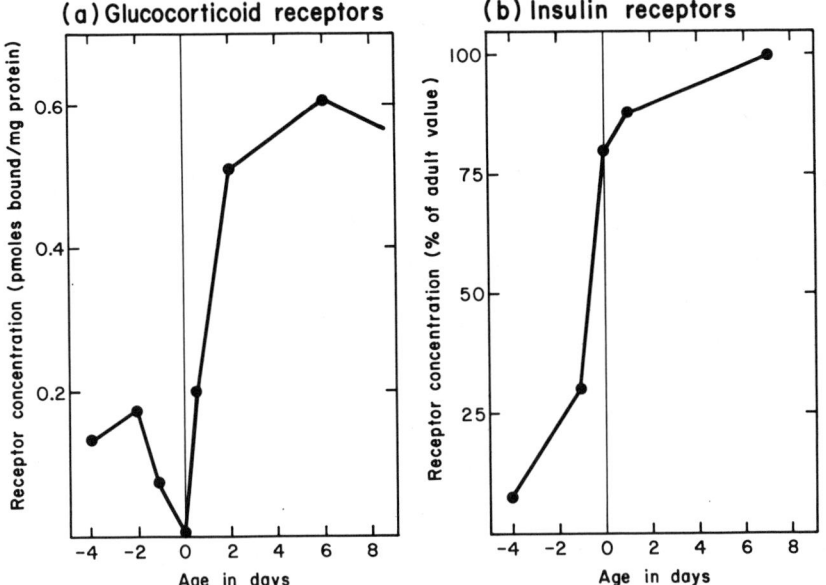

Fig. 14-12 Changes in the concentration of (a) glucocorticoid and (b) insulin receptors during liver development. The concentration of receptors is measured by the binding of radioactively labeled hormone and is subject to competition from endogenously produced hormones. [The data in (a) are reproduced from Fig. 2 of Giannopoulos (43) by permission of the American Society of Biological Chemists, Inc. The data in (b) are reproduced from Caliendo (45) with permission of *Federation Proceedings*.]

Although these are the only two receptors in liver for which detailed developmental profiles are available, the appearance of receptors during the course of tissue differentiation does not seem to be rate-limiting in the ability of the cells to respond to a hormone.

Adenylate Cyclase, cAMP, and Protein Kinases

Changes in the concentrations of adenylate cyclase, cAMP, and phosphodiesterase in liver have been measured from 5 days before birth to adulthood (Fig. 14-13) (46, 47). The enzymes, and a low concentration of cyclic AMP, are present 5 days before birth, and all three molecules increase to reach maximum values at about 5 days after birth, before declining to adult levels. Clearly, the ability to generate cAMP cannot be a rate-limiting step in the action of hormones on liver development.

Cyclic AMP-dependent protein kinase activity is also detectable in the liver 3–4 days before birth (48), and its concentration doubles between the last 2 days of fetal life and the first 2 days of neonatal existence. There have been no studies on the changes in the two isoenzymes of cAMP-dependent protein kinase, PK I and PK II, during development and no studies on the changes in the many

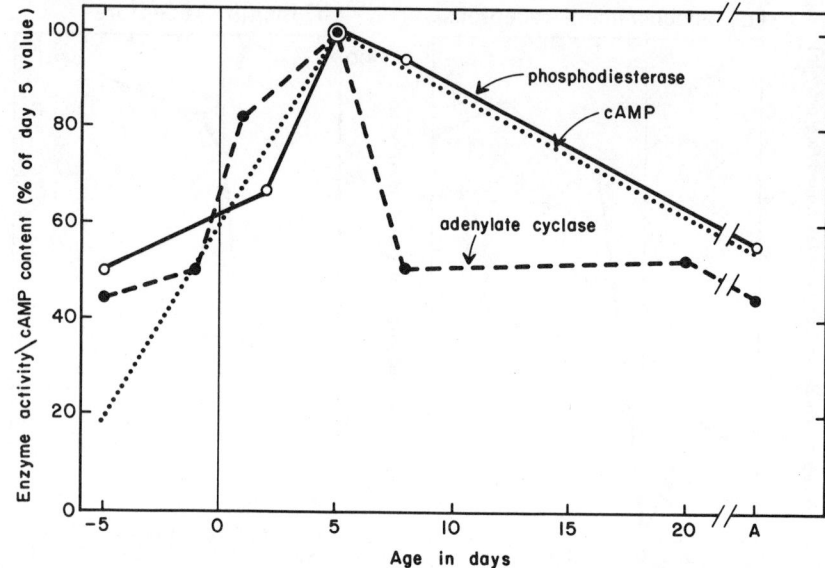

Fig. 14-13 Changes in the activities of adenylate cyclase and cAMP phosphodiesterase and the concentration of cAMP during liver development. [Reproduced from Figs. 7–9 of Christoffersen (46) with the permission of Elsevier/North-Holland Biomedical Press, and Fig. 1 of DiMarco and Oliver (47) with permission of the *European Journal of Biochemistry*.]

cyclic nucleotide-independent enzymes. However, at least in the case of cAMP-dependent protein kinase, this enzyme is present and capable of playing a role in signal transduction from several days before birth.

Hormone-Responsive Calcium Transport

Some hormones, particularly glucagon and catecholamines, mediate some of their effects by stimulating the release of calcium from intracellular stores (Fig. 11-14). The released calcium ions interact with calmodulin to generate an intracellular second messenger. Studies on the development of the capacity of this information-transfer system (49, 50) show that the liver cells first become able to respond to glucagon with a release of calcium about 2 days before birth (Fig. 14-14). The large, apparently glucagon-insensitive increase in calcium transport immediately after birth probably reflects endogenous stimulation of the system by either glucagon or adrenaline.

Summary of Changes in Information-Transfer Systems

Although far from complete, these studies indicate that receptors and the intracellular information-transfer or signal-transduction systems are all present in

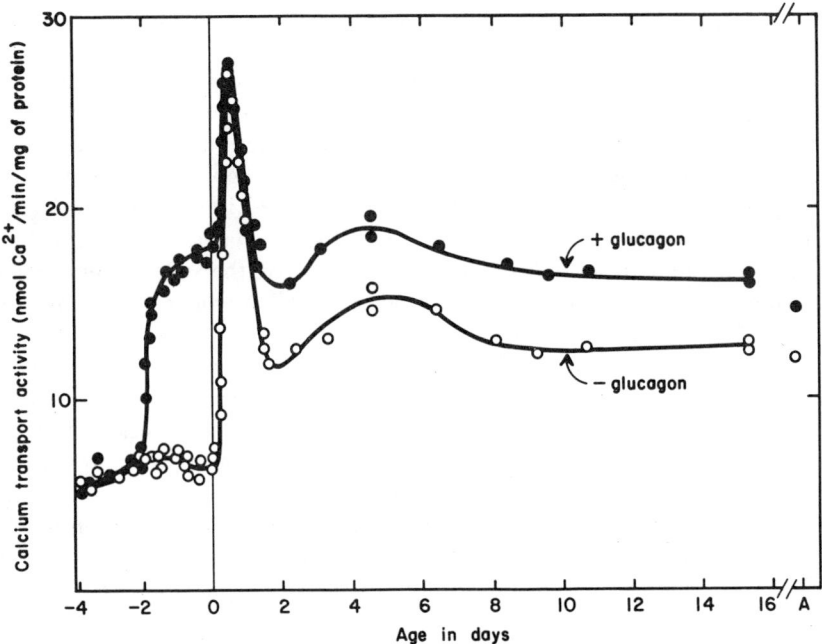

Fig. 14-14 Changes in the hormone-responsive, extramitochondrial calcium transport activity in the liver during development. [Reproduced from Reinhart and Bygrave (49) with the permission of the Biochemical Society.]

the liver cells before birth. However, since the changes in these components are often temporally related to the increase in the concentration of their hormone it is not possible to distinguish which may be rate-limiting. Differentiation is a highly coordinated series of events with many reactions occurring simultaneously. In general, these systems are formed 2 or 3 days before actually being used to mediate a change in gene expression in response to a hormone. Therefore, the regulation of the expression of genes during development is not limited by a lack of these molecules.

DEVELOPMENT OF DIURNAL RHYTHMS IN LIVER METABOLISM

About 15 days after birth the eyes of the neonatal rat open and the animal becomes aware, for the first time, of the cycles of light and darkness occurring each 24-h day. From this time onward there are profound changes in the physiology and biochemistry of the animal as it prepares for an independent existence in a world dominated by environmental rhythms. This section describes changes in the development of two of the components of the biological clock, the SCN, and the pineal gland, and then goes on to discuss the onset of diurnal rhythms of glucocorticoid secretion and diurnal rhythms of liver enzymes.

The Pineal Gland, SCN, and Glucocorticoid Rhythms

A diurnal rhythm of glucose uptake into the suprachiasmatic nucleus (SCN) is evident on the first day after birth (51), and this persists throughout neonatal development even though the eyes are closed and the retinohypothalamic pathways are not fully developed. This is of considerable interest since it implies that the rhythm is indeed endogenous because there appear to be no external cues. However, the possibilities of the persistence of some maternal signals or light perception through the eyelids or skull that could drive the SCN rhythm cannot be ruled out. In any case, the SCN is in a position to drive diurnal rhythms from a very early stage of development.

The pineal gland undergoes considerable development during the first 2 weeks of postnatal life (52), including innervation by the superior cervical ganglion that carries information from the SCN. (Thyroid hormones are thought to stimulate the development of the pineal gland.) The development of a neural connection between the SCN and the pineal gland leads to oscillations in the activity of serotonin N-acetyltransferase, which reach an "adult" amplitude by the eleventh day after birth. Thus enzyme synthesis (and melatonin production) exhibits a pronounced, SCN-driven diurnal rhythm *before* the eyes open on day 15.

A diurnal difference in the concentration of glucocorticoid in the blood is first observed 16–17 days after birth (53), which, in this particular strain of rats,

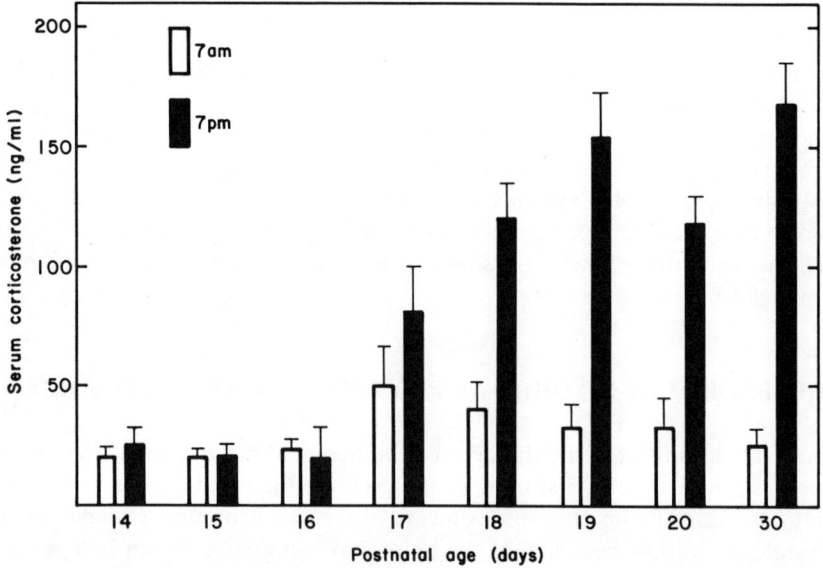

Fig. 14-15 The onset of diurnal differences in the concentration of glucocorticoids in the blood during development. An increased level at 7 PM (beginning of the dark period, Fig. 13-12) is indicative of a diurnal rhythm. [Reproduced from Fig. 1 of Poland et al. (53) with permission. Copyright 1981 The Endocrine Society.]

is 6–7 days before weaning (Fig. 14-15). The onset of the rhythm coincides with eye-opening and also with the appearance of enzymes in the liver such as glucokinase and malic enzyme (Figs. 14-6, 14-7), suggesting that some common mechanism may be responsible for the activation of all the components of the system for diurnal adaptability

Development of Diurnal Rhythms of Hepatic Enzymes

In the adult animal many aspects of the metabolism of the liver, and other tissues, are controlled by periodic changes in light, darkness, and food availability. The neonatal period, particularly the late-suckling period, is a time of transition from the fairly constant environment of the nest to the rhythmic environment of the independent adult. Consequently, this is a time of very complex changes in the physiology and behavior of the animal, which makes studies on the regulation of enzymes in any one tissue, such as liver, very difficult unless the influence of these changes is taken into account. It is this terminal phase of hepatocyte differentiation that gives the animal the capacity to metabolize dietary components on a diurnal basis.

The onset of rhythms for a number of liver functions has been studied (54) in young rats from 6 days after birth until 8 days after weaning (Fig. 14-16). In these animals eye-opening took place on day 15, and they were weaned onto the "8 + 16" feeding schedule (see Chapter 13) at the start of the dark period on day 22. In addition, the young were kept in cages that prevented them from gradually weaning themselves by nibbling the mother's food, and stomach weights were taken as an indication of the amount of milk or food consumed. Stomach weights were generally higher in the light period in the young, indicating that they suckled primarily at this time. There were no discernible rhythms in glycogen or TAT levels, and glucokinase was not detectable until the day before eye-opening. Between eye-opening and weaning the animals tended to suckle more in the dark period, and there was evidence for the onset of glycogen and TAT rhythms, indicating an influence of light and darkness.

Weaning the animals directly onto the "8 + 16" feeding schedule demonstrated that the animals are capable of responding to diurnal rhythms of food availability immediately. Thus TAT undergoes a pronounced rhythm and the tissue accumulates sufficient glycogen to provide the short-term energy store typical of the adult animal (since the diet used for weaning in these experiments was very high in carbohydrate content, glucokinase remains at its fully induced level rather than exhibiting a rhythm). It is clear that the animal has developed its *capacity* for diurnal fluctuations in metabolism prior to weaning.

The development of this ability to change the metabolism of the liver rhythmically in response to food availability is complete by day 14, as shown in Fig. 14-17 (54). This was demonstrated by adapting rats to a "16 + 8" suckling schedule that permitted the mother to feed in a separate cage for her 8-h feeding period (of the "8 + 16" feeding schedule) and then be returned to the young for the other 16 h of each 24-h day. Clear rhythms of TAT biosynthesis and

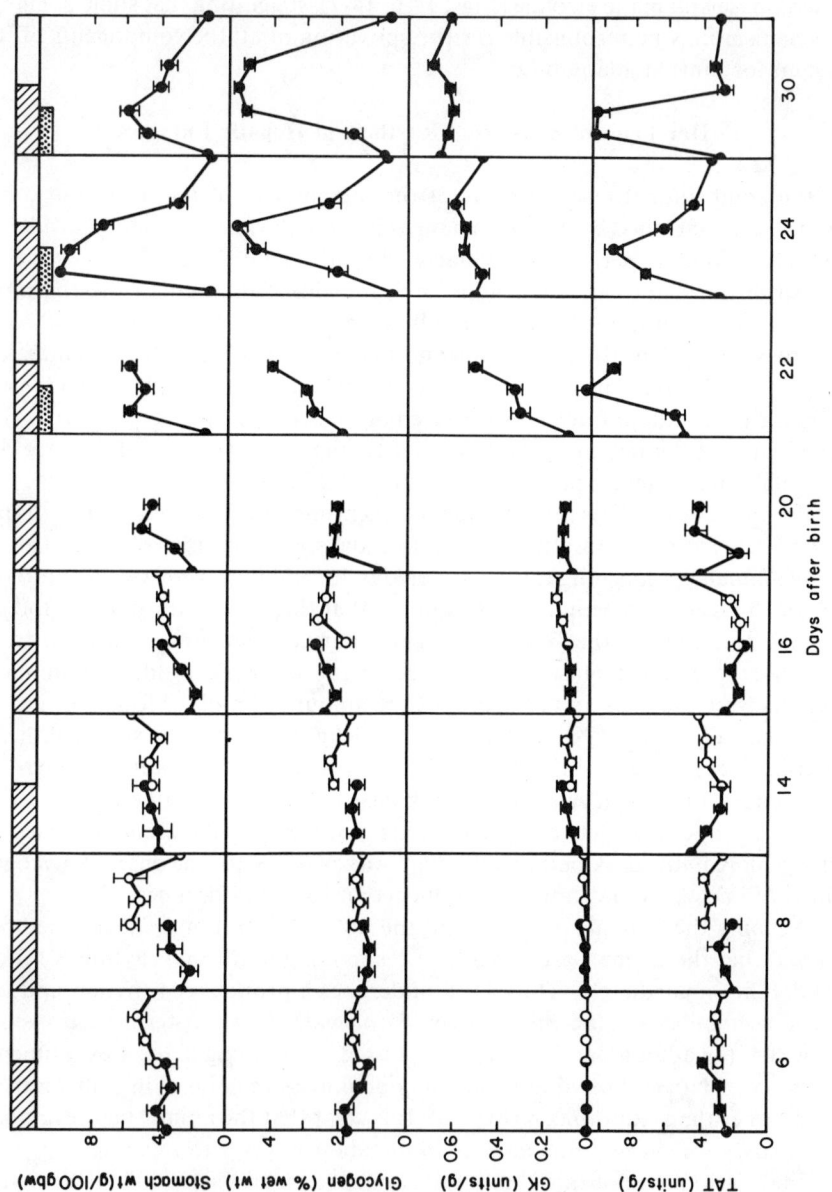

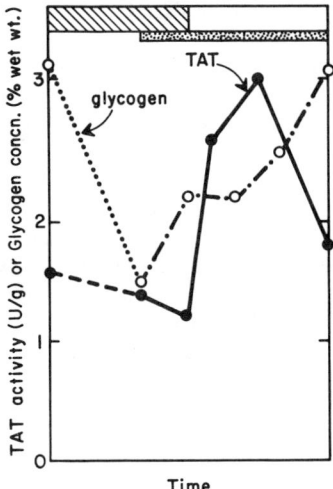

Fig. 14-17 The ability of 14-day-old neonatal animals to undergo food (mother's milk) related diurnal rhythms of glycogen accumulation and TAT activity. For these animals the mother was adapted to the "8 + 16" feeding schedule and fed for 8 h in a separate cage, before being returned to suckle the young for 16 h (stippled horizontal bar). [Reproduced from Fig. 2 of Ref. 54 with permission.]

glycogen accumulation and mobilization occur, and these are entrained to the intake of milk.

These studies clearly indicate that the developmental changes in the late-suckling period are not simply related to the shift in diet from the high-fat content of milk to the high-carbohydrate content of solid food. During the few days preceding weaning the animal develops a highly sophisticated, integrated system for dealing with the periodic nature of the environment upon which independent survival depends. The maturation of specific areas of the hypothalamus that form the basis of the biological clock(s) has a tremendous influence of the development of the animal at this time, including the changes in specific enzymes in the liver. An appreciation of the animals' sense of time is mandatory for the design of experimental protocols for the further study of the differentiation of the liver and other tissues at these times.

Fig. 14-16 Development of diurnal rhythms of glycogen concentration, glucokinase activity, and TAT activity. Animals from 6 days of age onward were kept with their mothers in special cages to prevent food nibbling until the start of the dark period on the 21st postnatal day, at which time they were placed on the "8 + 16" feeding schedule. Stomach weights were taken as a measure of milk or feed consumption. [Reproduced from Fig. 1 of Ref. 54 with permission.]

REFERENCES

1. O. Greengard (1970). The developmental formation of enzymes in rat liver. In G. Litwack (Ed.). *Biochemical Actions of Hormones*, Vol. 1. Academic Press, New York, pp. 53–87.
2. O. Greengard (1971). Enzymatic differentiation in mammalian tissues. *Essays Biochem.* **7**, 159–205.
3. O. Greengard (Ed.) (1973). *Biochemical Bases of the Development of Physiological Functions.* Karger, Basel.
4. E. Bailey and E. A. Lockwood (1973). Some aspects of fatty acid oxidation and ketone body formation and utilization during development of the rat. *Enzyme* **15**, 239–253.
5. E. H. Davidson (1976). *Gene Activity in Early Development.* Academic Press, New York.
6. C. C. Rider and C. B. Taylor (1980). *Isoenzymes.* Chapman and Hall, London.
7. C. B. Taylor, E. Bailey, and W. Bartley (1967). Changes in hepatic lipogenesis during development of the rat. *Biochem. J.* **105**, 717–722.
8. P. R. Walker (1974). Effect of controlled feeding on the isoenzymes of pyruvate kinase, hexokinase and aldolase during development. *Life Sci.* **15**, 1507–1514.
9. F. J. Ballard and R. W. Hanson (1967). Phosphoenolpyruvate carobxykinase and pyruvate carboxylase in developing rat liver. *Biochem. J.* **104**, 866–871.
10. D. Yeung and I. T. Oliver (1968). Factors affecting the premature induction of phosphopyruvate carboxylase in neonatal rat liver. *Biochem J.* **108**, 325–331.
11. D. Yeung and I. T. Oliver (1968). Induction of phosphopyruvate carboxylase in neonatal rat liver by adenosine 3′,5′-cyclic monophosphate. *Biochemistry* **7**, 3231–3239.
12. J. P. Ruiz, R. Ingram, and R. W. Hanson (1978). Changes in hepatic messenger RNA for phosphoenolpyruvate carboxykinase (GTP) during development. *Proc. Natl. Acad. Sci. USA* **75**, 4189–4193.
13. D. Mencher, D. Shouval, and L. Reshef (1979). Premature appearance of hepatic phosphoenolpyruvate carboxykinase in fetal rats, not mediated by adenosine 3′:5′-monophosphate. *Eur. J. Biochem.* **102**, 489–495.
14. J. G. Steele, M. C. McGrath, G. C. T. Yeoh, and I. T. Oliver (1980). Phosphoenolpyruvate carboxykinase in cultured foetal hepatocytes from the rat. *Eur. J. Biochem.* **104**, 91–99.
15. S. M. Andersson, N. C. R. Räihä, and J. J. Ohisalo (1980). Regulation of tyrosine aminotransferase in foetal rat liver. *Biochem. J.* **186**, 609–612.
16. A. H. Coufalik and C. Monder (1980). Regulation of the tyrosine oxidizing system in fetal rat liver. *Arch. Biochem. Biophys.* **199**, 67–76.
17. O. Greengard (1969). The hormonal regulation of enzymes in prenatal and postnatal rat liver. *Biochem. J.* **115**, 19–24.
18. A. V. Ghisalberti, J. E. Steele, M. H. Cake, M. C. McGrath, and I. T. Oliver (1980). Role of adrenaline and cyclic AMP in appearance of tyrosine aminotransferase in perinatal rat liver. *Biochem. J.* **190**, 685–690.
19. G. C. T. Yeoh, T. Arbuckle, and I. T. Oliver (1979). Tyrosine aminotransferase induction in hepatocytes cultured from rat foetuses treated with dexamethasone *in utero*. *Biochem. J.* **180**, 545–549.
20. K. K. W. Ho, M. H. Cake, G. C. T. Yeoh, and I. T. Oliver (1981). Insulin antagonism of glucocorticoid induction of tyrosine aminotransferase in cultured feotal hepatocytes. *Eur. J. Biochem.* **118**, 137–142.
21. C. Gautier, A. Husson, and R. Vaillant (1977). Effects des glucocorticostéroides sur l'activité des enzymes du cycle de l'urée dans le foie feotal de rat. *Biochimie (Paris)* **59**, 91–95.
22. Y. G. Yeung and D. C. K. Yeung (1980). Synthesis of threonine dehydratase in neonatal rat liver. *Eur. J. Biochem.* **111**, 389–393.

REFERENCES

23. F. J. Ballard and I. T. Oliver (1963). Glycogen metabolism in embryonic chick and neonatal rat liver. *Biochim. Biophys. Acta* **71**, 578–588.
24. W. Stalmans and M. Laloux (1979). Glucocorticoids and hepatic glycogen metabolism. In J. D. Baxter and G. G. Rousseau (Eds.). *Glucocorticoid Hormone Action*. Springer-Verlag, Berlin, pp. 517–533.
25. S. C. Jamdar and O. Greengard (1970). Premature formation of glucokinase in developing rat liver. *J. Biol. Chem.* **245**, 2779–2783.
26. N. C. Partridge, C. H. Hoh, P. K. Weaver, and I. T. Oliver (1975). Premature induction of glucokinase in the neonatal rat by thyroid hormone. *Eur. J. Biochem.* **51**, 49–54.
27. M. J. O. Wakelam, C. Aragon, C. Gimenez, N. B. Allen, and D. G. Walker (1979). Thyroid hormones and the precocious induction of hepatic glucokinase in the neonatal rat. *Eur. J. Biochem.* **100**, 467–475.
28. T. Nakamura, K. Aoyama, and A. Ichihara (1979). Precocious induction of glucokinase in primary cultures of postnatal rat hepatocytes. *Biochem. Biophys. Res. Commun.* **91**, 515–520.
29. M. J. O. Wakelam and D. G. Walker (1980). De Novo synthesis of glucokinase in hepatocytes isolated from neonatal rats. *FEBS Lett.* **111**, 115–119.
30. M. J. O. Wakelam and D. G. Walker (1981). The separate roles of glucose and insulin in the induction of glucokinase in hepatocytes isolated from neonatal rats. *Biochem. J.* **196**, 383–390.
31. P. R. Walker (1977). Regulation of liver glycogen metabolism by the portal blood glucose concentration in rats adapted to controlled feeding schedules. *Int. J. Biochem.* **8**, 555–556.
32. L. A. Killewich and P. Feigelson (1977). Developmental control of messenger RNA for hepatic tryptophan 2,3-dioxygenase. *Proc. Natl. Acad. Sci. USA* **74**, 5392–5396.
33. A. Cohen (1973). Plasma corticosterone concentration in the foetal rat. *Horm. Metab. Res.* **5**, 66.
34. G. C. Haltmeyer, V. H. Denenberg, J. Thatcher, and M. X. Zarrow (1966). Response of the adrenal cortex of the neonatal rat after subjection to stress. *Nature* **212**, 1371–1373.
35. C. E. Martin, M. H. Cake, P. E. Hartmann, and I. F. Cook (1977). Relationship between foetal corticosteroids, maternal progesterone and parturition in the rat. *Acta Endocrinol.* **84**, 167–171.
36. L. B. Rall, R. L. Pictet, and W. J. Rutter (1979). Synthesis and accumulation of proinsulin and insulin during development of the embryonic rat pancreas. *Endocrinol.* **105**, 835–841.
37. E. Blázquez, E. Montoya, and C. L. Quijada (1970). Relationship between insulin concentrations in plasma and pancreas of foetal and weaning rats. *J. Endocrinol.* **48**, 553–561.
38. J. R. Girard, G. S. Cuendet, E. B. Marliss, A. Kervran, M. Rieutori, and R. Assan (1973). Fuels, hormones, and liver metabolism at term and during the early postnatal period in the rat. *J. Clin. Invest.* **52**, 3190–3200.
39. P. N. DiMarco, A. V. Ghisalberti, C. E. Martin, and I. T. Oliver (1978). Perinatal changes in liver corticosterone, serum insulin and plasma glucagon and corticosterone in the rat. *Eur. J. Biochem.* **87**, 243–247.
40. D. van der Heide and M. P. Ende-Visser (1980). T_4, T_3 and reverse T_3 in the plasma of rats during the first 3 months of life. *Acta Endocrinol.* **93**, 448–454.
41. K. Pease, H. Shen, G. S. Acres, J. H. Rupnow, and J. E. Dixon (1980). Alterations in levels of thyrotropin releasing hormone and rates of hormone degradation of hypothalamic tissue of the developing rat. *J. Endocrinol.* **85**, 55–61.
42. D. Feldman (1974). Ontogeny of rat hepatic glucocorticoid receptors. *Endocrinol.* **95**, 1219–1227.
43. G. Giannopoulos (1975). Ontogeny of glucocorticoid receptors in rat liver. *J. Biol. Chem.* **250**, 5847–5851.
44. M. H. Cake, G. C. T. Yeoh, and I. T. Oliver (1981). Ontogeny of the glucocorticoid receptor and its relationship to tyrosine aminotransferase induction in cultured foetal hepatocytes. *Biochem. J.* **198**, 301–307.

45. A. M. Caliendo (1981). Development of insulin receptors in rat hepatic plasma membranes. *Fed. Proc.* **39**, 1877.
46. T. Christoffersen, J. Mørland, J. B. Osnes, and I. Øye (1973). Development of cyclic AMP metabolism in rat liver. *Biochim. Biophys. Acta* **313**, 338–349.
47. P. N. DiMarco and I. T. Oliver (1978). Adenosine 3':5'-monophosphate in perinatal rat liver. *Eur. J. Biochem.* **87**, 235–241.
48. P. C. Lee and R. A. Jungmann (1975). Ontogeny of cyclic AMP-dependent protein phosphokinase during hepatic development of the rat. *Biochim. Biophys. Acta* **399**, 265–276.
49. P. H. Reinhart and F. L. Bygrave (1981). Glucagon stimulation of ruthenium red-insensitive calcium ion transport in developing rat liver. *Biochem. J.* **194**, 541–549.
50. V. Prpič and F. L. Bygrave (1981). Maturation in liver mitochondria of ruthenium red-sensitive calcium-ion-transport activity and the influence of glucagon administration *in vivo* and *in utero*. *Biochem. J.* **196**, 207–216.
51. J. L. Fuchs and R. Y. Moore (1980). Development of circadian rhythmicity and light responsiveness in the rat suprachiasmatic nucleus: a study using the 2-deoxy [1-^{14}C] glucose method. *Proc. Natl. Acad. Sci. USA* **77**, 1204–1208.
52. D. C. Klein, M. A. A. Namboodiri, and D. A. Auerbach (1981). The melatonin rhythm generating system: developmental aspects. *Life Sci.* **28**, 1975–1986.
53. R. E. Poland, M. E. Weichsel, Jr., and R. T. Rubin (1981). Neonatal dexamethasone administration. I. Temporary delay of development of the circadian serum corticosterone rhythm in rats. *Endocrinol.* **108**, 1049–1054.
54. P. R. Walker, R. J. Bonney, and V. R. Potter (1974). Diurnal rhythms of hepatic carbohydrate metabolism during development of the rat. *Biochem. J.* **140**, 523–529.

CHAPTER FIFTEEN

Hormonal Control of Gene Expression During the Maturation of Reproductive Tissues

In mammals the final stage of development is the differentiation of the reproductive tissues to produce a sexually mature adult. Before puberty, these tissues are small and very poorly differentiated, but they each undergo a complex program of proliferation and differentiation under the control of the reproductive hormones, acting in concert with other hormones such as glucocorticoids and insulin. Since all these changes occur *ex utero* they are much easier to study and to manipulate; consequently, they have become particularly useful systems for the study of terminal differentiation. Indeed, much of our current knowledge of gene expression in eukaryotes has come from the study of these systems. For example, the ovalbumin and conalbumin genes of the oviduct were the first animal cell genes to be shown to contain intervening sequences. Because of their importance some of these model systems are described briefly in this chapter, even though most of the gene products are not enzymes. It is highly likely that mechanisms of gene expression discovered in these model systems will be applicable to changes in enzyme synthesis in other tissues.

HORMONAL CONTROL OF THE MATURATION OF THE CHICKEN OVIDUCT

Changes in Morphology

In the immature chick, the oviduct is lined with a mucosal layer of undifferentiated cells that, upon stimulation, proliferate extensively and differentiate into three main types:

1 Tubular gland cells that synthesize and secrete large quantities of the egg-white proteins ovalbumin and conalbumin, together with small amounts of lysozyme and ovomucoid.

2 Goblet cells that synthesize and secrete avidin.
3 Ciliated cells that are responsible for motility to facilitate movement of the egg.

In the mature bird the tubular cells account for approximately 90% of the total cell mass; these are the cells we focus upon in this section. *In vivo*, the differentiation of the oviduct is controlled by the increased circulatory levels of 17β-estradiol; but usually in the laboratory, the same program of differentiation is induced in younger animals by the administration of the estrogen analogue diethylstilbestrol (DES).

Within 24 h after injection of DES there are extensive morphological changes in the mucosal cells with an increased synthesis of ribosomes and the development of rough endoplasmic reticulum and golgi apparatus. These changes are typical of cells preparing for the biosynthesis of large quantities of secretory proteins. The secretion of ovalbumin and conalbumin (this is the iron-carrying protein also referred to as transferrin) commences 4 to 6 days after the first injection of DES, and the concentration of ovalbumin, for example, increases more than 300-fold during the next few days.

Virtually all of this program of differentiation is initiated by 17β-estradiol (or DES in the experimental situation) except for the synthesis of avidin by the goblet cells, which is stimulated by progesterone (a closely related steroid, see Fig. 10-4). The estrogen must be present continuously to maintain this differentiated state. If the hormone is removed there is a rapid decline in the biosynthetic capacity of the cells followed, eventually, by involution of the oviduct. However, if the hormone is administered again the synthesis of ovalbumin and the other proteins increases rapidly and within a matter of *hours* their levels are similar to those obtained over a period of *days* during primary stimulation. This rapid, secondary stimulation is the system most often studied experimentally as a model for steroid hormone control of gene expression. However, it must be remembered that these genes were previously activated, and even in the withdrawn state a small number of molecules of each mRNA is present owing to a basal level of transcription of their genes (1), as shown in Table 15-1.

Table 15-1 The Marked Contrast between the Numbers of Molecules per Cell of Egg-White Protein mRNAs in the Oviducts of Actively Laying Hens and Hormone-Withdrawn Chickens[a]

Hormonal status	Number of mRNA molecules/cell			
	ovalbumin	conalbumin	ovomucoid	lysozyme
Laying hen	77,000	11,000	27,000	29,000
Estrogen−withdrawn chicken	7	26	8	7

[a] Reproduced from Table 1 of Ref. 1 with permission of Elsevier/North-Holland Biomedical Press.

Hormone Interactions at the Transcriptional Level

When fully induced, ovalbumin synthesis accounts for 50–60% of total protein synthesis, conalbumin for about 10%, ovomucoid 8%, and lysozyme 2–3%. Since the rate of protein synthesis is directly proportional to the concentration of each individual mRNA, the concentrations of mRNA^{OV} and mRNA^{CON} are very high (Table 15-1). This facilitates their isolation, purification, and the construction of cDNA probes that can be used to measure changes in the transcriptional activation of their genes.

Changes in the rates of synthesis of mRNA^{OV} and mRNA^{CON} following secondary stimulation have been measured (Fig. 15-1) and related to the changes in the number of steroid receptors that are translocated to the nucleus (2, 3). If the time lag between the withdrawal of the hormone after primary stimulation and the administration of the secondary stimulus is short, the tubular cells retain a full complement of cytoplasmic steroid hormone receptors, and these are rapidly translocated in the presence of steroid to the nucleus. If the time lag is longer, the entry of hormone into the nucleus is limited by the concentration of cytoplasmic receptors (which decline in concentration as the interval between primary and secondary stimulation increases). In general, the rate of synthesis of mRNA^{CON} parallels, or follows closely behind, the entry of receptors into the nucleus, but the rate of synthesis of mRNA^{OV} always lags behind (Fig. 15-1).

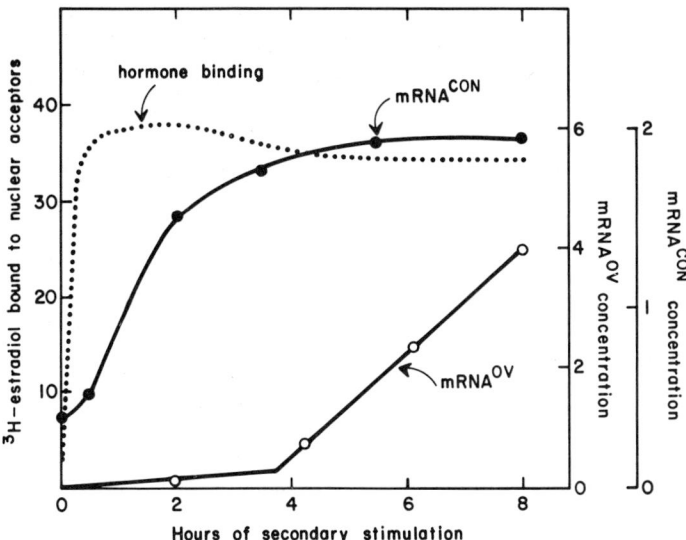

Fig. 15-1 Changes in estrogen binding to oviduct nuclear acceptors and the syntheses of ovalbumin and conalbumin mRNAs (mRNA^{OV} and mRNA^{CON}, respectively) in chicken oviducts during secondary stimulation with estrogen. [Reproduced from R. D. Palmiter, P. B. Moore, and E. R. Mulvihill (1976). *Cell 8* (1976), 557–572 with permission. Copyright 1976 Massachusetts Institute of Technology.]

Following a detailed, kinetic analysis of the synthesis of these two mRNAs, in comparison with the rates of accumulation of hormone–receptor complexes in the nucleus, Palmiter et al. (3) have recently concluded that even though the steroid activates both genes the mechanism of activation is somewhat different in the two cases. Since the synthesis of mRNA^{CON} is *directly* proportional to the concentration of receptors in the nucleus, it is believed that there is only one acceptor-binding site on chromatin, near the conalbumin gene, and when it is occupied the gene is expressed. This is similar to the activation of the tyrosine aminotransferase gene by glucocorticoids (Fig. 11-6). The relationship between the expression of the ovalbumin gene and receptor concentration in the nucleus is more complex, but the kinetics can be explained if, instead of one nuclear acceptor site, there are a total of five sites, all of which must be occupied for full expression of the gene. Moreover, the structural organization of the gene is such that if one of the five sites is occupied the gene can be expressed at about 15% of total capacity. However, occupancy of sites 2–4 does not produce any further increase in the rate of transcription, and full expression requires all five sites to be occupied. This explains the biphasic kinetics illustrated in Fig. 15-1. The nature of these binding sites is not known, but it is believed that hormone–receptor complexes for progesterone, androgens, and glucocorticoids can also bind to these sites and activate the genes, but with lower affinities.

Since cycloheximide (an inhibitor of protein synthesis) blocks steroid-hormone induction of these mRNAs, labile proteins are involved in the transduction of the hormonal signal into an increased frequency of transcription of the genes by RNA polymerase II. Moreover, inhibitors (butyrate, propionate) of deacetylation reactions also prevent expression of the genes, indicating that covalent modification of the histones of the nucleosomes or other nonhistone proteins is part of the activation or the transcription process. It should be noted that since changes in *mRNA* were taken as a measure of the activity of the gene, rather than changes in the production of nuclear *primary transcripts*, a post-transcriptional influence of the steroids cannot be ruled out.

There is a pronounced seasonal rhythm in the amount of mRNA^{OV} synthesized in response to steroid treatment (4). For example, the amount synthesized in December is only one third that in November (Fig. 15-2a), and in some months there is no response at all. When the acceptor activity of chromatin was examined on a seasonal basis over a 3-year period (5), the reason for the lack of responsiveness became evident (Fig. 15-2b). The birds had nuclear acceptor activity only during the summer and fall, and none in the winter and early spring. This is one of the few examples of a seasonal change in a molecule associated with the transcription of specific genes. The system has not yet been examined in terms of the influence of the pineal gland and other centers in the brain that may be responsible for timing this phenomenon. *However, it is clear that factors other than the steroids themselves control expression of these genes.*

Indeed, it now appears that polypeptide hormones, possibly of the somatomedin family, are required for the activation of the ovalbumin gene by steroid hormones (6). This effect is seen in oviduct explants *in vitro* when the tissue is removed from the influence of the high levels of circulating somatomedins.

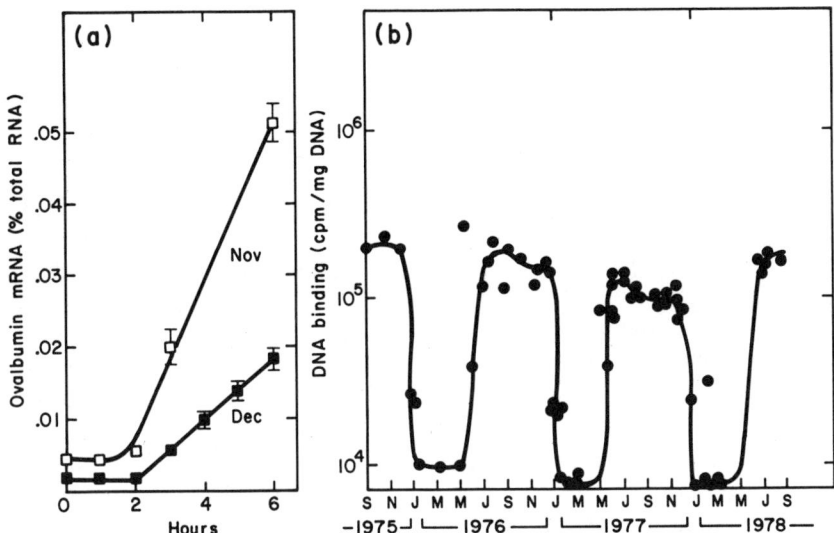

Fig. 15-2 Seasonal variations in (*a*) the synthesis of ovalbumin mRNA during secondary stimulation and (*b*) the binding of progesterone to nuclear acceptor sites in chicken oviduct. [The data in (*a*) are reprinted with permission from *Biochemistry* **19** (1980), 1410–1416. Copyright 1980 American Chemical Society. The data in (*b*) are reprinted from Fig. 3 of Ref. 5 with permission. Copyright © 1980 The Endocrine Society.]

These hormones are not yet completely characterized, but they are secreted under the control of the hypothalamus and the pituitary gland (possibly via growth hormone) and have a variety of growth-related effects. Insulin, which is much more readily available, can mimic the effect of the somatomedins (proinsulin also mimics this effect, ruling out a direct effect of insulin itself on the system).

In the explants, the synthesis of mRNA^{OV} is absolutely dependent upon the presence of insulin or other somatomedinlike hormones. The synthesis of mRNA^{CON}, which seems to be under less complex control, does not require the protein cofactor (the ovomucoid gene, like the ovalbumin gene *does* require the factor). Thus there is a requirement for a polypeptide hormone in the expression of at least two of these genes in response to steroids. The possibility that somatomedins are responsible for mediating the seasonal variation in gene responsiveness is currently being investigated because hypophysectomy is known to reduce the response of the cells of the oviduct to estrogens.

ESTROGEN REGULATION OF GENE EXPRESSION IN AVIAN AND AMPHIBIAN LIVER

In birds and amphibians (e.g., the frog, *Xenopus laevis*) the liver also functions in a reproductive capacity by synthesizing egg-yolk proteins. In response to stimulation by estrogen the liver synthesizes and secretes a large polypeptide, vitellogenin (M_r = 240,000). This molecule is taken up by the oocyte during formation of the egg-yolk, where it is cleaved to produce one molecule of lipovitellin (M_r = 130,000) and two molecules of phosvitin ($M_r \sim$ 32,000, the two phosvitins are slightly different). The hormonal regulation of vitellogenesis is being actively studied in avian and in amphibian species, and this work has recently been reviewed (7–9). The usual model system is the immature *male* animal that responds to an injection of estrogen in a similar way to the maturing female.

Synthesis of vitellogenin is more rapidly stimulated by a second injection of estrogen following withdrawal from a primary injection. Thus the liver is capable of being irreversibly "primed" by the steroid in a manner similar to that occurring during primary stimulation of oviduct. This is shown in Fig. 15-3a. The increase in the secretion of vitellogenin into the blood is preceded by an increase in the concentration of mRNA^{VIT} in the liver (Fig. 15-3b). In the unstimulated rooster liver there are less than 5 molecules of mRNA^{VIT} per cell, but 12 h after hormone injection this has increased to 2000 molecules per cell, and a maximum of 6000 molecules per cell is reached 3 days after hormone treatment—a greater than 1000-fold induction.

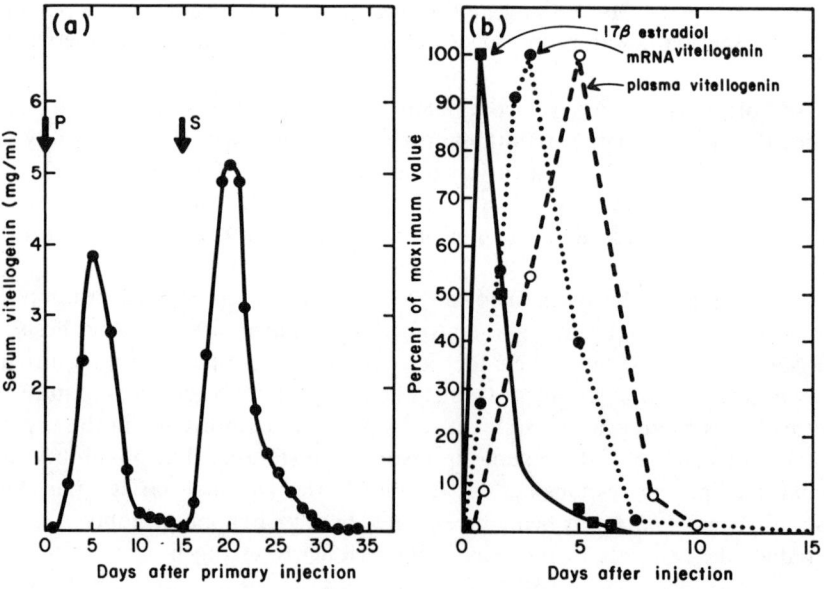

Fig. 15-3 Vitellogenesis in the rooster. (*a*) The levels of vitellogenin in the serum following primary (P) and secondary (S) injections of 17β-estradiol. The secondary injection was given 15 days after the primary injection. (*b*) Time course for the changes in the concentration of 17β-estradiol in plasma, the concentration of vitellogenin mRNA in the liver, and the concentration of vitellogenin in the plasma following secondary stimulation. [Reproduced from Figs. 5 and 6 of Goldberger and Deeley (7) with permission].

During primary stimulation the hormone induces a change in chromatin structure that makes the vitellogenin gene more sensitive to digestion by DNase-I (10). For 4 h after hormone injection the rate of mRNA^{VIT} synthesis during a primary stimulation is very low, but after this period there is a marked increase in the rate of synthesis of the mRNA. When animals undergo a secondary stimulation they commence mRNA^{VIT} synthesis at this higher rate immediately. The factors responsible for "priming" the liver cells are currently being studied because they represent important molecules in the regulation of the expression of previously unexpressed genes. It is likely that the reaction involved is one of the underlying mechanisms of cellular differentiation. The recent simulation of both primary and secondary responses in cultured liver cells from *Xenopus laevis* provides an excellent model system for the study of this phenomenon *in vitro*, particularly since the cells can be maintained in serum-free medium (11).

HORMONAL CONTROL OF α_{2u}-GLOBULIN PRODUCTION IN MALE RAT LIVER

α_{2u}-Globulin is a small polypeptide (M_r = 21,000) synthesized by the hepatocytes of sexually mature male animals. Even though the synthesis of this protein

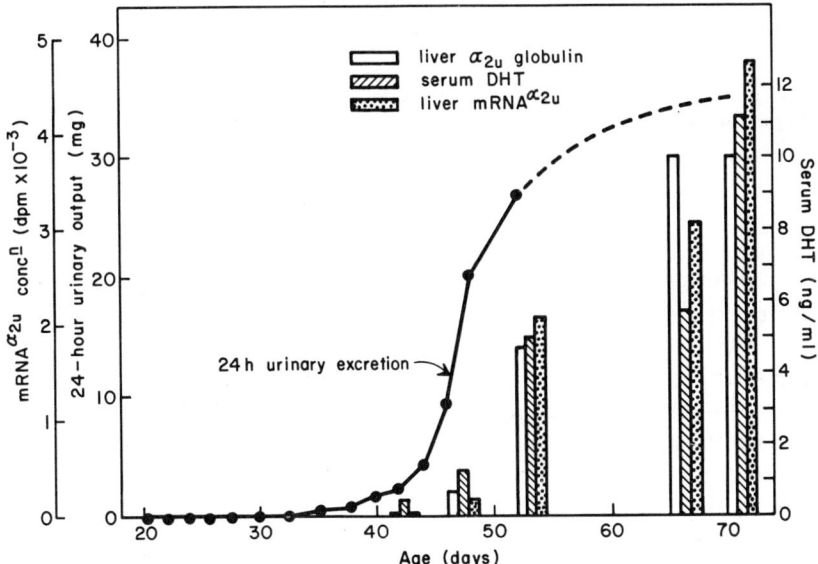

Fig. 15-4 The developmental changes in α_{2u}-globulin synthesis in male rat liver between 20 and 70 days after birth. The curve shows the amount of α_{2u}-globulin excreted into the urine during each 24-h period. The histograms give the concentration of dihydrotestosterone (DHT) in the serum and the concentration of α_{2u}-globulin protein and α_{2u}-globulin mRNA in the liver as indicated. [Reproduced from Roy et al. (12) with permission.]

accounts for 1% of total protein synthesis in the liver the function of the molecule, if any, is not known. Virtually all of it appears to be excreted (about 40 mg per day) in the urine. The control of the synthesis of this protein is quite complex and several hormones have been implicated in its regulation (see Refs. 12–14 for recent reviews).

α_{2u}-Globulin Synthesis during Development

The appearance of α_{2u}-globulin in the urine occurs between 5 and 6 weeks of age in parallel with the increase in the concentration of testosterone in the blood during sexual maturation (Fig. 15-4). Androgen receptors first appear in the liver cells at this time, and the concentration of both α_{2u}-globulin protein and α_{2u}-globulin mRNA also increases (Fig. 15-4). Injections of testosterone *before* the onset of puberty fail to preinduce the synthesis of α_{2u}-globulin, which suggests that some other developmental factors are involved in the regulation of the ability of the liver to synthesize this protein. Castration of the male prevents the appearance of the androgen receptor and α_{2u}-globulin synthesis in the liver, but castrated animals respond quickly to steroid injections.

Multihormonal Control in the Mature Animal

Androgens stimulate α_{2u}-globulin synthesis, and estrogens inhibit its synthesis. Thus castration of mature male rats reduces urinary output to about 10% of normal levels (12, 14). Daily injections of dihydrotestosterone restore α_{2u}-globulin production to normal levels. Moreover, dihydrotestosterone induces α_{2u}-globulin synthesis in the livers of spayed females (spaying reduces estrogen levels). Estrogen treatment of the dihydrotestosterone-treated spayed females or castrated males results in complete inhibition of α_{2u}-globulin production. All of these effects are accompanied by changes in the concentration of mRNA$^{\alpha 2u}$, indicating changes in gene transcription.

The inhibitory action of the estrogen on α_{2u}-globulin synthesis is mediated at two levels. It has been established that the estrogen competes with dihydrotestosterone for receptor binding when both hormones are present. During prolonged estrogen treatment the actual number of androgen receptors may decrease, resulting in a lack of responsiveness to dihydrotestosterone.

The effects of dihydrotestosterone and estrogen in castrated and normal males are summarized in Table 15-2, together with the effects of a number of other hormonal manipulations. Thyroidectomy abolishes α_{2u}-globulin synthesis, and this cannot be restored by injections of dihydrotestosterone. At least 10 days of thyroxine treatment are required to restore production to normal. The presence of thyroid hormone is necessary to maintain the concentration of mRNA$^{\alpha 2u}$ (Table 15-2), and since thyroidectomy does not affect the concentration of the androgen receptor in the cytoplasm the site of interaction between thyroid hormone and androgen must be in the nucleus. Adrenalectomy, hypophysectomy, and alloxan-induced diabetes all reduce the concentration of

Table 15-2 Effects of Endocrine Gland Ablation and Hormone-Replacement Therapy on the Synthesis of α_{2u}-Globulin and Its mRNA in the Livers of Mature Male Rats[a]

Endocrine Status	α2u-globulin concentration (percent of normal male)	mRNA$^{\alpha 2u}$ concentration (percent of normal male)
Thyroidectomized	0	0
" + T4 (4 days)	7.5	20.1
" + T4 (10 days)	97.5	105.0
" + DHT	0	0
Castrated male	N.D.	12.0
" " + DHT (4 days)	45.0	35.0
" " + DHT (8 days)	73.0	72.0
Intact male + estrogen (4 days)	35.0	34.0
" " + estrogen (8 days)	0	0
Adrenalectomized	N.D.	20.0
" + HC (10 days)	N.D.	32.0
Diabetic	24.0	27.0
" + Insulin	76.0	78.0

[a] Abbreviations: T_4 = thyroxine, DHT = dihydrotestosterone, HC = hydrocortisone (a glucocorticoid). The numbers in brackets refer to the number of days of hormone therapy. N.D. = not determined. Reproduced from Table 1 of Ref. 14 with the permission of John Wiley and Sons, Inc.

mRNA$^{\alpha 2u}$ to about 20% of normal levels. Insulin treatment can restore production in diabetic animals. However, the results of these hormonal manipulations are difficult to interpret because changing the concentration of one hormone, by either ablation or injection, in a mature animal is likely to result in compensatory changes in other hormones.

More recently, attempts have been made to reproduce these effects in hepatocytes in culture where the effects of individual hormones can be more easily studied (15). A mixture of dexamethasone, triiodothyronine, and growth hormone is required for optimal rates of production of α_{2u}-globulin in these cultured cells, indicating that all of them do, indeed, act directly on hepatocytes to maintain the α_{2u}-globulin gene in its fully active state. This is an unusually complex set of hormonal interactions, and elucidation of the role played by each component will contribute greatly to our understanding of the regulation of gene expression. A step toward this goal was made with the recent isolation of a cDNA fragment complementary to mRNA$^{\alpha 2u}$ (16). This cDNA will permit more direct studies on the transcriptional activity of the α_{2u}-globulin gene. Since mRNA$^{\alpha 2u}$ is only 1% of total liver mRNA (in comparison, mRNA^{OV} is 50–60% of total oviduct mRNA), its purification and the cloning of a cDNA fragment represent a fine technical achievement. It may be possible, by further refinement of the technology, to produce cDNA copies to the mRNAs for the enzymes discussed in Chapters 12–14, which often represent less than 0.1% of total cellular mRNA.

MULTIHORMONAL REGULATION OF THE DIFFERENTIATION OF THE MAMMARY GLAND

At least *eight* hormones are involved in the development of the mammary tissue during sexual maturation, pregnancy, and lactation to finally produce a fully differentiated tissue capable of secreting milk. Three steroids (17β-estradiol, progesterone, and glucocorticoids), two pituitary polypeptides (prolactin and growth hormone), two polypeptide hormones (insulin and placental lactogen), and thyroid hormones are all involved.

Morphological Changes

At birth, the mammary tissue is very poorly developed, consisting of cords of epithelial cells organized into primitive ducts (see Ref. 17 for a comprehensive review of the physiology and morphology of the tissue during development). There is little development of the tissue until the onset of sexual maturity, but if cells are isolated from a neonatal animal and challenged with steroids plus prolactin they are capable of synthesizing *in vitro* the milk proteins usually found only in fully differentiated lactating cells. Thus at birth the cells are already "programmed" to become active, secretory cells when appropriately stimulated.

During sexual maturation, which is 4–6 weeks after birth in the rat, there is extensive proliferation of the ductal cells, and this is probably stimulated and regulated by prolactin and estrogen (with a possible contribution from glucocorticoids). Growth of the tissue slows as the animal reaches maturity, and there are no further changes until the animal becomes pregnant.

The morphological, endocrinological, and biochemical changes that accompany the terminal differentiation of the mammary gland during pregnancy and lactation have been studied extensively and are reviewed in Refs. 18–20. There is extensive proliferation of the alveolar cells following the onset of pregnancy. These alveolar cells fill the interductal spaces to generate the typical lobuloalveolar structure of the mature tissue. Initially there is no synthesis of milk proteins or any of the enzymes associated with milk fat production, but toward the end of pregnancy some species synthesize and store many of these proteins. Other species, for example, the rat, do not commence synthesis of these molecules until after parturition. This is the most complex phase of differentiation of the mammary gland, and several hormones appear to be involved.

Changes in Hormone Levels During Pregnancy and Lactation in the Rat

Changes in the concentration in the blood of several hormones from the onset of pregnancy until the end of lactation are illustrated in Fig. 15-5. The concentration of prolactin declines during the first 5 or 6 days of pregnancy and remains at low levels until parturition. This drop in the level of prolactin is offset by an increase in the secretion of placental lactogen, which remains at a

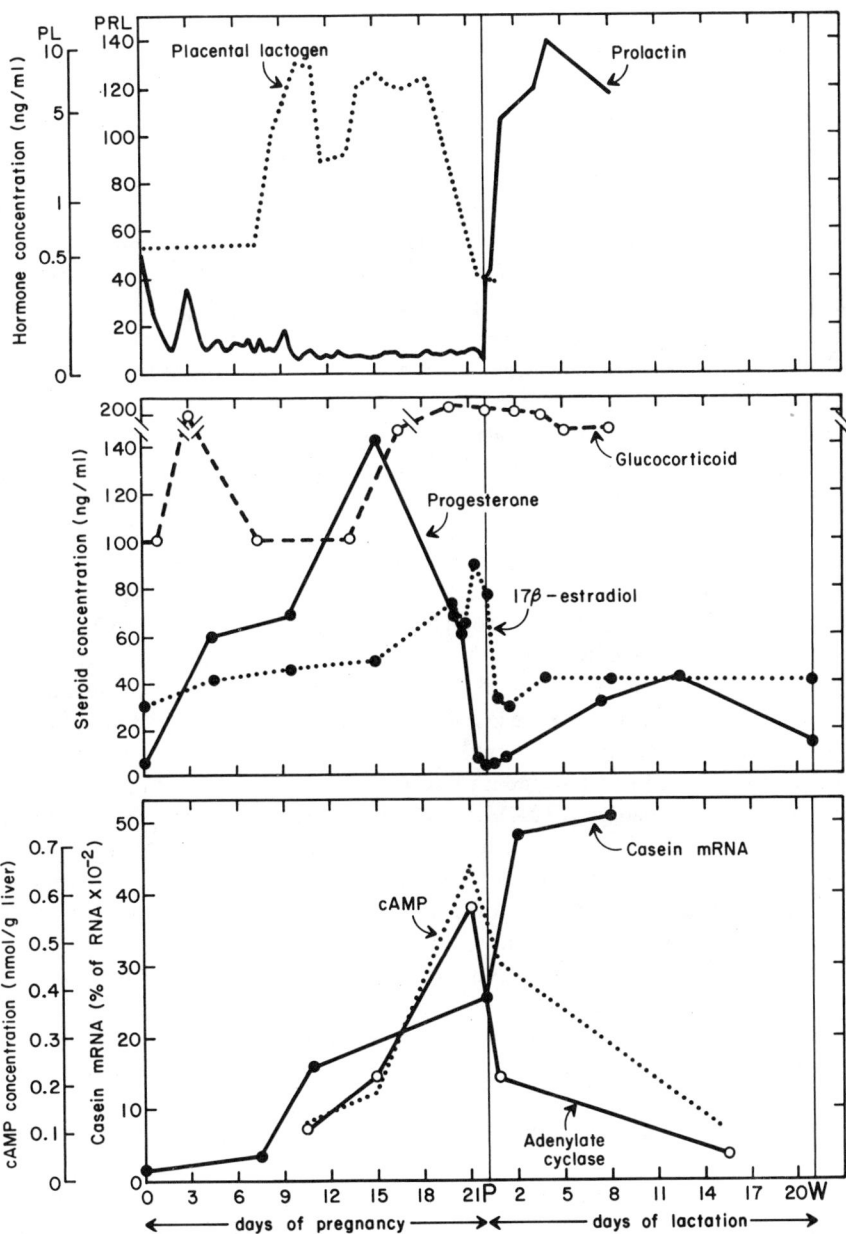

Fig. 15-5 Changes in the serum levels of hormones, and the concentrations of cAMP, adenylate cyclase, and casein mRNA in the mammary gland during pregnancy and lactation in rats. Parturition (P) occurred between 21 and 22 days after the start of pregnancy, and the animals were weaned (W) 21 days after birth. [Reproduced from Fig. 1 of Rosen et al. (18) with permission, Fig. 1 of Sapag-Hagar and Greenbaum (21) with permission of the *European Journal of Biochemistry*, and Table 1 of Sapag-Hagar and Greenbaum (33) with permission of the Federation of European Biochemical Societies.]

high level until just before parturition. At this time the concentration of placental lactogen falls, but there is a large increase in the concentration of prolactin. The concentration of prolactin remains high during lactation. Thus placental lactogen is the principal polypeptide hormone before birth, and prolactin is the principal polypeptide after birth.

Of the steroid hormones, progesterone shows the most dramatic changes: increasing throughout pregnancy to reach a peak 5–6 days before parturition. During the last 5 or 6 days before parturition, the concentration of the hormone drops and is at a very low level at the time of birth. This is followed by a smaller, more gradual increase in its concentration during lactation. In the 5–6-day period before parturition there are increases in the concentrations of 17β-estradiol and glucocorticoids. At parturition the concentration of 17β-estradiol falls abruptly, whereas glucocorticoids remain elevated during lactation (Fig. 15-5). It should be noted that none of these studies have taken into account the existence of diurnal rhythms of secretion for some of these hormones, and changes in vivo may be more complex than reported here.

Changes in Receptors and Second Messengers

Although neither steroids nor prolactin is thought to act via cAMP there are increases in the concentration of this cyclic nucleotide during late pregnancy (21), as shown in Fig. 15-5. There is a gradual increase in the concentration of the cyclic nucleotide during the last 8–10 days of pregnancy, which parallels the increase in the activity of adenylate cyclase in the lobuloalveolar cells. The concentrations of both enzyme and cyclic nucleotide fall after parturition. In addition to these changes in cAMP, there are also small changes in the concentration of cGMP (22) that appear to be opposite to those of cAMP.

Changes in the concentration of hormone receptors have not been particularly well-studied, despite their importance in determining tissue responsiveness to particular hormones. In one study (22) an increase in prolactin receptors on the surface of lobuloalveolar cells has been measured. In general, the change in receptors parallels the increase in hormone levels at parturition [it should be recalled (Table 11-5) that prolactin is one of the few hormones to stimulate the synthesis of its own receptors].

Changes in the Synthesis of Milk Constituents

The major constituents of milk are casein, fat, and lactose, and all three are produced in large quantities during lactogenesis. Casein is actually a mixture of closely related proteins (e.g., α-, β-, and κ-caseins; M_r = 20,000–28,000), and their synthesis accounts for 50–60% of total mRNA synthesis and protein production. Another milk protein, α-lactalbumin, accounts for an additional 20% of total mRNA and protein synthesis. Milk fat is mainly triglycerides containing predominantly C_8 and C_{10} chain-length fatty acids, and lactose is the disaccharide of glucose plus galactose.

Changes in the concentration of casein mRNA during pregnancy and lactation are shown in Fig. 15-5 (lower panel). The onset of the synthesis of this mRNA coincides with the increase in the circulating levels of placental lactogen midway through pregnancy, and a further increase takes place immediately after parturition, when there is a large increase in the concentration of prolactin. At the peak of milk production each alveolar cell contains approximately 79,000 casein mRNA molecules (20).

Synthesis of the enzymes of lipogenesis and lactose biosynthesis also commences during the second half of pregnancy, as shown in Fig. 15-6a. Changes in fatty acid synthetase and glucose 6-phosphate dehydrogenase are representative of changes in lipogenesis, and uridine diphosphoglucose pyrophosphorylase (UDPGPPase) typifies changes in the overall capacity to produce lactose. The concentrations of these enzymes continue to increase throughout lactation, but decrease dramatically once the young are weaned. The flow of carbon through the lipogenic pathway during pregnancy and lactation has been determined in rabbits using radioactively labeled acetate as precursor (Fig. 15-6b; note that pregnancy in the rabbit is 30 days compared to 21-22 days in the rat). There is

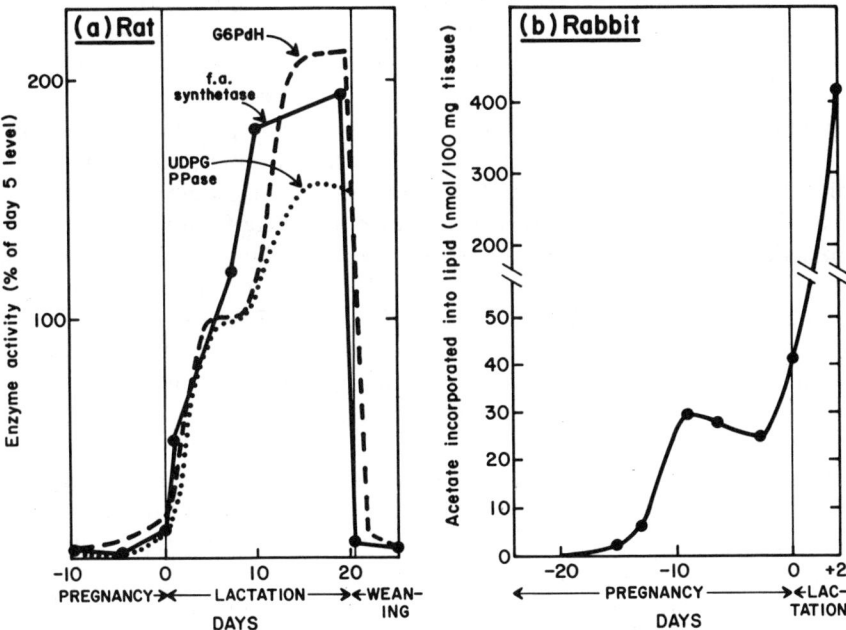

Fig. 15-6 Changes in lipogenesis in the mammary gland during lactation in (a) rats and (b) rabbits. (a) Changes in the concentration of two enzymes of lipogenesis (fatty acid synthetase and glucose 6-phosphate dehydrogenase) and one enzyme of lactose biosynthesis (uridine-diphosphoglucose pyrophosphorylase, UDPGPPase) are shown during late pregnancy, lactation, and after weaning. (b) Changes in the incorporation of radioactively labeled acetate into mammary gland lipids in the rabbit. [Data reprinted from Fig. 1 of Ref. 24 and from Ref. 19 with permission.]

some synthesis of milk fat before birth (which is stored in the cells and later released), but the major increase in milk fat production occurs immediately following parturition. More detailed accounts of these biochemical changes can be found in several reviews (19, 23–25).

Because of its abundance and the ease with which it can be studied most of the work on the regulation of mammary gland development has been carried out using casein as the model gene product. The role of the various hormones that control the synthesis of this protein is described below. It is not known if the expression of the genes for the enzymes of lipogenesis and lactogenesis is coordinately controlled by the same hormones that control casein production. Changes in the concentration of the mRNAs for these enzymes have not been studied. However, the recent successful isolation of fatty acid synthetase mRNA (24) should permit the synthesis of a cDNA probe that can be used to measure changes in the expression of the fatty acid synthetase gene and its control.

The Role of Prolactin

Most of the work on the role of prolactin, and other hormones, in the control of lactogenesis has been carried out on mammary gland explants in culture. Prolactin plays a major role in the activation of the casein and other genes, but it cannot achieve this alone. Moreover, the hormone appears to act at multiple levels within the cell (26–28).

Optimal rates of casein synthesis take place in the presence of prolactin, insulin, and glucocorticoids. Generally, the rate of transcription of the casein gene(s) is only increased 2–4-fold, whereas the rate of casein synthesis is increased 6–13-fold in the presence of prolactin. There is now evidence that prolactin also stabilizes the mRNA for casein by changing its half-life from approximately 5 h to greater than 90 h (26). Moreover, the hormone appears to stimulate the rate of *translation* of casein mRNA (27). Thus the hormone acts at the transcriptional, post-transcriptional, and translational levels to ensure that the rate of production of milk proteins can satisfy the demand.

The enzymes of lipogenesis are also induced by prolactin, but only when insulin and glucocorticoid are present. Figure 15-7a shows the effects of these hormones, individually and in combinations, on the induction of fatty acid synthetase biosynthesis (19, 25). Prolactin alone cannot even maintain basal levels of enzyme activity unless either insulin or glucocorticoid is present. When all three hormones are present the synthesis of synthetase is linear for at least 40 h in culture. In addition, this combination of hormones also induces acetyl CoA carboxylase and 6-phosphogluconate dehydrogenase, which are other key enzymes of lipogenesis (Fig. 12-1).

The Roles of Insulin and Glucocorticoids

In the presence of prolactin either insulin or glucocorticoid can stimulate a gradual accumulation of casein and the enzymes of lipogenesis (see Fig. 15-7a

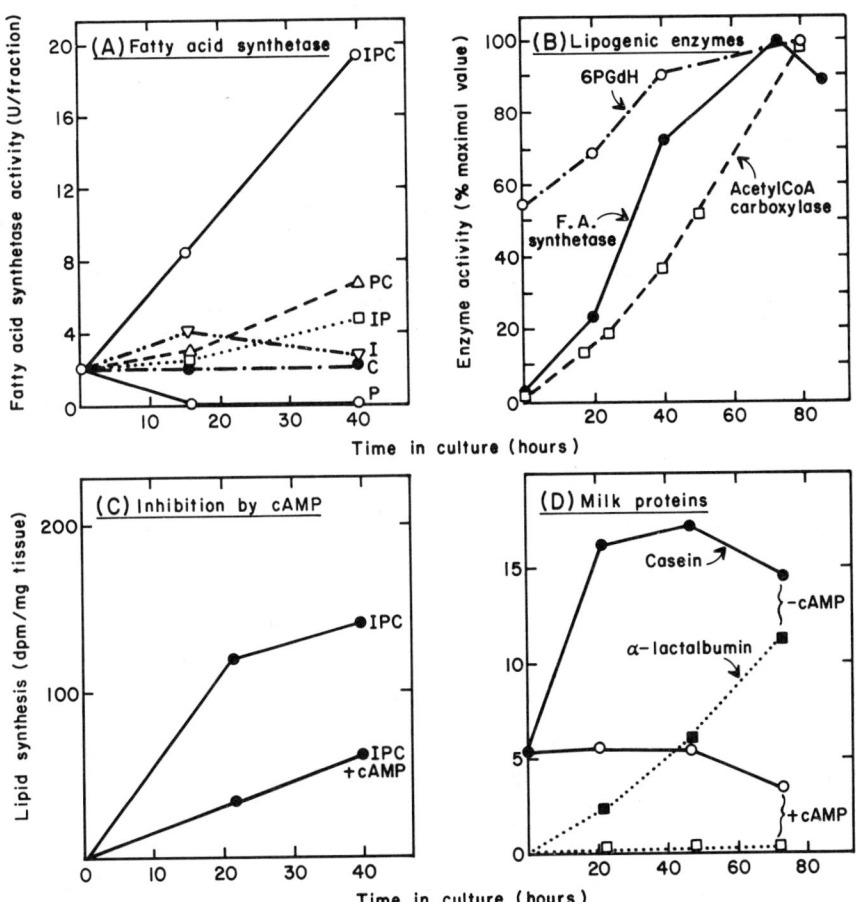

Fig. 15-7 Regulation of enzyme and milk protein synthesis in mammary gland explants. (A) The effects of prolactin (P), insulin (I), and cortisol (C), and combinations of the hormones on fatty acid synthetase production. (B) The effects of a mixture of prolactin, insulin, and cortisol on the synthesis of three enzymes of lipogenesis (6PGdH = 6-phosphogluconate dehydrogenase). (C) and (D), the inhibition of lipid synthesis and milk protein produced by cAMP. [Reproduced from Fig. 2 of Perry and Oka (32).]

for their effect on fatty acid synthetase) in explants of mammary gland from various species (28–31). However, the relative stimulation varies from species to species. For example, the role of insulin has been dismissed as being nonspecific (i.e., it only increases viability of the explant by stimulating glucose uptake) in some species, whereas in the mouse it has been shown to be essential and highly specific for the synthesis of casein (28). Proinsulin and somatomedin C were unable to substitute for insulin in these explants. Similarly, the role of glucocorticoids is not firmly established, but it appears to be a typical example of a steroid/polypeptide requirement for the full activation of the casein and other genes.

The Role of Progesterone

Despite the high levels of placental lactogen (functionally equivalent to prolactin), glucocorticoids, and presumably insulin in the pregnant rat, the amounts of casein and the enzymes of lipogenesis synthesized are quite small (Figs. 15-5 and 15-6). It is now thought that the increasing concentration of progesterone *represses* the activation of these genes until just before parturition. This effect can also be observed in explants, indicating that it is a direct effect of the steroid on the mammary gland cells. The site of action of the hormone is not yet known, but it may compete with glucocorticoids for the glucocorticoid receptor and prevent the formation of a functional glucocorticoid–receptor complex and its translocation to the nucleus. Progesterone also appears to inhibit the prolactin-induced increase in the translation of casein mRNA (27).

The Role of cAMP

The stimulation of the synthesis of milk fat and both casein and α-lactalbumin accumulation by a mixture of prolactin, insulin, and glucocorticoids is inhibited by cAMP (Figs. 15-7c, 15-7d) (32). Therefore, the cyclic nucleotide may play a role in the repression of these genes before parturition (see Fig. 15-5, lower panel, for time course of change). Similar results are obtained if explants are treated with either cholera toxin (an activator of adenylate cyclase) or isobutylmethylxanthine (an inhibitor of phosphodiesterase).

Since cAMP and progesterone both have the same effect, some effort has been expended in trying to establish a relationship between the two. This is not immediately obvious, since steroid hormones are not considered to exert any of their effects via second messengers (Chapter 11). However, it was shown some time ago (33) that progesterone and, more significantly, a mixture of progesterone and 17β-estradiol are effective inducers of the synthesis of adenylate cyclase in mammary tissue. Indeed, the concentration of 17β-estradiol during pregnancy increases in parallel with the increase in the activity of adenylate cyclase (Fig. 15-5). This is an interesting link between steroid hormones and cAMP. However, other factors must be involved since the newly synthesized adenylate cyclase molecules must be activated at the cell membrane by a hormone–receptor–cyclase interaction (Fig. 11-11).

Summary of the Control of Mammary Gland Differentiation

Prolactin is the most important hormone in the development of the mammary gland during pregnancy and lactation, but insulin and glucocorticoids are necessary to either coactivate or amplify some aspects of the developmental process. Of particular significance is the observation that the process of differentiation is regulated by *negative* factors such as progesterone and cAMP (and possibly estrogen). All of this makes the mammary gland an interesting experimental model for assessing the interactions of polypeptide and steroid hormones.

Differentiation of the mammary gland is determined by many simultaneous and sequential changes in specific mRNAs, some being expressed in just a few copies per cell and others in many thousands of copies per cell. The way in which the hormones interact to produce these differential changes in gene expression is an interesting subject for future research.

REFERENCES

1. A. E. Sippel, M. C. Nguyen-hus, W. Lindenmair, N. Blin, R. Lurz, H. Hauser, K. Giesecke, H. Land, M. Grez, and G. Schotz (1980). Mechanisms of induction of egg white proteins by steroid hormones. In M. Beato (Eds.). *Steroid Induced Uterine Proteins*. Elsevier/North-Holland, Amsterdam, pp. 297–314.
2. R. D. Palmiter, P. B. Moore, and E. R. Mulvihill (1976). A significant lag in the induction of ovalbumin messenger RNA by steroid hormones: a receptor translocation hypothesis. *Cell* **8**, 557–572.
3. R. D. Palmiter, E. R. Mulvihill, J. H. Shepherd, and G. S. McKnight (1981). Steroid hormone regulation of ovalbumin and conalbumin gene transcription. *J. Biol. Chem.* **256**, 7910–7916.
4. S. S. Seaver, D. C. Van Eys-Fuchs, J. F. Hoffman, and P. B. Coulson (1980). Ovalbumin messenger ribonucleic acid accumulation in the chick oviduct during secondary stimulation: Influence of combinations of steroid hormones and circannual rhythms. *Biochemistry* **19**, 1410–1416.
5. T. C. Spelsberg and F. Halberg (1980). Circannual rhythms in steroid receptor concentration and nuclear binding in the chick oviduct. *Endocrinol.* **107**, 1234–1248.
6. M. I. Evans, L. J. Hager, and G. S. McKnight (1981). A somatomedin-like peptide hormone is required during the estrogen-mediated induction of ovalbumin gene transcription. *Cell* **25**, 187–193.
7. R. F. Goldberger and R. G. Deeley (1978). The effect of estrogens on gene expression in avian liver. In A. K. Roy and J. H. Clark (Eds.). *Gene Regulation by Steroid Hormones*. Springer-Verlag, New York, pp. 32–57.
8. D. J. Shapiro and H. J. Baker (1979). Estrogen regulation of *Xenopus laevis* vitellogenin gene expression. In T. H. Hamilton, J. H. Clark, and W. A. Sadler (Eds.). *Ontogeny of Receptors and Reproductive Hormone Action*. Raven Press, New York, pp. 309–330.
9. J. R. Tata and D. F. Smith (1979). Vitellogenesis: A versatile model for hormonal regulation of gene expression. *Rec. Prog. Horm. Res.* **35**, 47–95.
10. S. Gerber-Huber, B. K. Felber, R. Weber, and G. U. Ryffel (1981). Estrogen induces tissue specific changes in the chromatin conformation of the vitellogenin genes in *Xenopus*. *Nuc. Acid Res.* **9**, 2475–2494.
11. P. F. Searle and J. R. Tata (1981). Vitellogenin gene expression in male Xenopus hepatocytes during primary and secondary stimulation with estrogen in cell cultures. *Cell* **23**, 741–746.
12. A. K. Roy, B. Chatterjee, and A. K. Deshpande (1980). Hormone-dependent expression of α_{2u}-globulin gene in rat liver. In A. K. Roy and J. H. Clark (Eds.). *Gene Regulation by Steroid Hormones*. Springer-Verlag, New York, pp. 230–246.
13. A. K. Roy (1979). Hormonal regulation of α_{2u}-globulin in rat liver. In G. Litwack (Ed.). *Biochemical Actions of Hormones*, Vol. 6. Academic Press, New York, pp. 487–517.
14. P. Feigelson and D. T. Kurtz (1978). Hormonal modulation of specific messenger RNA species in normal and neoplastic rat liver. *Adv. Enzymol.* **47**, 275–311.
15. C.-L. C. Chen and P. Feigelson (1979). Cycloheximide inhibition of hormonal induction of α_{2u}-globulin mRNA. *Proc. Natl. Acad. Sci. USA* **76**, 2669–2673.

16. B. Chatterjee and A. K. Roy (1980). Messenger RNA for α_{2u}-globulin of rat liver. *J. Biol. Chem.* **255**, 11, 607-11, 613.
17. Y. J. Topper and C. S. Freeman (1980). Multiple hormone interactions in the developmental biology of the mammary gland. *Physiol. Rev.* **60**, 1049-1106.
18. J. M. Rosen, D. A. Richards, W. Guyette, and R. J. Matusik (1978). Steroid-hormone modulation of prolactin action in the rat mammary gland. In A. K. Roy and J. H. Clark (Eds.). *Gene Regulation by Steroid Hormones*. Springer-Verlag, New York, pp. 58-77.
19. R. J. Mayer (1978). Hormonal factors in lipogenesis in mammary gland. *Vitamins and Hormones* **36**, 101-163.
20. J. M. Rosen, W. A. Guyette, and R. J. Matusik (1979). Hormonal regulation of casein gene expression in the mammary gland. In T. H. Hamilton, J. H. Clark, and W. A. Sadler (Eds.). *Ontogeny of Receptors and Reproductive Hormone Action*. Raven Press, New York, pp. 249-279.
21. M. Sapag-Hagar and A. L. Greenbaum (1974). The role of cyclic nucleotides in the development and function of rat mammary tissue. *FEBS Lett.* **46**, 180-183.
22. J. Dujiane, P. Durand, and P. A. Kelly (1977). Evolution of prolactin receptors in rabbit mammary gland during pregnancy and lactation. *Endocrinol.* **100**, 1348-1356.
23. B. L. Larson and V. R. Smith (Eds.) (1974). *Lactation: A Comprehensive Treatise*, Vol. 1. Academic Press, New York, Chaps. 1-4.
24. J. S. Mattick, Z. E. Zehner, M. A. Calabro, and S. J. Wakil (1981). The isolation and characterisation of fatty acid synthetase mRNA from rat mammary gland. *Eur. J. Biochem.* **114**, 643-651.
25. B. L. Baldwin and Y. T. Yang (1974). Enzymatic and metabolic changes in the development of lactation. In B. L. Larson and Y. T. Yang (Eds.). *Lactation: A Comprehensive Treatise*, Vol. 1. Academic Press, New York, pp. 349-411.
26. W. A. Guyette, R. J. Matusik, and J. M. Rosen (1979). Prolactin-mediated transcriptional and posttranscriptional control of casein gene expression. *Cell* **17**, 1013-1023.
27. B. Teyssot and L.-M. Houdebine (1981). Induction of casein synthesis by prolactin and inhibition by progesterone in the pseudopregnant rabbit treated by colchicine without any simultaneous variations of casein mRNA concentration. *Eur. J. Biochem.* **117**, 563-568.
28. F. B. Bolande, Jr., K. R. Nicholas, J. J. Van Wyk, and Y. J. Topper (1981). Insulin is essential for accumulation of casein for mRNA in mouse mammary epithelial cells. *Proc. Natl. Acad. Sci. USA* **78**, 5682-5684.
29. M. Ono and T. Oka (1980). The differential actions of cortisol on the accumulation of α-lactalbumin and casein in midpregnant mouse mammary gland in culture. *Cell* **19**, 473-480.
30. B. Teyssot and L.-M. Houdebine (1981). Role of progesterone and glucocorticoids in the transcription of the β-casein and 28S ribosomal genes in the rabbit mammary gland. *Eur. J. Biochem.* **114**, 597-608.
31. L. J. Burditt, D. Parker, R. K. Craig, T. Getora, and P. N. Campbell (1981). Differential expression of α-lactalbumin and casein genes during the onset of lactation in the guinea-pig mammary gland. *Biochem. J.* **194**, 999-1006.
32. J. W. Perry and T. Oka (1980). Cyclic AMP as a negative regulator of hormonally induced lactogenesis in mouse mammary gland organ culture. *Proc. Natl. Acad. Sci. USA* **77**, 2093-2097.
33. M. Sapag-Hagar and A. L. Greenbaum (1974). Adenosine 3':5'-monophosphate and hormone interrelationships in the mammary gland of the rat during pregnancy and lactation. *Eur. J. Biochem.* **47**, 303-312.

INDEX

Acetylation:
   of histones, 80–82
   of other proteins, 153
   role in transcription, 77, 80–82
AcetylCoA-carboxylase:
   regulation of induction, 282
   role in metabolism, 264–265
Adaptation, 1
   characteristics of adaptive enzymes, 264–265
   differences between pro- and eukaryotes, 263
   of gluconeogenic enzymes, 268–279
   nutritional, hormone basis of, 266–268
   summary of regulation of gluconeogenic enzymes, 277–279
Adenylate-cyclase:
   ADP-ribosylation, 81, 151
   changes during pregnancy, 365
   developmental pattern, 345–347
   mechanism of activation, 244–246
   properties of, 245
Adrenaline, function of, 224
Adrenocorticotrophin, 213
Albumin, induction by glucocorticoids, 286
Aldolase, changes in isozymes during development, 328–331
Alkaline-phosphatase, induction by cAMP, 290
Amino-acids:
   biosynthetic pathways in bacteria, 182
   covalent modifications of residues, 147–156
   diurnal rhythm of transport in liver, 309
   involved in protein binding to DNA, 19
Aminoacyl-tRNA, role in attenuation, 186
Androgens, metabolic functions of, 221
Arabinose, pathway of catabolism, 170
Arabinose operon:
   control of, 169–171
   structure of, 170
Arginase, induction of, 274
Arginine, biosynthesis of, 187, 188
Argininosuccinase, 274
Arginino-succinate lyase, 274
Arg operon:
   regulation of, 188
   structure of, 188

Attenuation:
   attenuator structure, 31
   general model of, 196–201
   of *his* operon, 195
   of *ilv* operon, 192
   leader transcript, consensus sequences and, 199
   of *leu* operon, 192
   mechanism of, 29–34
   of *phe* operon, 186
   preemptor, structure of, 32
   regulatory site, on operons, 29
   ribosome binding sequence, 199
   RNA secondary structure and, 31–34
   role of aminoacyl tRNA molecules, 200
   role of DNA sequence, 196
   role of leader transcript, 31–34
   role of ribosome, 199
   of *thr* operon, 190
   of *trp* operon, 29–34, 184–186
Attenuator, role in attenuation, 29–34
Autogenous gene expression, 171
Auxic growth, 163–165

Bacterial cells:
   biosynthetic pathways of, 182–184
   catabolic pathways of, 165
Base-pairs:
   in DNA double helix, 18
   interaction with proteins, 19
Biological-clock:
   nature of, 298
   role of pineal gland, 316
   role of suprachiasmatic nuclei, 316
Bio-operon:
   regulation of, 196–198
   structure of, 197
Biotin, pathway of biosynthesis, 197
Brain, higher-centers and control of rhythms, 315
Branched-chain amino acids, biosynthesis of, 190–192

Calcium binding proteins and information transfer, 250–252

INDEX

Calcium ions:
  cytoplasmic concentration, 250
  development of second-messenger capability, 346
  intracellular stores, 250–251
  role in information transfer, 250–252
Calmodulin, role in information transfer, 250
cAMP:
  biosynthesis of, 243
  changes during pregnancy, 365–366
  changes in liver, in response to diet, 266
  concentration, and insulin/glucagon ratio, 267
  and control of energy metabolism, 150
  developmental pattern, in liver, 346
  discovery of, 173–242
  and glucose levels in bacteria, 174
  induction of lactate dehydrogenase, 288
  induction of tyrosine hydroxylase, 289
  as intracellular second messenger, 242
  kinetics of information transfer, 243
  mechanism of action in enzyme induction, 290–292
  and mRNA-TAT levels, 271
  and phosphoenolpyruvate carboxykinase induction, 275–277
  and protein kinases, 246
  role in:
    catabolite repression, 173–174
    enzyme synthesis in development, 332
    lactogenesis, 369, 370
    N-acetyltransferase rhythm, 303
    TAT biosynthesis in development, 334
    and transcriptional activation, 247
    tryptophan oxygenase induction, 273
    and tyrosine aminotransferase induction, 269–272
    and tyrosine aminotransferase rhythm, 315
CAP:
  DNA, mechanism of binding to, 27
  interaction with cAMP, 27, 174
  promoter activation, 27, 174
CAP-cAMP:
  and catabolite repression, 173
  control of arabinose operon, 170
  control of *gal* operon, 168
  control of *lac* operon, 27, 168
  mechanism of action, 176
Carbohydrate metabolism, controlled by phosphorylation, 150
Carboxykinases A and B, 145
Carboxymethyl-transferase, 152
Casein:
  changes in mRNA levels, 356
  changes in pregnancy and lactation, 366–368
Catabolite activator protein (CAP), 27, 174

Catabolite repression:
  and CAP-cAMP, 173
  escape from, 178
  and glutamine synthetase, 187
Catecholamines:
  and calcium fluxes, 250
  pathway of biosynthesis, 289
  properties of, 223
  surface receptors for, 253
cGMP:
  changes during pregnancy, 366
  and information transfer, 249
  role in glucokinase induction, 281
Chromatin:
  amount dispersed by steroids, 239
  amount dispersed by thyroid hormones, 231
  association of steroids with, 235
  association of thyroid hormone with, 230
  classes of, 39
  definition of, 39
  digestion studies on, 49–50, 62–66
  diurnal rhythm of template activity, 307
  euchromatin, 39
  heterochromatin, 39
  higher orders of structure, 55–57
  and 10 nm fibre, 49, 62
  and 20–30 nm fibre, 49, 62
  nuclease sensitivity of active sequences, 62–66
Chromosomes:
  and chromatin, 39
  DNA content of, 38
  in mammalian cells, 38
Chymotrypsin, 145
Chymotrypsinogen, 145
Circadian oscillator, 298, 316
Circadian rhythms, *see* Rhythms of enzyme synthesis
Codons, usage of, 126
Constitutive enzyme, 1
Controlled feeding schedules, 297–320
Core particle, of nucleosome, 49–52
Cot value, 40–42
Cyclic AMP receptor protein (CAP), 27, 174

Degradation of enzymes, 156–158
Deoxyribonucleic acid, *see* DNA
Differentiation:
  enzyme changes during, 325
  of reproductive tissues, 355
Diurnal rhythms, *see* Rhythms of enzyme synthesis
DNA:
  A and B structures of, 17
  amount in bacterial cells, 13, 15, 38

INDEX

amount in mammalian cells, 13, 38
amount transcribed in mammalian cells, 39–40, 42
base-pairing in helix, 18
base-pairs and protein interaction, 19
binding of thyroid hormone receptors to, 230
breathing and polymerase binding, 71
breathing and structure, 71
nuclease sensitivity, and DNA binding proteins, 65
of nucleosomes, 49
organization into chromosomes, 38
organization on nucleosomes, 52
rhythms of synthesis, 307
satellite, 42
selfish, 44
sequence complexity of, 40–44
sequence complexity and structure, 42–44
specific protein binding, 19–21
split genes, 94–96
structural organization of fibers, 48, 49
structure in nucleoid, 15–18
DNAse-I, digestion of chromatin, 63
DNA-sequences:
  of *bio* regulatory region, 197
  highly repetitive in mammals, 41
  of *his* regulatory region, 195
  of *ilv* regulatory region, 193
  of *lac* regulatory region, 24
  of *leu* regulatory region, 192
  moderately repetitive in mammals, 41
  of *phe* regulatory region, 187
  of *thr* regulatory region, 190
  of *trp* regulatory region, 30
  unique sequences in mammals, 40–42
Down-regulation of receptors, 254

Embden-Meyerhof glycolysis, 165
Enterokinase, 145
Enzymes:
  acetylation of, 152
  adaptive, 1
  adaptive, in mammalian cells, 264–265
  adaptive enzymes of gluconeogenesis, 268–279
  adenylylation and uridylylation of, 151
  ADP-ribosylation of, 151
  of arginine biosynthesis, 188
  of branched chain amino acid biosynthesis, 191
  cascades, 54
  constitutive, 1
  degradation, kinetics of, 5
  degradation of, 156–158
  development of rhythmic capacity, 347–351
  discontinuous synthesis in development, 327
  induction, 2
  of enzymes of NADPH production, 283
  kinetics of, 3
  mechanism of translocation across membranes, 145
  methylation of, 152
  patterns of change in liver development, 328–330
  phenylalanine biosynthesis, of, 184
  phosphoramidation of, 148
  phosphorylation of, 147–151
  redox changes, 152
  relationship between hormone and enzyme rhythms, 312–316
  repression, 2
  rhythms of activity, 296
  of RNA processing in bacterials cells, 113–116
  role of higher centers of brain, in rhythms, 315
  summary of induction of lipogenic, 284–286
  summary of regulation of gluconeogenic, 277–279
  synthesis, techniques for measuring, 8
  threonine, biosynthesis of, 189
  tryptophan biosynthesis of, 185
  turnover, kinetics of, 3–8
Estrogen:
  changes during pregnancy, 365
  control of oviduct maturation, 356
  metabolic function, 221
  and vitellogenin synthesis, 359–361

Fatty-acid synthetase:
  changes in mRNA levels, 283
  kinetics of induction, 282–283
  role in metabolism, 264–265
Follicle-stimulating hormone, 215

Galactose, pathway of catabolism, 169
β-Galactosidase, 3, 163, 164
Galactoside permease, 166
Galactoside transacetylase, 166
Gal-operon:
  and colonic acid formation, 169
  control by repression, 168
  dual promoters, 169
  organization of, 169
Genes:
  amylase gene, 95, 108
  globin gene, 95
  nuclease sensitivity of active sequences, 62
  number in bacterial cells, 13
  number in mammalian cells, 13

ovalbumin gene, 95
regulation of expression of 5S-RNA, 75–77
RNA polymerase binding to, in mammals, 70–71
structure of histone gene, 64
structure of ribosomal gene in mammals, 43, 72
structure of 5S-RNA gene, 74–77
α_{2u}-Globulin:
  developmental profile of, 361–362
  synthesis in liver, 361
  testosterone and biosynthesis, 361
Glucagon:
  and calcium fluxes, 251
  changes during development, 341
  changes during starvation/refeeding, 266
  induction of urea-cycle enzymes, 273
  insulin/glucagon ratio and cAMP, 267
  and PEPCK changes in development, 334
  and phosphoenolpyruvate carboxykinase induction, 275–277
  role in:
    glucokinase induction, 280
    tryptophan oxygenase induction, 273
    tyrosine aminotransferase induction, 268–272, 314
  structure and function, 223
  and TAT synthesis in development, 334
  and transcriptional activation, 247
Glucocorticoids:
  and activation of TAT gene, 238
  changes during neonatal development, 340–341
  changes during pregnancy, 365
  changes in response to diet, 266
  development of rhythm of secretion, 348
  and glucokinase synthesis during development, 336–339
  induction of nonhepatic enzymes, 286–287
  metabolic functions of, 218–221
  role in:
    enzyme synthesis in development, 330
    glucokinase induction, 280
    lactogenesis, 368
    phosphoenolpyruvate carboxykinase induction, 277
    TAT rhythm, 299
    tryptophan oxygenase induction, 272
    tyrosine aminotransferase induction, 269–272
  and urea-cycle enzymes, 273
Glucokinase:
  and portal blood glucose, 279
  changes in development in liver, 330, 336–339
  development of rhythm, 349–352

  diurnal rhythm of, 300
  hormonal regulation during development, 336–339
  kinetics of induction, 280–282
  role in metabolism, 264–265
Gluconeogenesis:
  adaptive enzymes of, 268–279
  summary of enzyme adaptation, 277–279
Glucose 6-phosphate dehydrogenase:
  changes during pregnancy and lactation, 361
  role in metabolism, 264–265
Glutamine synthetase:
  and catabolite repression, 178
  control of, 151
  induction by glucocorticoids, 274
  role in metabolism, in mammals, 264–265
Glycogen:
  development of diurnal rhythm, 349–352
  diurnal rhythm of, 300–301
  effect of feeding schedule on rhythm, 301
  hormones, and rhythm in liver, 313
  levels during development, 335–336
  pathways of metabolism, 279–281
Glycogen-phosphorylase:
  diurnal rhythm of, 300
  levels during development, 335–336
Glycogen-synthetase:
  changes in development in liver, 330
  diurnal rhythm of, 300
  levels during development, 335–336
Glycosylation of proteins, 146
Growth-hormone:
  induction by glucocorticoids, 287
  induction by thyroid hormone, 287
  lack of second messenger, 249
  properties of, 217
Guanylate cyclase, 249, 281

Heterochromatin, 39
High mobility group proteins, 57–58
His operon:
  regulation of, 194–196
  role of histidyl tRNA, 195
  structure of, 194
Histidine:
  catabolic pathway, 171
  pathway of biosynthesis, 201
  regulation of biosynthesis and catabolism, 201–202
Histones:
  association with thyroid receptors, 230
  H1 and nucleosomes, 49
  H1 phosphorylation, 78–79, 248
  modification:
    by acetylation, 77, 80–82
    and cell cycle, 79

by phosphorylation, 78–79
of nucleosome core, 49
in nucleus, 45
other modifications, 77–78, 81
organization of histone genes, 64
sequence conservation of, 51
structure of, 49, 52
structure of nucleosome core, 52
HnRNA:
    organization into particles, 97
    and perichromatin granules, 97
    sequence complexity, 96
HnRNP-particles, composition of, 98
Hormone-receptors:
    receptors, steroid acceptor binding to DNA, 84
    receptors, thyroid receptor binding to DNA, 84
    surface receptors for polypeptides, 240
Hormone rhythms:
    adrenocorticosterone, 309
    corticosterone, 309
    glucocorticoids, 309
    insulin and glucagon, 312
    melatonin, 310
    pituitary hormones, 311
    thyroid hormones, 312
Hormones:
    hierarchy of control, 209–212
    role in information transfer, 209
    translocation to nucleus, 227–228
Hut-operon, structure of, 172
Hydroxymethylglutarylcoa reductase, diurnal rhythm of, 302
Hypothalamus:
    control of hormone secretion, 209
    peptides released by, 212
    structure of, 211

Ilv-operon:
    regulation of, 192–194
    structure of, 191
Insulin:
    changes during development, 341
    changes during starvation/refeeding, 266
    and glycogen rhythm, 313
    insulin/glucagon ratio, and cAMP, 267
    lack of second messenger, 248
    role in:
        glucokinase induction, 280
        in development, 338
        lactogenesis, 368
        malic enzyme induction, 284
        tyrosine aminotransferase induction, 269–272
    structure and function, 223

Insulin-glucagon ratio, changes during development, 342
Intervening sequences, 93–96
Introns, 93–96
Isoenzymes, and enzyme adaptation, 265
Isoleucine, biosynthesis of, 191

Lac-operon:
    mechanism of control, 166–168
    organisation of, 167
Lactate Dehydrogenase:
    changes in mRNA, 288
    induction by cAMP, 288
Lactate dehydrogenase, isoenzymes and induction, 289
Lactogenesis, 368–371
Lactose:
    as milk constituent, 366
    pathway of catabolism in bacteria, 165–167
Lateral-hypothalamic nuclei, and enzyme rhythms, 315
Leucine, biosynthesis of, 191
Leu-operon:
    regulation of, 190–192
    structure of, 191
Lipogenesis:
    and adaptive enzymes, 279–281
    changes during pregnancy and lactation, 367
    diurnal rhythms of, 308
Lipovitellin, 360
Luteinizing hormone, 215

Malic-enzyme:
    changes in mRNA levels, 284
    developmental pattern, 339
    role in metabolism, 264–265
Mammary gland:
    development of, 364–371
    morphological changes, 364
    summary of differentiation, 370
Matrix, *see* nuclear matrix
Melanocyte stimulating hormone, 217
Messenger RNA, *see* mRNA
Micrococcal nuclease, and digestion studies, 48, 63
mRNA:
    degradation of, 118
    kinetics of accumulation, in oviduct, 357–359
    levels in hypo- and euthyroid animals, 230
    levels in oviduct, 356
    mRNA-TAT and diurnal rhythm, 298
    and poly(A)tail, 102
    processing of, 98–112
    and protein synthesis, 126–129
    sequences involved in translation, 127
    specific increases with steroid hormones, 237

stability of, 93
structure of particles, and translational control, 139–141
mRNP particles in cytoplasm, 117

N-acetyltransferase:
  diurnal rhythm of, 305
  effect of light/dark on rhythm, 303
  rhythm in cultured pinealocytes, 305–306
  role in melatonin biosynthesis, 303–304
Nitrogen catabolism in bacteria, 178–180
Nonhistone proteins:
  and gene expression, 82–85
  modification of, 83
Noradrenaline, function of, 224
Nuclear matrix:
  association with nucleic acids, 46–47
  composition of, 46
  role in replication, 47
  role in transcription, 47
  and steroid action, 239
Nuclear membrane, 48
Nuclear proteins:
  histones, 45
  nonhistone (NHP), 45
  structural, 47
Nucleosomes:
  DNA and histones of, 49
  interaction with RNA polymerase, 66–67
  molecular weight of, 49
  and organization of DNA, 52
  organization into polynucleosomes, 53
  phasing, 63
  as product of digestion, 49
  sites of action of nucleases, 65
  sites of HMG protein binding, 57–58
  and steroid hormone binding, 236
  structure of, 49
  and thyroid receptors, 230
Nucleus:
  DNA content of, 44
  protein content of, 44
  RNA content of, 44
  and steroid acceptor sites, 236
  structure of, 44–45
  and thyroid hormone binding, 230

Operator, and operon control, 23
Operon, 23
  and other regulatory proteins, 34
Ornithine aminotransferase, adaptation to diet, 318–320
Ornithine decarboxylase:
  diurnal rhythm of, 302
  role of rhythm in adaptation, 318–320

Ovalbumin:
  estrogen, and mRNA levels, 357, 359
  synthesis in oviduct, 356
Oviduct:
  induction of differentiation of, 356
  morphological changes during maturation, 355
  mRNA levels during maturation, 356
Oxytocin, 217

Pentose-phosphate pathway, 165
Phenylalanine, pathway of biosynthesis, 184
Phe operon, regulation of, 186, 187
phosphodiesterase, changes in development, 346
phosphoenolpyruvate carboxykinase:
  changes in mRNA during induction, 276
  control of synthesis near birth, 330–334
  diurnal rhythm of, 302
  role in metabolism, 264–265, 275
Phosphoramidation of histones, 77
Phosphorylation:
  and changes in chromatin structure, 77–79
  and control of energy metabolism, 150
  of eIF-2, 137
  of gluconeogenic enzymes, 150
  of glycolytic enzymes, 150
  of histones, 77–79
  mechanism of, 147
  regulation of information transfer, 246
  regulation of translation, 137–139
  and thyroid hormone action, 232
Phosvitin, 360
Pineal gland:
  biosynthesis of melatonin, 303–306
  function as clock, 318
  innervation by superior cervical ganglion, 303–304
  relationship with SCN:
    in chickens, 305
    in rats, 303
  rhythm of N-acetyltransferase, 303–306
  role in development of rhythmic capacity of liver, 348
Pinealocytes, N-acetyltransferase rhythm in culture, 305
Pituitary gland:
  and control of hormone secretion, 209–211
  hormones released by, 213–218
  structure of, 211
Placental-lactogen, changes during pregnancy, 365
Plasma-membrane receptors for polypeptide hormones, 241
Polynucleosomes:
  and histone H1, 53
  structure of, 53, 54
  and 20–30 nm fibre, 54

Polypeptide hormones:
  and adenylate cyclase activation, 244–246
  and cAMP as second messenger, 242–248
  cell-surface receptors, 240–242
  fractional occupancy of receptors, 242
Polysomes, membrane-bound, 146
Post-translational modifications involving proteolysis, 144–145
Preemptor, role in attenuation, 32
Procarboxykinases A and B, 145
Progesterone:
  changes during pregnancy, 365
  role in lactogenesis, 370
  role in oviduct maturation, 359
  seasonal rhythm of secretion, 359
Prolactin:
  changes during pregnancy and lactation, 365
  and control of lactogenesis, 368
  nature of second messenger, 249
  and receptor concentration, 255
Promoter, 23
  activation by CAP, 27
  and polymerase binding, 23
Promoters, and polymerase binding in mammals, 69–77
Prostaglandins:
  general properties of, 253
  role in information transfer, 253
Protein kinases:
  activation by calcium-calmodulin, 252
  activation by proteolysis, 145
  activation of RNA polymerase, 77
  cAMP-dependent, 150
  cGMP-activated, 249
  and control energy metabolism, 150
  HnRNP-associated, 98
  involvement in enzyme phosphorylation, 149–151
  involvement in regulation of translation, 137–139
  possible involvement in thyroid hormone action, 232
  role in information transfer, 246
  types of cAMP-dependent, 246–247
Proteins:
  covalent modification of, 147–156
  DNA binding, kinetics of, 21
    nature of interaction, 20
    sequence specificity of, 21
  glycosylation of, 146
  mechanism of synthesis, 122–136
  regulation of synthesis, 134–141
Protein synthesis:
  mechanisms of elongation and termination, 133
  mechanism of initiation in bacterial cells, 129–131
  mechanism of initiation in eukaryotes, 131–133
  translational control in bacterial, 134–136
  translational control in eukaryotes, 136–141
Pyruvate kinase, changes in isozymes in development, 328–331

Receptors:
  activation by steroid, 232
  α-adrenergic, 250
  β-adrenergic, 251
  association of thyroid receptors with chromatin, 230
  for catecholamines, 253
  changes in liver in development, 344
  cytoplasmic, steroid (src), 232
  and down-regulation, 254
  kinetics of steroid hormone action, 237–240
  movement on cell surface, 244
  nuclear, for estrogen in oviduct, 358
  number of thyroid hormone receptors in tissues, 230
  polypeptide, and cAMP, 242
  surface:
    and adenylate cyclase activation, 244–246
    for polypeptides, 240–242
  translocation of SRC to nucleus, 234
Regulon, 188
Repressor, 2
  ara, 170
  arg, 188
  gal, 168
  hut, 171
  lac, 167
  obstruction of promoter, 28
  trp repressor, 185
Rhythms, biological, 296–297
Rhythms of enzyme synthesis:
  N-acetyltransferase, in pineal gland, 303–306, 315
  development of capability for, 347–351
  enzymes of glycogen metabolism, 300
  hepatic enzymes, 302
  in other tissues, 306
  regulation by higher centers of brain, 315
  tyrosine aminotransferase rhythm, 297–300, 312
Ribosomal RNA gene, structure of, 43
Ribosomes:
  important sequences of, 124
  involvement in attenuation, 199
  proteins of, 123
  as sites of protein synthesis, 123
Rifampicin-challenge-assay, 231, 239

RNA-heterogeneous nuclear, 96, 97
RNA polymerase:
  in bacterial cells, 22
  biphasic changes with steroids, 240
  in mammalian cells, binding to active genes, 66
  and sequence recognition in mammals, 69
  sites of initiation, in mammals, 70–71
  thyroid hormones, and activity, 230
  types, in mammalian cells, 67–69
RNA processing:
  in bacterial cells, 113–116
  capping, 98
  in eukaryotes, 98–112
  methylation of bases, 100
  polyadenylation, 100
  summary of, 110–112
RNA splicing:
  models of, 103–106
  and new gene products, 106–110
  regulation of, 106
RNA-transport from nucleus to cytoplasm, 116
RNA synthesis, rhythms of, 307

SCN, see Suprachiasmatic nuclei
Second-messengers:
  calcium ions, 250
  cAMP, 242–248
  cGMP, 249
Serine dehydratase, adaptation to diet, 318–320
Sigma factor, 22
Signal hypothesis, 146
Split genes, 94–96
Steroid hormones:
  cytoplasmic receptors for, 232–235
  mechanism of receptor activation, 233–235
  mechanism of transcriptional activation and, 237–240
  nuclear acceptor sites, 234–236
  and receptor down-regulation, 255
  translocation to the nucleus, 234
Steroids, chemistry of, 218
Stringent factor, 136
Supercoils of DNA, 15, 52
Suprachiasmatic nuclei:
  connection with pineal gland, 303
  function as biological clock, 316
  role in development of rhythmic capacity, 348

Terminator sequences:
  in animal cells, 72
  in bacterial cells, 25
Testosterone, and α_{2u}-globulin synthesis, 361–363

Threonine operon:
  regulation of, 189
  structure of 189
Thyroid-hormone:
  changes in development, 342
  changes in response to diet, 266
  general functions, 222
  and glucokinase synthesis in development, 337–339
  and growth hormone induction, 287
  and induction of lipogenic enzymes, 282
  induction of malic enzyme, 284
  mechanism of transcriptional activation by, 230–232
  nuclear receptors for, 229–230
  role in glucokinase induction, 280
  stability in blood, 229
Thyroid stimulating hormone, 215
Thyroxine, transport in blood, 227–229
Transcription:
  activation by phosphorylation, 247
  changes in oviduct, 358
  definition of mechanism, 21
  diurnal rhythms of template activity, 307
  DNA template and, 22–25
  histone modification, and regulation, 77–82
  of ribosomal DNA, 72
  role of labile proteins, 358
  of small RNA molecules, 72–74
  and specific sequences in mammals, 69–77
  and steroid hormones, 237
  termination of, 25–27
  termination in mammals, 71–72
  and thyroid hormones, 230
Transcripton, 110
  and new gene products, 106–110
Transferrin (Conalbumin):
  estrogen and mRNA levels, 357–359
  induction by glucocorticoids, 286
  synthesis in oviduct, 356
Transfer RNA in protein synthesis, 125
Translation:
  control and eIF-2, 137–139
  control of elongation, 139
  mechanism of, 122–134
  and mRNA structure, 139–140
  regulation in bacteria, 134–136
  regulation in eukaryotes, 136–141
Tricarboxylic acid cycle, 165
Triiodothyronine, 227–232
Trp-operon:
  attenuation of, 29–34
  regulation of, 184–186
  structure of, 184–186
Trypsin, 145

Trypsinogen, 145
Tryptophan:
  pathway of biosynthesis, 184
  stabilisation of tryptophan oxygenase, 272
Tryptophan oxygenase:
  developmental pattern, 339
  mechanism of induction, 3, 4, 7, 268–273
  mRNA, developmental pattern, 339
  role in metabolism, 264–265
Tyrosine aminotransferase:
  changes in mRNA during induction, 270–272
  development of rhythm, 349–352
  diurnal rhythm of, 297–300
  effect of diet on rhythm, 298–300
  feeding activity and rhythm, 297
  and histone H1 phosphorylation, 248
  kinetics of gene activation, 238
  mechanism of induction, 268–273
  regulation of synthesis in development, 334
  rhythm of:
    and glucagon, 315
    and glucocorticoids, 313
    rhythms and adaptation, 318–320
    role in metabolism, 264–265
Tyrosine pathway of biosynthesis, 184

Ubiquitin, 58
Urea-cycle enzymes:
  kinetics of induction, 274
  role in metabolism, 264–265
  urea cycle, 274

Valine, biosynthesis of, 191
Vasopressin, 217
Ventro-media hypothalamic nuclei, and enzyme rhythms, 315–317
Vitellogenin:
  induction by estrogen, 360
  mRNA during induction, 360
  synthesis in liver, 359–361

Zymogens, 154